QUANTUM COWBOY

For Heather + Byron

David L Brown

Copyright © 2010 David L. Brown

ISBN 978-1-60910-465-8

All rights reserved. No part of this publication may be reproduced, stored in a retrieval system, or transmitted in any form or by any means, electronic, mechanical, recording or otherwise, without the prior written permission of the author.

Printed in the United States of America.

The characters and events in this book are fictitious. Any similarity to real persons, living or dead, is coincidental and not intended by the author.

Booklocker.com, Inc.
2010

www.dlbrown-inc.com

QUANTUM COWBOY

A Novel

By David L. Brown

Chapter 1

Quantum Center, Los Alamos, New Mexico
Saturday, 7:32 a.m. MST

Tommy Dawson leaned back from the keyboard and raised his arms above his head to stretch tired muscles. The extra-large Aeron chair creaked under his considerable weight. Gazing at the ceiling he emitted an expansive yawn. He and his team had been at work for several hours.

"Hey! Need some coffee here!" he yelled to no one in particular. Although several heads turned briefly in his direction, no one responded. Dawson grunted, acknowledging the usual indifference to his demands.

"Get it myself!" he announced to his computer screen. Grumbling he rose and wandered toward the coffee machine in a far corner.

Tommy Dawson, Ph.D., Nobel Laureate and senior fellow of the New Mexico Institute, was deeply respected by all—but this spacious research center held enough accomplished physicists and engineers to staff a first-class university and none were prepared to become Dawson's personal errand boy or girl.

Still mumbling to himself Dawson seized an unwashed mug. He used his fingers to rub lipstick stains from the rim then held it under the spout. The coffee had festered in the urn for several hours. Even with three packets of creamer it remained the color of molten lead. Dawson tore open four packets of sugar, dumped them into the cup and stirred with a swizzle stick before turning back to the room. He raised the cup to his lips and drank with a loud slurping sound. Grunting in satisfaction he let his eyes travel around the scene.

The center was big, almost the size of an aircraft hanger. It was a picture of chaos, filled with scattered workstations, strange equipment and busy people. Heavy cables snaked across the floor, connecting workstations and various machines. Near the far end where Dawson had begun his coffee quest was a raised platform on which stood an open U-shaped control center topped by an array of over-sized flat-screen monitors.

Opposite him and rising almost to the 20-foot ceiling were racks containing several hundred 12-core servers. They were configured in parallel as a single super computer running on an advanced Unix-based operating system. Atop two rolling ladder stairs, several technicians were busy attending to the bank of machines.

For a long moment Dawson continued to stare around the room. He was a round man. Round head, round body, round like the Michelin Man. His face was round, too, with sagging jowls and at least two double chins, the result of too many sugar-laden drinks and donuts. His cheeks were colored by rosacea and although he shunned alcohol his nose was cherry red.

Dawson's dark hair was shaggy and striped with gray. He wore a custom-made baseball cap featuring a stylized atom and the words "Quantum Cowboy." The greasy hair on the back of his head had been threaded over the adjustment strap of the cap to create a sad approximation of a ponytail.

Dawson wore a stained T-shirt imprinted with a picture of a cat. There were actually two images of the cat, superimposed. In one the cat was sitting up attentively, in the other it was lying with its legs sticking up in the air, apparently dead. A legend in old-style Germanic script read: "Schrödinger's Cat." There was a large coffee stain on the front of the shirt. His casual appearance was completed with a pair of faded jeans and suede roper boots.

Tales still circulated about Dawson's attendance at the Nobel awards program in Stockholm. He had famously shown up for the strictly black-tie event wearing a tan cowboy-style suit with a turquoise-and-silver bolo tie, two-toned tan and red Tony Lama boots with matching belt, and a dove gray size seven and seven-eighths Stetson 10X beaver hat. Had he not, after all, been an important guest of honor he would surely have been asked to leave.

Snatching two stale donuts and stuffing one into his mouth Dawson wandered toward the center of the lab where four desks were positioned around a square table to create a big workspace in the style of a Greek cross. Several engineers were standing at one side examining large format schematics. The printouts were visualizations of program code produced by advanced software running on the server array. Graphic visualization of software was a breakthrough method that made it easier to "see" how code strings interacted. Programming bugs and awkward pieces of code appeared as broken or erratic lines, making it easy to identify and correct mistakes in millions of lines of complex software.

Nearby an eight-foot wide plotter was industriously churning out another schematic visualization. Dawson eyed it suspiciously. The engineers were studying a drawing that had been printed as multi-layered acetate overlays. Dawson sidled up behind them, munching on the second donut and taking a loud slurp of coffee.

One of the men picked up a grease pencil and marked a circle on the drawing.

"We need to take another look at this area," he murmured. "It should work, but there has to be a more elegant solution."

"Yeah, I see that," another engineer agreed. He leaned over to take a closer look. "This was probably written as a temporary work-around, not as final code." He glanced over to a female engineer who was standing to his left. "Can you get on this right away?"

"Sure, Boss," the programmer replied. She picked up a folder full of documents and stepped away to her nearby workstation.

Seeing Dawson, the project manager nodded a greeting. Dawson said nothing. He stepped closer and gazed at the printouts. He was still sipping from the grungy coffee cup, unintentionally spilling some of the nasty brew onto the layout. No one said anything but a young engineer let out an impatient breath and rolled his eyes briefly at the ceiling. He picked up a shop rag and wiped up the spill.

Dawson, his attention on the schematic, noticed none of this. Inhaling the last bite of donut he wiped his greasy hand on his pants leg.

"What's the problem?" he grunted abruptly.

Although brilliant, Dawson suffered from high-functioning autism and his social skills were marginal at best. His associates were accustomed to his hard edges and usually took in stride what would be considered unacceptable rudeness in anyone else.

"Sure," replied the project manager, Stuart Reeves. He smiled and beckoned Dawson to come nearer. Reeves was slender with thinning blond hair trimmed short. He wore gray Dockers and a blue short-sleeved pilot-style shirt with two pockets and shoulder tabs, no tie and black penny loafers with white socks. He drew Dawson's attention to the schematic the engineers had been studying.

"This is some of the software for the interface between the input console and the processing bank." He waved at the wall of servers. "As you know, our main engineering challenge is to link with the processor bank and receive data back."

"Yeah, yeah," Dawson muttered impatiently. Reeves took a half step back and tried again.

"We're pretty happy with the software. Just need to work out some wrinkles. Of course, it's the processor bank that actually provides the link to the quantum device."

He was referring to the heart of the quantum computer they were preparing to test. It resided in a tank holding a cloud of molecules serving as quantum binary digits, or qubits. It was located in a nearby cryolab dubbed the Qubit Farm, equipped to maintain the atoms in deep chill near absolute zero.

"Yeah, yeah, I know," grumbled Dawson, leaning one elbow against the desktop and slopping more coffee on the printouts. "I designed the damned thing you know. Just tell me what's the problem."

The young engineer with the shop rag reached around him to wipe up the spill, once again favoring the ceiling with an irritated glance. He groaned as Dawson leaned his hand on the printout, leaving a blob of sugar and grease.

Reeves began to describe the minor details of the interface program, running an index finger across the acetate drawing to explain each point. Dawson eyed the diagram skeptically. When the project engineer had finished the physicist just grunted. He tipped up and drained the last of his coffee, set down the cup with its donut residue on a stack of documents and turned away without a word. The young engineer snatched up the filthy cup and grimaced.

"Well, shit, he's got the manners of a pig," he muttered to himself.

"Settle down," Reeves told him. "You know Tommy's different from the rest of us, and he does the best he can. And besides, if not for him we'd all be doing something else, something a lot less interesting."

"Oh, sure," the young engineer admitted, wiping off the offending cup. "I really do understand. I knew a kid in school who was autistic, and he was sure strange in some ways but pretty much okay. Smart, too, but not like Doctor Dawson."

He turned to follow Dawson with his eyes as the physicist meandered through the lab, stopping to stare over people's shoulders or poke at documents or machines.

"It's just that sometimes the pressure gets to me and I forget to remember to be nice," he added, smiling wanly at the coffee cup in his hands.

Reeves slapped him on the shoulder then glanced at his watch.

"I have to get going," he announced. "I've got a few things to take care of and then later I have an appointment with a science writer. You know how those journalists are—she'll be waiting at my office, sharpening her claws."

The other engineers chuckled sympathetically as the meeting broke up.

Chapter 2

Palo Alto, California
Saturday, 7:34 a.m. PST

Chris Fisher clenched the butt of a Heckler & Koch 45-caliber pistol in one sweaty hand. Half-blinded, he slid in a fresh magazine and chambered a round. He was surrounded by a cloud of dust and sand, swept up by the rotors of the Pave Hawk rescue helicopter that was descending above him. He had lost touch with the rest of his team and now the extraction mission was in serious trouble. His M-4 carbine was out of ammo and the fight wasn't over.

From somewhere upslope came the crackle of gunfire, the distinctive *clap-clap-clap* of AK-47s firing in three-round bursts. Bullets zipped through the air like swarms of hornets. Reacting to the fire the helo banked sharply away as jacketed rounds began to strike it. Fisher saw star-shaped cracks appear in the craft's windscreen as the pilot struggled to control the Pave Hawk.

It rocked away and the starboard gunner opened up at the insurgents with a 7.62mm mini-gun. Firing several thousand rounds per minute from its rotating barrels, the gun poured forth a torrent of expended brass shell casings that rattled and bounced like a hailstorm among the rocks near where Fisher was pinned down.

It was not his first visit to this place. It was not real, but a distorted dream-memory from his past. As he tossed uneasily in his bed he was not the Stanford professor of physics, but once again a

Navy lieutenant leading a special ops unit of SEAL Special Forces commandos. In his mind he was back in Afghanistan, engaging al-Qaeda fighters across the desiccated foothills in the rain shadow of the Hindu Kush mountain range.

The setting of the dream was vaguely familiar, for it was loosely based on the events of an actual mission. That real firefight had ended in success, but in the nightmare version everything possible seemed to go wrong. It always did, somehow morphing into a worst-case scenario.

The Pave Hawk banked away. Fisher could hear the *ulu-ulu-ulu* battle cries of Islamic warriors, drawing close. More bursts of gunfire raked at his position among the rocks. He ducked and waited for them to come nearer, ready to fire as they broke over the rocky ridge.

The sounds of the approaching enemies grew in intensity. The dust began to clear as the Pave Hawk's racketing sound faded. Fisher tensed for the final confrontation…

…and everything began to dissolve into a mist-like fog.

Chris slowly rose from the dream state. His right hand was clutched into a fist, but it held no pistol. The rocks, which in his dream had been cold and hard, grew soft and became his mattress, a foam pillow, and damp sheets that were twisted around his legs.

And there was another sound.

It was not the rattle of AK-47s or the chatter of the mini-gun and thump of rotors. Nor was it the ceaseless, unforgiving wind of the Afghan foothills. For an instant he struggled to identify the sound before snapping fully awake.

Down the hall in the living room of his apartment the phone was ringing.

Chris tried to sit up, then had to lie back and untangle his legs from the sheet. The phone rang again.

Grabbing a bathrobe and slippers, he leapt through the door and down the hall dressed only in his briefs. The phone rang.

Chris stumbled into the living room, dodged around the couch and dove across the coffee table. The phone rang. Performing a rolling dive he grabbed the handset from its charger base just as it stopped in mid-ring.

He brought the portable to his ear in time to hear a faint, metallic hum, followed by the click of disconnection and a dial tone.

With an exclamation Chris put the phone back into its charger, climbed to his feet and sank onto the couch. Leaning forward with elbows on knees he rubbed his face to wipe away the last remnants of the dream.

It visited him less frequently now as the real events in Afghanistan receded with each passing year. But when it did return, it still had power. He could sense the adrenaline surging through his body. His heart thumped in response to the fight-or-flight reaction. His recurring dream was a secret he had shared with no one. He didn't believe it was a sign of post-traumatic stress syndrome, for he felt exhilaration, not fear when his mind took him back to a time and place of danger.

He took five slow, deep breaths and felt his pulse rate slow as the oxygen reached his bloodstream. He looked at the clock on the desk across the room. It was still early, only 7:48.

Who would be calling at this hour?

He shrugged, stood up and went into the kitchen to brew a double espresso, his stimulant of choice. As the machine whirred to grind fresh beans he pulled on the dressing gown and slipped his feet into the lambskin slippers he'd snatched up from beside his bed.

He was a ruggedly built man but more leopard than lion, still carrying most of the lean muscle of a SEAL leader. No glutton for punishment he didn't frequent gyms or work out on exercise machines. He kept his edge by running a few miles three or four times a week, riding his mountain bike, taking strenuous hikes in the nearby Santa Cruz Mountains and enjoying the occasional scuba diving adventure.

He stood two inches over six feet and weighed a solid 210 pounds. He wore his light brown hair short in almost a military style, keeping it neat with standing weekly appointments at a nearby barbershop owned by a former chief petty officer. Snipping with his scissors the old sailor spouted endless tales of his years at sea. Chris sometimes spoke of his own experiences, but generally relaxed in the chair and watched himself in the mirror as the chief prattled on about long-ago events in the Pacific and South China Sea.

The buzzcut went well with his rugged face. He had a square chin that would give envy to a Humvee and his nose showed signs of having been broken more than once. Each time the damage had been repaired in a workmanlike manner but not up to the standards of civilian cosmetic surgery. A hairline scar, souvenir of a Taliban firefight, marked his left cheek.

His brown eyes held a glint of gold and seemed unusually alert. At 36 his vision was still excellent. During SEAL training he had topped his class as a marksman, able to spot his hits on targets up to 50 yards away without resorting to a scope. In combat his keen eyesight and powers of observation had helped earn him a reputation as an outstanding leader.

He sat down at the little kitchenette table and sipped the bitter espresso. It was Saturday and he had thought of taking a bike ride along the Pacific Coast, perhaps following scenic back roads across to Highway One and looping up to Half Moon Bay, maybe stopping at Sam's Chowder House. One of Sam's famous blue crab sandwiches and a cold bottle of Anchor Steam would go down well and he could watch the gulls squabble over crumbs.

He rose and set to work preparing a simple breakfast. As his espresso machine ground the beans for a second cup he reached into a drawer and pulled out a well-honed combat knife. He sliced an English muffin, dropped the two halves into the toaster, wiped the blade on a towel and placed it back in the drawer. Opening the refrigerator he drew out butter, a crock of plum jam and a container of orange juice and set them on the table.

As the toaster and espresso machine worked he went to the front door for the morning paper. In a few minutes he was chewing and sipping contentedly, reading about unlikely events that had allegedly taken place somewhere in the crazy, out-of-control world, or perhaps only in the imagination of the writers.

Later as he shaved, showered and dressed in his cycling jersey and pants his mind wandered idly back to the dream and the early phone call that had ended it. He tried to think who might have called. No one from the university, surely, for the physics department shut down on weekends. All the faculty members and staff would be enjoying their own pursuits. Nor could it have been his sometimes girlfriend, for she was leading an archaeological dig in the remote Andes Mountains of Chile, far from the nearest phone.

Chris had recently been granted tenure as full professor at Stanford University, one of the youngest ever to attain that rank. Following undergraduate studies at Virginia Tech on a Navy ROTC scholarship and six years of active service he had used his military benefits and student loans to squeeze both a masters and doctorate into three years of concentrated effort. Now, in just seven years at Stanford he had climbed to the top of his profession.

Ready to leave, Chris had his hand on the doorknob when the phone rang again. He glanced at the dial of his dive watch. It was 8:32. For a moment he considered ignoring the call, then shrugged. He turned back into the room and picked up the phone.

"Hello, this is Chris."

There was a faint click, then the low humming sound he had heard earlier. After a pause a slightly distorted voice spoke.

"Dr. Fisher," it stated. Not a question.

"Yes, this is Dr. Fisher," Chris replied. His eyebrows rose with curiosity. "Who is this please?"

There was another pause before the stranger spoke again.

"This call concerns national security. I'm using a secure phone but yours is not safe. We need to meet, soon."

Chris scowled. "What's this about?"

The man on the other end cleared his throat impatiently. "Yours is not a secure phone," the voice repeated. "We must meet, soon."

Chris paused for a moment before replying. *What is this?* He had no idea why he would be involved in a matter of national security.

"Okay," he said at last. "I was just leaving for a bike ride. We could meet later…"

"No," the voice said sharply. "This is not secure. Stay there. Someone will pick you up. Fifteen minutes. Be ready."

The humming sound was replaced by a dial tone as the caller abruptly ended the connection.

Chapter 3

Los Alamos, New Mexico
Saturday, 10:47 a.m. MST

Stuart Reeves pulled his anthracite gray Jeep Grand Cherokee into a reserved parking space in the multistory parking garage adjacent to the main administration building of the Los Alamos National Laboratory. Grabbing his briefcase he jogged to the elevator and was soon striding down the hallway toward his corner office on the top floor. As he approached the small waiting area a tall brunette woman stood up to meet him, setting aside a silver Macintosh laptop.

"Sorry I'm running late," Reeves puffed, a little breathless after his fast trip from the parking garage. "You're Kate Elliot, right? From *Science Today* magazine? Hi, Ms. Elliott, I'm Stuart Reeves." He stopped to take a deep breath then offered his hand.

The woman chuckled at Reeves' slightly manic greeting and took his hand. "Pleased to meet you," she replied. "Yes, I'm a senior editor at *Science Today*. You can call me Kate."

"Okay, Kate," he responded. "And I'm Stuart. Look, come in and sit down." He ushered her into his office and gestured to a seat in front of the desk.

He hovered for a moment as she sat down, asking if she would like something to drink. She shook her head and he rounded the desk to sit down facing her. The Sun was shining brightly through the south-facing window. She squinted and he hurriedly rose to lower a translucent shade.

"There, is that better?"

"Oh, yes, absolutely," Kate replied, smoothing her skirt and leaning toward the heavy glass top of the desk. She was an attractive woman in her early 30s. Her dark hair was done in a conservative style, with short bangs and ringlets next to each ear. She smiled and set her laptop on the desk to one side.

"Well, how was your trip in from Boston?" Reeves inquired, playing the small talk gambit that usually served him well.

Kate grinned. "Actually, I didn't come from Boston," she replied. "In fact, I haven't been in my office for more than two weeks. I just flew in from Beijing to LAX yesterday. Over-nighted there and caught a morning flight into Albuquerque today." She smiled. "Science happens everywhere, Stuart, so I spend a lot of time tracking down stories. Sometimes I wonder if I'll every have a chance to use all my frequent flyer miles."

"I bet you have a lot of them," Reeves said inconsequentially, straightening some papers on his desk and setting them to one side. He sat back in his executive swivel chair and smiled at her uneasily, wondering why the Department of the Navy had recommended her magazine be given confidential press access to the secret project.

"Okay, fair enough," he said at last. "You asked for an interview about our quantum computing experiment. As I told your publisher when he first called, there are security issues…"

He reached into a drawer and pulled out a document. "Here's the release I mentioned." He shoved the document across the desk and Kate reached for it. "By signing you agree that everything you learn here is for background only and that you won't publish until we give you clearance to do so." He paused as she began to read the document. "Also, and I realize this is somewhat unusual, it requires you to submit anything you write for fact checking. You agreed to these terms verbally, so this just memorializes it. There's a copy there for you."

Frowning in concentration Kate read the agreement, pausing several times to go back over a paragraph. Finally she fumbled in her Coach messenger bag, brought out a pen and signed the document with a flourish. She shoved the signed copy back to Stuart and stuffed the second copy into the bag. She pulled out a steno pad and small digital video recorder.

"I plan to take notes by hand, but if it's okay with you I'd like to record the interview. The recording is password encrypted for security and only I know the password." Seeing some hesitation, she added. "It's a real bastard, 18 letters and digits having nothing to do with my birthday or the name of my favorite movie star." Reeves smiled and nodded in agreement.

Kate set up the tiny video camera to one side, turning its LCD screen so that she could monitor the view. She pressed a button to start the recording and spoke toward the camera's microphone.

"This is an interview made on August twenty-seventh at"...she glanced at her watch... "Eleven-oh-four in the morning. We're in the office of Doctor Stuart Reeves at the Los Alamos National Laboratory. He's the deputy director of the lab and also project manager for a classified program to develop a new kind of super computer."

She turned to Reeves and smiled. "There, that takes care of the preliminaries. Now, tell me about this quantum computer—and remember, I'm a science journalist, not a scientist, so you're going to have to explain things in plain language or you'll lose me."

"I hardly think that'll be a problem," he said with a smile. "I've read some of your work, including your excellent book on alternative energy. Your description of why fusion power will always be thirty years in the future was one of the best treatments of the subject I've seen anywhere."

Kate accepted the compliment with a nod then shook her head in denial.

"You have no idea how hard it was for me to understand all of that, much less write it all out in words that non-scientists could understand. Everyone I interviewed had to repeat everything at least six times and draw pictures for me. I've got a whole file cabinet full of notes, diagrams, recordings, and even sketches on bar napkins from almost everywhere that fusion scientists hang out."

Reeves chuckled in appreciation, then fidgeted with the stack of papers before sitting back in his chair and beginning to speak.

"What we're working toward is to build the first truly practical quantum computer," he said. "I trust you've done your homework and know at least the general outline of what quantum computing is all about?"

"I have, but presume that I never heard of it," Kate prompted.

"Okay," he agreed. "First you need to know a little about what ordinary computers do. They operate with a large number of tiny relay switches, microscopic transistors. Each switch can be in only one of two positions, on or off, in the same way that a light switch can make the room light or dark."

To demonstrate he reached over to a desk lamp and switched it on, then off again.

"Each of the transistors in a computer registers information in the form of numbers. The off position equals zero, and the on position equals one. The information is recorded in the form of what we call "bits," which is short for 'binary digits.' We say they're binary because they are expressed in a base-two numerical system, instead of the base-ten system we commonly use. Computers can only recognize zeros and ones, but by combining those in binary notation they can express any value. For example, the binary numeral system goes like this..."

He reached for a writing pad and picked up a pen. Holding the pad so that Kate and her video camera could see it, he wrote: 00, 01, 10, 11, 100, 101, 111, 1000...

"You see, that is binary notation for the numbers zero through seven. A computer puts bits together to express whatever number it needs."

"But what about other values, such as letters or pictures?" Kate asked.

"Good question," Reeves responded, putting down the pad. "The answer is simple: The computer sees everything as binary numbers. For example, when you use your computer to compose an article or an email you may think you're typing letters of the alphabet, but the computer only sees numbers that correspond to the letters. As you are probably aware, your computer keyboard uses a system called ASCII, which is simply a set of numbers to identify characters."

"Okay, I think I understand that," Kate said. "You've told me about bits, but what are these 'bytes' I hear about?"

"Oh, yes," Reeves chuckled, "you do hear about bytes and they're the next step up in computer science. Bits are such tiny things that they need to be organized. If they were just floating around by themselves the computer couldn't make much of them. So the bits are grouped into batches, originally containing eight bits. A group of bits is called a byte. The word actually was coined by an IBM computer scientist back in the fifties, and he defined it as the smallest unit that a computer could 'bite' at one time."

He paused to let Kate catch up on her notes before continuing: "We usually don't deal with bytes individually, but once again by grouping them. You are of course familiar with the term 'kilobyte,' and while you may think that means one thousand bytes that is not quite right. A kilobyte is actually one thousand and twenty-four bytes. That's because eight-bit bytes don't divide evenly into 1000, which is technically what the prefix 'kilo' means.

"Same with a megabyte, which is the nearest the quantity comes to a million, but is actually a bit more. And of course these days we're working with even larger numbers of bytes—gigabytes, terabytes and even petabytes, which is a lot of bytes."

Kate scribbled in her notebook for a moment before looking up.

"So computers use these bits and bytes to process all information, no matter what kind. For example, to a computer there isn't really any difference between, say, a text document or an image file from the Hubble Space Telescope, right?"

"That's correct. All the computer sees are strings of zeros and ones, packed together in bytes. Incidentally, as computers become more powerful the standard eight-bit form is being replaced by larger bytes, for example 16-bit, 32-bit, and even encryption as high as 256 bits is coming on the scene. The more bits you have in a byte, the faster the computer can perform operations."

"Okay," Kate mused, wrinkling her brow in concentration, "so we have computers using just two numbers to do all the things they do. You know it's really hard to see how you can get so much from so little. For example, this Mac of mine can hold hundreds of picture files from my digital camera, all of the articles I write, reference materials that I've downloaded from the internet, and even digital video streams. I have to admit that seems like a lot of work for a computer that only weighs about three pounds."

"Actually, the computer itself is only a tiny processor chip on the motherboard," Reeves explained. "It only weighs a fraction of an ounce. All the rest is there just to support the processor, things like the hard drive, batteries and so forth. There's no doubt we've come a long, long way in a very short time. For example, take a look at that."

He pointed to the wall behind Kate. She turned in her chair to see an old sepia-toned photograph in a black frame.

"That's a picture of the first commercial computer ever made," he told her. "It was called Univac and the first one was installed in 1952 at the Pentagon."

Kate got up to inspect the photograph.

"Why it's a whole room full of big machines," she exclaimed. She picked up the video camera and pointed it to capture the image. "What are all those things that look like juke boxes or something, and with those movie reels on the front?"

Reeves got up and stepped over to the picture, pointing with a ballpoint pen.

"Those are tape drives that were used to record data. That was long before the days of hard drives and everything had to be recorded on analog tape, sometimes even punch cards or perforated paper tape. To operate, Univac had to run those reels back and forth to find the data it needed to process. This massive unit over here is the actual computer itself, and those workstations are the controls."

"Gosh, it makes my little Mac look pretty small," remarked Kate, looking down at her laptop.

"Small, but powerful," Reeves told her. "The Univac weighed twenty-nine thousand pounds and operated with fifty-two hundred vacuum tubes—remember this was before transistors. Its main memory consisted of just one thousand 'words' of twelve characters.

It needed a hundred and twenty five kilowatts of power to operate, enough to run about six average houses."

He pointed at her laptop.

"Your Mac is about ten thousand times faster and more powerful than Univac," he added. "And the cost difference is almost beyond belief. A Univac sold for more than a million dollars, and that was back when a dollar was actually worth something. Your Mac probably cost less than three grand, and it's ten thousand times better than Univac. Plus, you can carry it around in your briefcase and it only needs a little twelve volt power to run."

Kate looked at her laptop with new respect.

"Well, I'm sure glad I don't have to carry a twenty-nine thousand pound computer around with me," she exclaimed. "Think of the overweight fees the airlines would charge!"

"Somehow I don't think they'd let you on the plane," Reeves responded with a hearty laugh. He glanced at a clock. "It's nearly noon. Why don't we take a break and get some lunch. Do you like Greek food? There's also a little bistro that isn't bad."

"The bistro sounds great," Kate said, turning off the camera and putting it back in her bag along with the Mac. "Then we need to get to the real questions I have, about what makes quantum computing so special."

Chapter 4

Palo Alto, California
8:36 a.m. PST

Bemused, Chris Fisher set down the phone. *Is this a joke? Somebody pulling my leg?* But it sounded serious. Credible somehow. Shrugging, he decided to see what would happen.

He quickly changed out of the cycling clothes, pulling on a pair of stone washed Levis, a black t-shirt, white athletic socks and a pair of running shoes. Pocketing his keys, wallet and cell phone he stepped out the front door. His three-bedroom unit was on the top floor, two flights up from the landscaped area between the building and the street.

Chris jogged down the steps and paused to look around before slipping through the passageway that led to the parking lot in the back. He glanced left and right, then emerged and walked directly to his Wrangler. The extended 4-door version of the classic Jeep was forest green with gold pin-stripes and dark gray upholstery. He checked over the vehicle then walked the perimeter of the parking lot alert for anything that was out of place, an old habit from his military days. Seeing nothing amiss, he returned to the building and walked back through the passageway to the front. He took up a position by the open stairway, able to watch the street side of the building and also see back through the passageway to the rear.

Everything was quiet. His neighbors were either sleeping in or had already left for their day's activities. Falling into the familiar pattern of recon patrol duty, Chris slouched against the staircase, his observant eyes sweeping from left to right.

After about ten minutes he heard the rumble of a powerful motorcycle approaching on the main street. He listened as it slowed and turned into the complex, circling around to the parking lot. Curious, he walked to the end of the passageway and looked out just as the rider pulled into the loading zone.

The matte black Kawasaki Ninja had the deadly lines of a shark. Its rider wore matching leathers and helmet. Chris stepped back as the man swung the bike around in a sliding turn directly in front of him, dropped the six-speed transmission into neutral and blipped the throttle. The engine roared, then dropped to idle speed.

The rider turned his head without lifting the face-plate of his helmet. Chris could see his own distorted reflection in the insect-like mask. "Get on," the man barked, gesturing behind him. His voice was muffled. Chris flashed on an image from Star Wars. The cyclist blipped the throttle impatiently.

His days as a Navy SEAL had prepared Chris to make fast decisions. He didn't know what this was about, but figured he would find out soon enough. Taking three quick steps forward he vaulted onto the tiny passenger seat behind the rider and seized the grab bars.

Immediately the rider popped the cycle into gear and released the clutch. The bike lurched into motion, leaving behind a little cloud of exhaust and a strip of tire rubber. He swung the Kawasaki expertly in a slewing turn and ran it in second gear across the parking lot and down the entry drive. Signaling, he made a sedate left turn onto the street. Shifting up the biker cruised for several blocks well within the speed limit then made a wide right turn onto a through boulevard.

Two blocks further on he suddenly punched the accelerator. The machine surged forward as he ran it up through the gears, engine screaming. Within a few seconds they were traveling at nearly 80 miles per hour.

"Easy," Chris cried out, but with the engine roaring beneath him and wind whistling in his ears Chris knew the man couldn't hear him.

Now the cyclist was weaving in and out of traffic. Throttling down as they approached an on-ramp to the Bayshore Freeway, the biker threaded his way around several slow moving cars, passed a bus on the right and shot onto the northbound ramp, the machine's 4-cylinder engine screaming.

Alarmed, Chris held on tighter than ever as the bike continued to accelerate, dodging and weaving through the Saturday morning traffic at what must be more than a hundred miles an hour. Somewhere behind them a siren began to wail, but no police car could ever catch them, not in this traffic.

An exit ramp was coming up and the rider shifted down. He careened onto the shoulder, blasted past a van full of surfers and shot down the ramp. As he reached the bottom he suddenly slowed to a normal speed, flicked his turn signal and sedately turned west on the cross street. He continued at a calm and steady pace. *All as innocent as a little lamb,* Chris thought.

At the next intersection the biker turned right, traveled calmly about another block then coasted toward a dark maroon Lincoln Town Car that was idling in a bus zone. As they coasted to a stop next to the car the left rear door swung open.

"Get in the car," the biker advised him.

A bit cramped from clinging to the motorcycle during the wild ride, Chris stood up and stepped away. The biker immediately rode off. Bending down, Chris looked into the car. A man was sitting on the far side of the leather bench seat. He gestured Chris forward.

Instead he stood up and checked the street. Nothing seemed out of order. He looked back at the man in the Town Car, who was impatiently waving at him. Chris smiled, climbed into the seat and shut the door. The vehicle was set up for livery service with a smoked glass panel separating the front seats from the passenger area.

"Well, Mr. National Security, can we talk now?" he asked.

The stranger merely looked at him and said nothing. He leaned forward and tapped on the glass. The driver, visible only as a silhouette, put the car into gear and eased away from the curb. Like the divider, the outside windows of the big car were heavily filtered, dimming the morning California light. Chris took off his aviators and hung them by one earpiece from the neckband of his t-shirt.

"Thanks for the exciting ride," he remarked, half turning toward the stranger beside him. "I haven't had a thrill like that since I can't remember."

The man chuckled.

"I'll bet you haven't," he replied. "Maybe not since Afghanistan that time you took a Humvee down the side of a mountain and through the back door of an al-Qaeda camp."

Chris was startled. This was secret information. SEAL special ops were classified, details known only to the participants and those in their immediate chain of command. Special ops reports were redacted to replace actual names with noms de guerre and kept separate from each sailor's personnel jacket. Or at least, that's what Chris had been told.

He decided to play it cool, test the guy without giving anything away.

"Yeah," he said, affecting a bored tone. "Something like that—but I felt a lot safer in the Humvee." That brought a little chuckle from the stranger. "Actually I think the last time I felt such a thrill was the time I had to do a HALO drop into the Persian Gulf to place a limpet mine on an Iraqi gunship…"

"That's bullshit," the man stated. "You never did that. That was a SEAL named Perry and you damn well know it. He was a buddy of yours from back at Coronado. Don't try to shit me; I know all about you."

Chris sat back in the leather seat. What the man said was true. Leonard Perry had been one of his SEAL commandos. It was Perry who had done the high approach, low opening parachute drop into an Iraqi port.

They rode in silence for a few moments. At last Chris nodded and turned to his inquisitor.

"Okay, I was just testing you. Was that you on the phone, or is there another layer to this thing?"

The man relaxed and fixed his face in what might be the way a sheepish grin would appear on the face of a wolf.

"Yeah, that was me," he admitted. "We needed to get you out of there without any chance of your being followed. That's the reason for all this crap."

He held out his right hand.

"My name's Jones, Jeff Jones. Pleased to meet you, Doctor. Or should I say Lieutenant-Commander Fisher. Congratulations on your promotion. You're back in the Navy, sailor."

Oh shit! Scowling, Chris gave Jones's hand two limp shakes, sighed and turned away to look out the window. The ordinary world of California was passing by as the Town Car carried him toward some uncertain and perhaps dangerous future. Oddly, he realized it felt kind of good. Being a professor was okay—but something had been missing from his life.

Lieutenant-Commander, huh? He sighed again, leaned back against the supple leather and crossed his legs. With a faint smile he watched the mundane world drifting past beyond the smoked glass.

Chapter 5

Los Alamos, New Mexico
Saturday, 12:20 p.m. MST

Stuart Reeves leaned back in the straight wooden chair at a corner table in the trendy downtown bistro and looked across at Kate Elliott. She was demolishing a plate of fish tacos with sides of black beans and slaw. Having finished his crab cakes, Reeves was sipping a second glass of herbal iced tea.

Kate was an attractive woman, he observed, but not in the spectacular way of a model or beauty queen. Her body was feminine yet firm with the muscle of a long distance runner. Her face projected a serious, determined look that conveyed a sense of pride in her place in life.

"Yum, that was good," she said, setting down her fork and meeting Reeves's eyes. "I love fish tacos, ever since I discovered them down in Cabo San Lucas. You don't often see them on menus outside of San Diego and down the Mexican coast."

Reeves nodded in agreement. "I like them, too. When were you in Cabo?"

"Oh, I've been there several times. One summer when I was in college four of us drove a van down the Baja, camping on the beach, surfing, and pigging out on seafood." She laughed, a musical, two-toned sound. "We lived like hippies for six weeks. It was great.

Later my ex- and I spent our honeymoon there, and I was back again two years ago to cover a conference on marine biology."

She flipped back an unruly lock of hair, tucking it behind her ear. In the backlight from a window Stuart noticed that she was a natural brunette with a mix of fine light brown strands mingled with darker brown ones. The simple wavy style complemented her tanned complexion and made a nice contrast to her serious green eyes. She wore no jewelry other than the barrette and a simple silver necklace that he recognized as Zuni work.

"I gather from your pendant you've been in these parts before," he commented.

"Actually I have," Kate replied. "I've been covering an archeological dig off and on over the past couple of years. It's not too far from here, down in the San Mateo Mountains northeast of Grants. It's a recently discovered ruin, part of a pueblo that was associated with the Chaco Culture. I picked up this piece from an artist at the Zuni Pueblo south of Gallup."

She fingered the necklace, displaying the pendant with inlaid mother of pearl, jet and turquoise forming the image of a stylized bird.

"This is Thunderbird, a symbol of great power," she told him with mock seriousness. "He's the bringer of rain, or so the Zuni say. When he flies his wings make the thunder and the flash of his eyes makes the lightning."

She let the pendant fall back to her blouse. Stuart smiled encouragingly and set his glass down carefully on the tile-inlaid tabletop.

"It's beautiful," he said. "I've spent some time in Indian Country so I know how powerful their myths can be. The Zuni, the Navajo, the Hopi —all the Native tribes in fact — are closer to nature than we are. Those myths are their way of understanding their place in the world, aren't they?"

"Yes, myth is at the center of Native American life," Kate replied. She reached over to pick up the dessert menu. "Do you want to share something?" She glanced at the offerings. "They have Tiramisu."

Reeves nodded and Kate waved the waiter over. "One Tiramisu, two plates and two forks," she instructed him.

As they waited for the dessert and coffee to be served Kate fingered her necklace pensively. Then she let go of it and looked away out of the window, slightly embarrassed.

"I feel some connection with those mythic traditions," she said. "My grandmother was one-half Cherokee, so I'm part Native American myself."

"That's great," Reeves replied. "I never thought about it much, as you obviously have, but I've got some Indian blood, too, at least according to the family history. I had a great-grandfather who was pure Algonquin."

She turned back to look at him, peering sharply at his face.

"Yes, I think I can see a little bit of him in you. It's hard to see, but around the eyes, and your nose of course. Just a little, but I think I can see it."

Reeves chuckled self-consciously. "No one ever noticed that before," he said as the waiter approached. "I'm afraid I'm mostly Scots-Irish and English. Oh, here's our dessert."

In moments they were sharing the rich Italian treat. Later as the waiter poured cups of black coffee Kate reached into her messenger bag and pulled out the steno pad. She folded it back and laid it on the table next to her coffee cup, then placed her ballpoint on the pad and straightened it neatly.

"All right, here comes the hard part. What is this 'quantum computing' you're working on? You've explained about bits and bytes and told me how powerful regular computers are. What makes a quantum computer, whatever it is, so much better?"

Reeves took a sip from his coffee cup and thought for a moment. Setting down the cup he clasped his fingers together and waggled his thumbs for a moment.

"We'll get into more detail after we get back to my office," he said softly. "Remember that this is a secret project so I can't say much about it in public. But I can give you a sort of Cliff's Notes version." He chuckled as Kate picked up her pen and prepared to write.

"I've explained how an electronic computer works, by using thousands of little switches that can be either on or off, registering either a zero or a one. Now to understand quantum computing you must know a few things about quantum mechanics—the laws that apply on very small scales, at the level of the atom and beyond. At quantum scales, which are more than a million times smaller than the smallest thing that you can see, the usual laws of nature don't apply. I'm talking about gravity, momentum, things that we take for granted in the ordinary world. At the quantum level those common-sense rules just don't apply at all. What does happen seems, well, very strange to us."

He paused and took another sip from his coffee cup, then gestured to the waiter for a refill.

"When we study atoms and subatomic particles we find that they're quite remarkable things," he continued as the server topped up their cups and laid the check on the corner of the table. "For one thing, quantum particles such as atoms and photons can be in more than one place at once, that is, they can be in many different states simultaneously."

Looking up, Kate raised an eyebrow.

"You know, I remember hearing that but I find it hard to believe," she remarked.

"Well, it's not at all easy for us to visualize but it's been demonstrated to the satisfaction of science. I can show you some diagrams and formulas when we get back to the office. The classic demonstration is a simple experiment using two slits in front of a screen. When a single photon is sent through one of the slits, a diffraction pattern appears on the screen just as if photons had passed through both slits."

"Okay, that background material will help but you need to explain it a bit more to this non-physicist."

Reeves picked up a bar napkin and took out a pen. He drew two small dots on opposite sides of the napkin, marked one with an A and the other with a B and pointed to them in turn.

"Let's say that these dots represent a quantum particle, and that it appears to be in positions A and B," he said. "I've drawn them as if they were particles, but one of the strange things about quantum mechanics is that sometimes things such as electrons or photons behave like waves. In quantum mechanics we actually interpret particles as waves."

He drew a series of curved lines across the area between the two dots.

"When seen as waves, a quantum particle can not only be in two states, it can actually be in all the states in between, all at the same time."

"Um, okay. I don't really follow you yet." Kate furrowed her brow and concentrated on her notes for a moment. She circled a word and looked up again. "Can you tell me a bit more about the diffraction pattern you mentioned? What's that about?"

"Well, imagine that you're standing on the shore watching ocean waves. Some waves are smaller, some are larger, and every so often there will be a really big one. That's because there are actually different wave patterns in the water. Some may be due to a storm somewhere miles away, others from wind blowing somewhere else, and yet others due to the interaction of currents. These different patterns are all mixed up together," Reeves explained.

"Sometimes the different waves cancel each other, like when a wave peak from one pattern and a wave trough from another arrive at the same time. But sometimes they add to each other, and then you get a big wave. In quantum physics we call that 'superposition,' where different wave fronts add or subtract from one another."

He drew a series of black dots across the napkin.

"In simple terms that means that a quantum particle can not only appear to be in two states, but in every state in between A and B, simultaneously. But there's another characteristic of quantum particles, and that's that they can become entangled."

Kate jotted a few words on her steno pad then looked up again.

"Entangled? You mean, like when I get my connector cables out of my bag and they're all tangled together?"

"Well, kind of like that," Reeves said with a chuckle. "In fact, someone recently proposed a theorem that goes something like this:

Anything that can become entangled, will. It's a corollary of Murphy's Law."

They both laughed as Kate slid the check toward her and pulled out a credit card.

"Actually, that isn't quite what entanglement means," he continued. "When two quantum particles become entangled they're linked in a common state. That is, if one is spinning in a certain direction, the other is spinning in exactly the same direction. If you were to separate them by a distance they still maintain their entangled state. And here's where something strange happens.

"Let's assume that you apply a force, say a laser beam, to change the state of one of the entangled atoms. To make it spin in the opposite direction, for example. Lo and behold, even though it may be far away the other atom will also change its direction of spin. It's as if there were an invisible bond between the two, and that's the mysterious force that we call entanglement."

Kate set down her pen and stared at Reeves. "That's really weird," she exclaimed. "So, I presume that if you take the atoms far enough apart eventually they'll lose that mutual attraction?"

"No," Reeves said. "As far as we know, changes between entangled particles take place instantaneously no matter how far apart they are. We could send one of them to Alpha Centauri and the moment we changed the atom here on Earth, the other one would change too."

Kate looked at him with astonishment.

"You can't be serious,' she said after a moment. "I thought nothing could move faster than the speed of light, so how could the atom at Alpha Centauri instantly know to change its spin? It doesn't make sense."

"Not to our usual way of thinking," Reeves admitted. "Even Albert Einstein didn't understand how entanglement could work. He called it 'Spooky action at a distance,' and that's about as good a description as anyone has come up with."

The waiter returned with the bill. Kate scribbled a signature on the form, took her card and receipt and placed the charge slip inside the leather folder.

"Another thing about quantum dynamics is that no action can be predicted except by the rules of chance. Every possibility is, well, possible," Reeves said with a sheepish grin. "You've probably heard of the Uncertainty Principal?"

Kate nodded but waved to indicate he should continue.

"Werner Heisenberg developed the concept back in the 1920s. His theorem states that we can't know both the position and the momentum of a particle at the same time. The simple fact of measuring one changes the other, so even as we determine one state, we can never be certain about the other.

"For a long time even Einstein couldn't accept the quantum theory because of the Uncertainty Principal. As he put it, 'God doesn't play dice.' Even though he himself had replaced the mechanical laws of Newton with his model of relativistic space-time, Einstein couldn't accept that quantum events are purely ruled by chance."

He waited as Kate scribbled some notes. He noticed that she drew an arrow at the last entry and placed several question marks in the margin. After a moment she paused, looked back at her notes and shrugged.

"Okay," she said at last. "So quantum particles can be in more than one place at the same time; they can become linked together so that no matter how far apart they are anything that changes one will immediately affect the other; and everything is uncertain." She set down her pen and sat back in the chair with an ironic smile. "I can tell you that I really get that uncertainty part."

Reeves laughed out loud. Kate chuckled then arranged her face in a comic scowl. In a fair imitation of John Wayne she said: "You've got a lot of 'splainin' to do, Pilgrim."

"Yeah, I know," Reeves acknowledged. "But now you're speaking cowboy? Makes me wonder, when you were a little girl and you went to see those western movies, whose side were you on?"

"Ha, that's easy," she replied with a smirk. "I was cheering for both the cowboys and the Indians." She stopped short and a quizzical expression came over her face. "Hey, I guess that makes me kind of like one of those quantum thingamajigs, doesn't it? Uncertain, for sure, and able to be on both sides at once."

They were still laughing as they walked out to the parking lot and got into Reeves's Grand Cherokee for the drive back to his office. They hadn't noticed two men in a black Ford Expedition that pulled out of the parking lot behind them.

Chapter 6

Palo Alto, California
Saturday, 10:35 a.m. PST

Chris watched as the maroon Town Car glided smoothly south on Page Mill Road and entered Skyline Ridge Park. Jones had said no more after his startling announcement and Chris remained silent, contemplating what this might mean. The driver turned into a pull-off and stopped the car. He set the parking brake and stopped the engine. Jones opened the door on his side of the car and stepped out. After a brief pause Chris followed suit and stood looking at the open land and mountain views. Without a word Jones began to walk away from the car. Chris followed him. They walked together for several minutes before Jones stopped and turned.

"For a guy who just got some interesting news, you're pretty quiet," he said. "I figured you'd have a lot of questions."

"Oh, I do," Chris responded, "but I figured you'd tell me when you were ready, so there wasn't any point in making a fuss."

"Good move," Jones said approvingly. "I like a man with patience. Well, why don't you ask me some of those questions now?"

Chris looked across the landscape for a moment, collecting his thoughts.

"Okay, let's start with this: Who are you? And don't tell me that Jeff Jones is your real name."

Jones chuckled and reached inside his jacket to pull out a leather case. He flipped it open to display a gold badge and oversized identification card.

"Says here I am," he stated. Chris leaned forward to read the information on the card. At the top was the shield and eagle symbol bearing the words "Central Intelligence Agency." It identified the bearer as "Jeff Jones, Senior Case Officer" and showed an address in Virginia.

Chris stepped back and looked Jones in the eye.

"Now I know you're a fake," he said. "The Company doesn't issue ID badges and cards to its staff—they always use fronts, claiming to be with the FBI or military intelligence, never CIA. Now tell me who you really are."

Jones smiled, flipped the case shut and put it back into his inside coat pocket. As he did so his jacket gaped slightly and Chris caught a glimpse of a compact pistol in a concealed holster under the man's left armpit. It looked like a Beretta 9mm.

"You're right, and I'm surprised you knew that," Jones admitted. "That phony spook ID impresses most people. But the CIA isn't the only outfit to engage in a bit of misinformation. Who do you think I am, in reality?"

Chris looked him up and down, checked the cut of his suit, observed the well-shined black brogues.

"First, I'd say you're military. I say that from the way you hold yourself, the tailoring of your clothes, the way your shoes are polished. I'd bet your suit was made by a military tailor, with special fitting to cover that Beretta."

Jones smiled and patted his armpit. "Go on," he said.

Chris walked around the man, thinking.

"Now listening to the way you talk, you sound like you grew up somewhere out East, but not Boston. Connecticut, maybe, or Long Island. You probably attended an Ivy League university where I'd guess you majored in political science or philosophy, maybe economics. Your sports were either wrestling or the rowing team."

Jones looked impressed.

"I crewed at Yale," he admitted. "What gave you that?"

"The shoulders," Chris told him. "Those skinny legs and broad shoulders. So, Yale eh? Did you make it into Skull and Bones?"

"What makes you ask that?" Jones sounded curious.

"You're the type," Chris said without further explanation. "And there's something else, a sort of faraway look in your eyes that I've seen before in senior Navy officers. Comes from spending too much time at sea, I think. If I had to guess, I'd bet you're with Naval Intelligence. Either that, or you're a damned good fake."

Jones broke out laughing then turned to look across the hills to the sparkling water of San Francisco Bay on the far horizon. Unconsciously he clasped his hands in the small of his back in the stance of an officer on the deck of a ship.

"You're as good as they said you were," he said without turning around. "I'm a career Navy man, served at sea for a dozen years before I got into the secrets business." He paused for a moment before turning back. "Know much about us sailor spooks?"

"A little," Chris replied. "We worked with some of your guys in Afghanistan. As I recall the headquarters is somewhere in Maryland.

"Yes, the Office of Naval Intelligence is based in Suitland," Jones confirmed. "What more can you tell me about myself?"

"That's it, except that your name isn't Jones unless you're Davy's great grandson," Chris told him. "I understand that using common names is considered safe cover, but when spies call themselves Jones, Smith and so forth it always seemed a little obvious to me."

"Hmm, you have a point there," Jones said. "What if I told you that I have more than a half dozen different names, with ID to go with them. You just met my CIA legend. Would you like to meet the FBI version of me, or the Secret Service one? I can be an ordinary businessman, too, when it suits my needs. I even played a legend as a bum once, hanging around the Reflecting Pool in Washington, just across from the White House. That wasn't my favorite assignment."

"A man of many faces," Chris mocked. He gave the man another look, taking in his short haircut, jug ears, and piercing blue eyes. His neck was thick, the shirt collar snug with a small roll of flesh protruding above the knotted rep tie. Jones was about six feet tall and weighed a little over 200 pounds.

"All right, enough about you," he said. "What's the idea of kidnapping me, trying to press-gang me back into the Navy? I served my time and I've got a brand-new career now, as you very well know. I've got responsibilities, classes to teach, a project at SLAC."

The Stanford Linear Accelerator Center is one of the world's oldest research laboratories devoted to high-energy particle research. It was there that Chris was working as part of a team of physicists and engineers to develop the Linac Coherent Light Source, a two-mile long linear accelerator designed to produce high-energy X-ray beams that could create clear images of molecules and even individual atoms.

"I can't just walk in and tell them I've decided to run away and join the Navy." Chris held out his hands palms up and raised his eyebrows.

"That isn't going to be a problem," Jones told him. "This afternoon you'll call your department head and request an immediate leave of absence. You'll do the same with the project manager at SLAC. I have their private numbers here." Jones opened a notebook, tore out a page and handed it to Chris.

"Yeah, sure," Chris said, taking the sheet with an ironic grimace. "Look, even though I have tenure, I can't just walk away on a moment's notice."

"Yes you can," Jones replied. "Do you know what they're going to say? They're going to say, 'How long do you need?' and you're going to reply, 'At least three months but possibly more,' and they're going to say 'Okay, just let us know when we can expect you back.'

"And, oh yes, they're going to make sure no one else is within hearing distance of their phone conversation, congratulate you on your important special assignment from the Department of the Navy, and wish you the civilian version of Godspeed and fair winds," Jones concluded.

"Oh, they are huh?" Chris responded. "These must be a couple of different guys from the ones I know."

"Not at all, not at all," Jones assured him. "You see, they've received personal and very confidential phone calls from someone in Washington, someone very high up in Washington, who has told them that your country needs to borrow you for a while. I assure you, there will be no problem at all. In fact, this will actually enhance your reputation.

"We've even arranged for you to continue to draw full pay and benefits from Stanford during your leave of absence," Jones continued, "and you'll be glad to know that you'll also draw your Navy pay. At your new oh-four rating it amounts to about five grand a month, and we're even adding hazardous duty pay to the package."

"Ah, that's what I was afraid of," Chris said with a sour expression. "Back in my day, you had to volunteer for hazardous duty instead of being kidnapped."

"Who said anything about it actually being hazardous," Jones retorted. "I just mentioned that we added the extra pay. It's a gift to you from your government."

"Oh, yes, that same government that staffs the Internal Revenue Service with the meanest bunch of sharks outside of the Pacific Ocean to go around giving gifts to we lucky citizens." Chris kicked at a clod of dirt and looked up at the sky. "All right, I suppose I'm volunteering, just mark it down that it was involuntary. Now what have I just involuntarily volunteered for?"

"No, no, nothing involuntary about it. Just hear me out and I can assure you that you're going to be happy to accept this assignment."

"All right, let's hear it—and it better be good."

"Oh, it is. You're probably wondering 'why me?' and now I'm going to tell you. We need someone to represent the Navy in a sensitive matter and the ideal candidate would be a former active Navy officer who also happens to be a particle physicist. How many of those do you think there are in this vast world of ours?"

"Um, I don't know. Five? Six?"

"Well, you're probably right, but one of those is a veteran of World War Two, two more served during the Korean affair, and two others were in Vietnam. We need someone a bit younger, someone exactly like you."

"So you're saying you want me for my knowledge of particle physics, not my SEAL training?"

"Now you've got it. You're just the man we need because of your Navy rating as well as your academic position. We're not going to send you into combat!" Jones chuckled and scratched his right earlobe. "In fact, I think you're going to enjoy this."

"I hope so," Chris told him earnestly. "So what are my orders?"

Jones began to wander across the grass, moving further away from the Town Car. Chris glanced back. The driver was leaning against the hood smoking a cigarette. He looked attentive, his eyes switching up and down the nearby roadway.

They approached a fallen tree and Jones sat down on the horizontal trunk. He patted the log to his left and Chris eased himself down beside the agent. They were near the top of a ridge and the view was expansive. Chris could see across Sunnyvale and Santa Clara to the mountains beyond.

"There's a secret project underway at the Los Alamos lab in New Mexico," Jones began. "They're attempting to build and operate the first practical quantum computer. I think you know what such a computer could do."

"It would revolutionize computer science," Chris replied. "A truly functional quantum computer would be ten thousand times more powerful than our fastest super computers."

"Yes, at least. This could be a very important project. And if you're wondering what it has to do with the Navy, we've funded the lion's share of the project, directly and through some fronts. It's very hush-hush."

"I think I heard some rumors floating around at a conference at CERN in Geneva last year. There was some hallway gossip. I didn't pay much attention to it. Most of us believe that advances in quantum computing are nowhere near the point of anything but small proof-of-concept machines. So this is the real thing?"

"It could be," Jones affirmed. "You never know until the eggs actually hatch. But the project is the brainchild of Tommy Dawson and he's put together a crack team out there. All indications are that they may be on the right track."

"Ah, then it is serious," Chris exclaimed. "Dawson doesn't make many mistakes. I worked with him for a while as a graduate student at Cal Tech, before he moved to the New Mexico Institute."

"Yes, we're aware of that," Jones pointed out. "That's another reason why you're the right man for this assignment. The Navy, well we sailor spooks, want to have a pair of eyes and ears on the ground there at Los Alamos, someone who knows what to look for, the right questions to ask, that sort of thing. That's the assignment. So now what do you say? Wanna volunteer?"

"Sure," Chris said without hesitation. "Sounds really interesting, a new challenge. But I have to wonder why one of those older men wouldn't do just as well. Hell, old man Nichols down at Cal Tech commanded a destroyer during Vietnam and made full captain in the reserves. He's forgotten more about particle physics than I ever knew."

"Oh I doubt that, but remember that Professor Nichols is 69 years old and has pretty nasty arthritis in his legs. He's not up to very much activity these days."

"Yeah, that's true," Chris admitted. "As you probably know he was the advisor on my doctorate. Last time I saw him he was leaning on a cane. Okay, so I guess I am the natural guy to pick. But I have one more question, and that's why is this assignment coming from the Intelligence branch. Why not the Naval Research Lab? That would make more sense."

Jones sighed and shifted his butt on the fallen tree trunk, trying to find a more comfortable position. Failing, he stood up and stretched, raising his arms above his head and yawning.

"Well, now that you mention it," he said, avoiding Chris's eyes, "there is a bit more to it than just being a watcher. It seems we're not the only ones with interest in this quantum computing stuff. We've picked up some signals chatter about it. From Iran, to be specific. The Ayatollahs would love to get their hands on this. Anyway, we think the project is in danger of being infiltrated, perhaps just to spy but maybe with intent to sabotage."

"Oh, now I get it," Chris exclaimed. "You want me to be a Navy spook in physicists clothing, right?"

Jones regarded him for a moment then shrugged and curled his lips in a wry smile.

"Yeah," he said. "You got it correcto-mundo, bang-on first time. We're gonna turn you into an amateur sailor spook. Just for a little while."

"Um, so I see," mused Chris. "But why all the secrecy today, the high-speed extraction and our meeting here? Am I already being watched or what?"

Jones chuckled and waved the question away with an airy gesture. "Nah, that was just for effect. I wanted to make sure you'd take me seriously. And you did. We spooks are into psychology don'tcha know? And that reminds me, you were wrong when you tried to guess my university major. I got my Master's in social psychology. Handy stuff to know, real handy."

Chapter 7

Los Alamos, New Mexico
Saturday, 1:25 p.m. MST

Stuart Reeves and Kate Elliott were back in his corner office. While Kate set up her mini video camera to record the conclusion of her interview Reeves dug into a file drawer and pulled out some documents.

"I don't have a lot more time," he said, glancing at a clock on the wall near his photograph of Univac, the 1950s computer. "Here's some background material that will be of help to you. I'll have my secretary make you some copies." He picked up the phone and spoke briefly. In a moment a middle-aged woman wearing a tan pants suit and comfortable looking shoes came in, smiled at Kate, took the papers and left.

"All right," Kate said as the door closed. "You've given me a little background on how standard computers work, and some idea about the strangeness of quantum mechanics. Now how can all that be put together to create a quantum computer, and what are the advantages of doing so?"

Reeves thought for a moment then picked up a paperweight from his desk. It was a lump of petrified wood that had been ground to a two-inch sphere and polished to bring out the colorful bands of agate. He held the paperweight up for Kate's inspection.

"Imagine this is an quantum particle," he told her. "In fact, this stone actually contains billions if not trillions of atoms. That's how small they are. But let me use this to demonstrate my points.

"If this were an actual subatomic particle it could be in different states simultaneously, thanks to the Uncertainty Principal."

"Okay, I've got that, although I don't really understand it."

"Well you're not alone. I don't think anyone can grasp quantum effects in any rational way. What we're talking about can really only be expressed in the language of mathematics."

"Right. Perhaps before you go on you should tell me more about this term 'quantum' and why it's important."

"Oh, good point. The word itself is from the Latin. As you know, a simplified representation of an atom looks like a little solar system, with a center core made up of protons and neutrons and with little electrons spinning around it. You probably remember from your chemistry classes that electrons remain in certain specific distances from the nucleus, that is, the center of the atom. Those orbits are called 'shells'. If an electron is excited, that is kicked to a higher energy level, it can jump from a lower shell to a higher one."

"Yes, I do recall that," Kate said. She picked up the paperweight and turned it in her hands.

"Well only a certain force will cause an electron to jump, a very specific wavelength of energy. That is, the electron can only respond to packets of energy in certain precise amounts, depending on the starting state of the electron and other factors. That amount of force that will act on a particle is described by the word 'quantum," which simply means 'a certain amount,' as in the word 'quantity'.

Kate scribbled in her steno pad for a moment, and Stuart noticed her make a quick sketch of an atom. She drew an arrow to mark an electron and labeled the arrow "quantum force." She looked up and signaled him to proceed.

"Okay, now we know where the 'quantum' comes from, now let's see how that can be used to do computing. It's been learned that under certain conditions we can control the state of atoms by beaming laser light at them. If the light's at the correct frequency, that is, the correct quantum wavelength, we can make the atom change its state by jumping an electron to another level. That amounts to the same thing as flipping a transistor switch in an electronic computer. You can say that the atom is registering a zero, and when it flips to the other state it's registering a one. You've created a bit."

"Ah," Kate breathed, "I think I begin to see where you're going. But that seems like an awfully hard way to do computing. Aren't atoms really hard to control?"

"Yes, they are, but we're learning their secrets," Reeves replied. "And the reason why quantum computing is so appealing is that atoms actually don't create bits, they create 'qubits'—quantum bits. And qubits are quite versatile things. You see, because of the Uncertainty Principal an atom can be both a one and a zero at the same time."

Kate looked skeptical. "That just makes no sense at all," she remarked. "Wouldn't it cancel itself out?"

"No, not at all," the engineer said. "In fact, what it means is that in a quantum computer a single atom can perform more than one calculation at the same time. Remember that in an electronic computer the bits can only be in one state at a time, on or off. In a quantum computer things can happen simultaneously at every level."

He reached over the desk to pick up the paperweight again.

"Then there's the fact that an atom can change in other ways, too. Besides changing its energy state we can make it spin clockwise, counterclockwise, or in every direction possible in space-time, all at once. And, we can entangle it with other atoms. That means that when you change the state of an entangled atom, you change the state of other atoms with which it's entangled. It's all very technical but the end result is that a quantum computer can operate many times faster than ordinary computers. Thousands of times faster."

He glanced at the clock again.

"And speaking of time, I'm sorry but I need to get back to the center. As you know we're planning our first test run in just a few days so we're working around the clock. As I promised when we first spoke on the phone I've requested a security clearance for you to observe our startup. When you have the clearance I'll show you the actual hardware we've built and it will all be a little easier to understand."

"Yes, that's great. I'm looking forward to it. Meanwhile, I've got some reading up to do to understand what you've told me." Kate folded her steno pad, turned off the video camera and began stuffing everything into her messenger bag.

"I guess I see that a quantum computer would be pretty powerful for high-end scientific computations. Things like high-resolution meteorology models come to mind. But it wouldn't mean much for most ordinary situations would it?"

"Oh, I think you're wrong there," Reeves replied. "We don't really know how powerful our system will be, but there are some people who are worried about what quantum computing would mean. For example, the products of large prime numbers are used as cryptography keys to protect financial data. It works because it would take a conventional computer a long time to break the code. I'm talking about years, even centuries to crack a sufficiently large cryptography key. A fully functional quantum computer could break the code much quicker."

"Oh, I see," Kate said with a note of surprise in her voice. "You mean it might be able to do in days what could take a regular computer years?"

"Possibly just hours or even minutes."

"Oh, that would be something important, wouldn't it?"

"Perhaps. We really won't know until we find out whether our experiment is a success. It's all still a bit theoretical right now, but we think it could be important."

He stood up and walked around the desk to open the door. Just then his secretary appeared and handed him a stack of documents.

"Ah, here are your copies," he said, passing them to Kate. "Get in touch with me in a couple of days to confirm your security clearance. Are you planning to stay here in Los Alamos until the experiment starts?"

"No, but I'll be in the area," Kate told him, stuffing the papers into her already bulging bag. "I plan to drive down to the archeological site I was telling you about and interview some of the researchers about their progress. They're about to shut down the dig for the season and I want to see what new discoveries they've made. You have my cell number, so give me a call whenever you have news."

"Great," Reeves replied. "Have fun and we'll see you in a few days. My secretary will escort you back to your car." He turned back into his office with a wave, then stopped and turned back. "I enjoyed meeting you Kate, and I'm pleased to know that we share that little common bond of Native American heritage. I hope we have a chance to talk about that some more."

Kate smiled and fingered her necklace, holding up the Zuni pendant.

"Thunderbird and I look forward to that," she said.

Stuart smiled and shook her hand. "I'll see you both in a few days," he told her with a grin.

Chapter 8

Durango, Colorado
Sunday, 12:10 p.m. MST

Chris Fisher had packed and hit the road soon after the man who called himself Jeff Jones dropped him back at his apartment. He had been informed that further instructions would await him in Los Alamos.

He'd driven most of the night, pausing only for a few hours sleep at a motel in eastern Nevada. Now he was making a quick lunch stop. After wolfing down a hot beef sandwich and apple pie with ice cream he walked back to his Jeep and checked the road atlas. Another three hours to Los Alamos. Before continuing his drive he stretched and looked around the spectacular mountain scenery. On the high slopes the aspens were beginning to show the first hint of fall color, their leaves turning from pale green to vivid yellow. Smiling, Chris climbed back in the Jeep for the final leg of his trip.

Jones had given him only a quick briefing and Chris had hardly had time to ponder even those sketchy details of his assignment. The hard drive across the Mojave Desert and this morning's push through Slick Rock Country had kept his mind occupied. Now, with most of the trip behind him, he began to think about what he had learned.

Chris knew something about the theories of quantum computing and was aware that simple demonstrations had proved the general concept. But the idea of a full-fledged quantum computer, if what Jones said was right, opened an entire world of possibilities.

He passed through the towns of Aztec and Bloomfield, New Mexico and entered a long, straight stretch of U.S. 550, a divided four-lane highway leading southeast. He was making excellent time now, pushing the Jeep five over the 70 mph speed limit. The whine of the off-road tires on pavement and the rush of air provided a white noise background to his thoughts.

According to Jones, Chris would be acting completely in the open in his putative role as the Navy's observer to the project. No clever disguise, no false identification papers. That was a relief. But the second, less visible part of his assignment made him wonder what was in store. Jones had only hinted at the suspected spying activity that might be taking place, or what he was expected to do about it.

As a Navy SEAL in Afghanistan Chris had dealt with a few CIA case officers but knew little about the black world of spycraft. He had read a lot of thrillers and was a special fan of Tom Clancy's books but it was his impression that the actuality was far less exciting than what was portrayed in fiction. In fact, he decided, the assignment would probably be fairly routine. He could think of this as a well-deserved vacation from the years of hard work invested in earning his doctorate and tenured post at Stanford.

No sweat, he thought as he turned onto highway 126 just south of Cuba and headed East into the Santa Fe National Forest.

About an hour later Chris pulled up in front of the motel where he had a reservation. It turned out to be an upscale bed-and-breakfast with individual outside entrances for each room and plenty of parking. The reception area had the ambience of a fine home. Off to one side was a comfortable dining room with tables already set with crystal and silver for the next morning's breakfast service.

After checking in Chris hefted his suit carryall and duffel up a short flight of stairs from the parking lot and opened the door to a comfortable mini-suite. He hung the carryall in the closet, dumped the duffel on a chair and grabbed a cold bottle of spring water from the fridge. Popping the top he took a long drink, then walked to the window and opened the drapes, what he and his SEALs used to call "checking their six". He saw only a bare lawn with a lonely gazebo. Further away were the stony peaks of nearby mountains, scarred by recent forest fires. Far to the Northeast he could see the gleam of snow on higher peaks. Above it all was an incredibly blue sky punctuated by a few white clouds.

Finishing the water Chris stripped and stepped into the shower. Five minutes of hot water scrubbing was followed by a two-minute cold rinse. Shivering, he toweled off then rummaged in his duffel for a razor.

Wearing a pair of Dockers and a knit golf shirt he lounged barefooted on the bed, hands clasped behind his head. Taking several deep breaths he stared at the ceiling. It was Sunday afternoon, quiet time after a long drive. Relaxing, he wondered what would happen next.

Ten minutes later the phone rang. Chris grabbed the handset and was informed that a package had been delivered for him at the front desk. *Aha! Time to go to work.*

Slipping his feet into a pair of loafers he jogged down to the office and was handed a large brown envelope. In bold black ink it was marked "Lieutenant-Commander Christopher Fisher". There was no return address and no stamps or courier documents to indicate how it had been delivered. When he inquired the desk clerk shrugged and told him the envelope had appeared on the front desk while he was on a rest break. No, he hadn't seen who dropped it off.

Another mystery, Chris thought. *Is it just coincidence, or am I under surveillance?*

He thanked the clerk and walked back to his room, weighing the packet in his hand. It was an ordinary nine-by-twelve brown envelope sealed with both a glued flap and bent metal prongs. A standard office supply item.

He closed and locked the door, found his combat knife in the bottom of his duffel and slit open the envelope. He pulled out a folder containing documents. As he did so something fell out on the floor with a little tinkle. *What the heck?* He stooped down to retrieve it. It was a key attached to a cheap plastic key fob marked with the number 42. He shrugged and slipped it into his pocket.

Sitting down at the little desk he turned on and adjusted the compact anglepoise lamp. He spread the folder open and prepared to examine the contents.

On top was a hand-written note on a full-sized piece of plain white paper. It had neither a salutation nor a signature and its succinct message was: "Please read the enclosed carefully. All will become clear."

Chris turned the note over and set it aside. The first document contained his official orders on a letterhead of the Office of Naval Research, Arlington, Virginia, signed by a rear admiral. In sparse numbered paragraphs it described Chris's assignment at the Los Alamos National Laboratory. He set it aside.

The second document was a letter of introduction on the same letterhead addressed to Paul Bonner, the LANL director of security. It confirmed that Chris was an official representative of the Navy and requested the cooperation of the Los Alamos staff during his assignment there. It was signed by the same rear admiral.

Stapled to the letter was a certificate confirming Chris's Top Secret clearance from the Department of Homeland Security. That gave him access to most classified material on a need-to-know basis. He had been given the TS status since beginning work on a classified project at SLAC several years before.

Next was a customized map of the area, with several sites clearly marked. The LANL security office was highlighted in yellow. So were two buildings marked as "Quantum Center" and "Qubits Farm". Chris surmised these were the locations of the quantum computer project. Also marked were the inn where he was staying and several other locations including a nearby restaurant.

Setting aside these documents Chris discovered a multi-page report bearing a bold red stamp at the top. The imprint read: "Classified/Top Secret." The subject line was: "Internal Security Problems, Quantum Computing Project." He set it aside to read later and examined the last document. It was a handwritten invitation to dinner at a Mexican restaurant, the one marked on his map. He was asked to arrive at 7:30 p.m. and to give his name to the hostess. The note did not say whom Chris was to meet. *Another mystery.* He set the invitation aside and stood up to stretch. Then he picked up the classified report, settled in a nearby easy chair, propped up the footrest and began to read.

Chris approached the rendezvous a little before 7:30. At first he drove past, eyeballing the parking lot. He turned left at the next street and proceeded along an alley, discovering more parking behind the restaurant. He reversed into a spot toward the back next to a line of bushes. This area seemed to be reserved for staff and overflow parking with main customer parking in front.

He shut off the engine, set the parking brake and sat quietly to scope out the area, a habit learned in the uncertain war in the Afghan mountains. After a moment he picked up a thin leather case from the seat beside him and checked its contents. It held the documents he'd received that afternoon as well as a legal pad, micro recorder and compact digital camera.

Chris stepped out of the Jeep and locked the door. He was dressed comfortably in a blue oxford cloth shirt, open at the collar, a soft camel hair jacket over brown slacks, and tan leather Wellington boots. Tucking the portfolio folder under his left elbow he paused to take another look around. He could see smoke rising from a kitchen chimney and the pungent odor of green chilies was in the air. A busboy emerged from a rear door carrying a pail. He turned its contents into a dumpster and went back into the building without glancing up.

Chris strolled across the parking lot and around the side of the building. There were about 20 vehicles parked in front. He noted a security vehicle, a black Durango SUV with low-profile light bar. There were several vehicles displaying LANL stickers, but that was nothing unusual in this town.

He stepped up to the door and bent forward to peek through the diamond shaped glass panels. The walls of the foyer were decorated with bullfight posters, several brightly colored serapes and a couple of oversized Mexican straw hats. A reception lectern was at the left side next to the entrance to the dining room.

Chris stepped inside and waited for a moment. Presently a middle-aged woman dressed in a long peasant-style dress stepped out from the dining room. "Dr. Fisher?" she inquired with a broad smile.

A bit surprised, he nodded.

"Right this way. Your guests are waiting," the woman said, leading him across the dining room and opening a door to reveal a small private room. She waved him inside and closed the door behind him.

Two men sat at the large rectangular table in the middle of the room. Chris had never seen either of them. They looked up at him expectantly. A florid-faced man got to his feet and held out his hand.

"Dr. Fisher, welcome," he said in a raspy voice. "Thank you for coming. I'm Paul Bonner."

"Ah, the lab's security director. Pleased to meet you." Chris shook his hand.

Bonner turned to the other man, also now standing.

"This is Stuart Reeves, our deputy director. Chris stepped forward to shake hands. The two nodded in recognition of mutual respect.

"Well, hello," Chris said, a bit uneasily. "I'm glad we could get together. Mr. Bonner, I have some documents for you." He opened his leather portfolio case and handed over the letter of introduction and security clearance. "I understand you'll issue me a pass."

Bonner nodded curtly, examined the papers briefly and handed them back.

"Call me Paul," he said. "Present your credentials at the main gate first thing tomorrow and they'll issue your ID. Meantime, Stuart and I thought it a good idea to get to know you. Welcome to Los Alamos Lieutenant-Commander. Would you like to join us in a drink?"

"You can call me Chris, and yes that would be nice. A bottle of Corona please." As Bonner left to get the drinks Chris turned to

Reeves. "I recognize your name. You're heading up the work I'm here to observe, isn't that right?"

"Yes, I'm doing double-duty as the project manager. We can speak openly about it here. Paul's security staff swept this room for listening devices less than an hour ago."

Chris considered this as Paul Bonner returned holding two bottles of Corona longnecks in one meaty hand and a glass of red wine in the other. Each beer bottle had the traditional wedge of lime stuck in the top. Bonner handed the wine to Stuart and set down the beers. Chris picked one up, tipped it in salute, removed the lime and took a sip while examining the security officer. Bonner picked up his own beer and returned the salute.

The security chief was a husky man with a size 19 neck and broad shoulders. His graying hair was cut short in the style favored by police and military men. He wore a western-style gabardine suit in a light fawn color. A two-inch wide leather belt with a brass rodeo buckle encircled his waist, complemented by highly polished calfskin roper boots. Around his neck was a bolo tie with a turquoise slide in the form of the Sacred Bear, a traditional Southwest Native American symbol. Chris noticed a Stetson on a nearby hat rack that was surely part of Bonner's outfit.

"I was just telling Chris that we can speak freely here," Reeves said. "As you know he's here officially only as an observer, but he has a second purpose which is to help us with security issues." Shifting his eyes back to Chris he continued, "Only Paul and I know that about you. Even Doctor Dawson is unaware. I understand you've worked with Tommy?"

"Yes, at Cal-Tech," Chris responded. "I was one of his graduate assistants for a while before he came here to the New Mexico Institute. He's a brilliant man."

"Yes, he truly is, and the work he's doing now will be his crowning achievement. I understand you've been told something about our project?"

"Yes, but only enough to make me very curious. Are you actually on the verge of creating a fully functional quantum computer?"

"Yeah that's exactly what we're doing." Reeves grinned and raised both hands with crossed index and middle fingers. "At least, that's what we hope we're doing. We may know in a couple of days when we make the first trial run. But in the meantime, Paul here tells us that we need to be aware that some outsiders are interested in what we're doing."

"So I understand. I read the briefing document this afternoon. It was helpful but not very specific. Frankly I'm not sure I understand why there's so much concern. I could see the need for strict security back when the atomic bomb was being invented here, but why all the interest in this new kind of computer?"

Elbows on the table, Reeves held his wine glass poised in both hands, like a priest about to offer the sacrament. He thought for a moment before replying.

"As you'll learn, quantum computing might be a world-changing technology. It wouldn't just be a better computer, but light years ahead of present technology. It could perform calculations that are completely impossible now, or that would take even the fastest super computer years or even centuries to perform. Quantum computing has the potential to be to standard computers as a jumbo jet is to a box kite."

He took a sip of wine and Paul Bonner picked up the thread.

"If the secrets of quantum computing should fall into the wrong hands, it could give other countries advantages that we don't want them to have. I'm talking about nations like Iran and North Korea, and even the Chinese. Quantum computers could break codes, design nuclear weapons, and perform all kinds of tricks that could be put to use against us."

"Okay, I can see that," Chris murmured.

"That's why we're worried there may be spies attempting to infiltrate our project," Reeves pointed out. "It's particularly difficult because this project has brought nearly a hundred people together from all over the country and even a couple from friendly nations. We just can't be sure they're all on the up and up."

"I hate to admit it, but our own security organization has serious limitations," Bonner interjected, "Most of my staff are just glorified security guards. We can call on the FBI as needed, but even the smartest federal agents stand out like sore thumbs among the rocket scientists and super geniuses that are working on this. About ten seconds after a federal agent opens his mouth he might as well be wearing a big sign with flashing lights."

"Yes, I can see that would be a problem," Chris said. "So you think I can be of help because I can poke around and ask intelligent questions in my role as an official observer."

"You got it," Bonner said. "We need someone like you to mix and mingle, act curious, and keep your antennae up for anything that smells wrong. That's all, just be a smart pair of eyes and ears and report anything suspicious to me. Stuart here will introduce you around and help you get started."

"Okay, that's pretty much what I was told to expect," Chris responded. "It sounds like an interesting project. I've read the background report and I'm looking forward to getting started. When does the lab open tomorrow?"

Reeves emitted a short laugh. "We've been working twenty-four seven for the past six weeks," he said. "In shifts, of course. I'll be there at six tomorrow. Why don't you plan to meet me then? I'll be pretty busy, but you can learn a lot by tagging along."

Chris smiled. "I think the way Yogi Berra put it was, 'you can see a lot just by observing.' I'll keep my eyes open."

Reeves chuckled and glanced at Bonner. "Can you get his papers ready and have someone brief him on security measures if he drops by your office about 5:30?" Bonner nodded. "Okay, so that's set. Let's order some food and we can talk some more while we eat. They have great enchiladas here, and the chilies relleno are to die for, as my 12-year-old daughter would say."

Chapter 9

Los Alamos, New Mexico
Monday, 6:45 p.m. MST

Doctor Harvey Pearson lived a quiet life. His neighbors thought of him as a pleasant man who sometimes stopped to chat when they encountered him on the street or at the nearby market. Yet no one really knew him. He seemed to value his privacy far more than most. He lived in a modest two-bedroom house near the edge of Los Alamos. Known for his frugality, he drove a nine-year-old Volvo with a dent in the left rear door and a faded Kerry/Edwards sticker on the rear bumper.

Pearson was a nuclear physicist employed in the high explosive research lab at LANL. He had been on the staff for more than a decade and although no standout his work had always been deemed adequate.

He was an average sort of man, which helped to explain why he attracted so little attention. About five foot nine and weighing no more than 165 pounds he often wore jeans and penny loafers to work. As far as anyone knew he had no social life to mention and was never seen in public with others. When invited to parties, cookouts or private dinners he always had an excuse not to attend. Consequently, for several years no one had bothered to extend an invitation to him. On rare occasions he'd been seen at a performance of the Santa Fe Opera or dining alone in a restaurant. No one took much notice and he was pretty much left alone.

Pearson had regular features, light brown hair, a narrow nose and wide-spaced blue eyes. His complexion hinted at Mediterranean roots. His lean body carried more muscle than his modest weight might suggest. He kept that condition by performing a daily thirty-minute regimen of exercises each morning before showering and dressing.

His security file said that he was the son of Jonathan and Eleanor Pearson, late of Seattle. According to the records his parents had died in an automobile accident when Harvey was an 18-year-old freshman at M.I.T. His father had been a successful investor, or so the records said, and as their only child Harvey inherited a sizeable estate bolstered by a million dollar insurance policy that had paid double indemnity for accidental death.

His file recorded that he had graduated from M.I.T. with top honors, stayed on to earn a Master's in nuclear physics then completed his Ph.D. at the Cavendish Laboratory, Cambridge University, England. He joined the LANL staff as a research associate at the age of 28 and had advanced to assistant project manager. He was now 39 years old.

Almost all of that was false. His entire history up until the time at Cambridge was a fabrication. It was only then that "Harvey Pearson" first appeared. The true history was quite different.

His birth name was Sanjar Jalaly. He was born not in Seattle but in the Iranian holy city of Qom. His father, still very much alive, was a highly-placed Shiite mullah, a member of the top echelon in the theocracy that ruled Iran. A fanatic, the elder Jalaly had dedicated the life of his son as a holy warrior dedicated to the destruction of the West.

From his first years as a toddler Sanjar had been exposed to American language and customs. He was cared for by a series of American nannies. Prior to the fall of the Shah, American tutors saw to his education.

For nearly a dozen years Jalaly had led a double life at LANL. To his fellow scientists and around his neighborhood he projected the image of an ordinary middle class American. But behind the façade of privacy he was an expert spy devoted to his nation and Islam. For years Sanjar Jalaly had passed vital contributions to Iran's effort to obtain nuclear bombs, weapons to bring the Great Satan and the evil Zionists to their knees.

This day he was worried. He had received a signal calling for a face-to-face meeting with his handler, a spymaster through whom he passed stolen secrets. They had seldom met, and not since more than a year earlier. Their usual contact was through clandestine drops, carefully camouflaged by innocent covering activities. Most recently had been a handoff to an intermediary in the busy parking lot at the conclusion of a performance of the Santa Fe Opera.

The call for a direct meeting was most unusual. The danger of discovery was too great to discount. The signal had come as an innocuous want ad in the *Santa Fe New Mexican*. It offered for sale certain ordinary items, using certain ordinary words that by prearrangement had special meanings. The simple code instructed him of the time and place to meet.

It was nearly seven p.m. Jalaly stepped into the small basement room of his house, equipped as a home office. On the back wall was a floor-to-ceiling bookcase filled with technical manuals, textbooks and reports. He had no taste for fiction or history and his personal library reflected the dedication of a hard-working physicist. Searchers would find no hint of his Muslim roots, and in fact a faked Pearson family Bible held a prominent place on the shelf.

He lifted the Bible to release a hidden pressure switch while activating a concealed button at the side of the bookcase. One section swung back to reveal a low tunnel. He stepped inside and the hidden door swung closed behind him. Jalaly entered the world of his third life.

Hunching over he duck walked down the four-foot-high tunnel and entered a small underground room. Here he paused. On the floor was an antique prayer rug, a fine hand-woven Persian piece worth several thousand dollars. It was carefully aligned to face Mecca. For a few moments he waited, checking the time for *Salaah*, evening prayers. When the moment came Jalaly knelt in submission and began the Muslim prayer in Arabic, the language of the Koran.

"Allahu Akbar...," he intoned in Arabic. *In the name of Allah, The Most Compassionate, the Most Merciful. All praise belongs to Allah...*

Finishing his prayers he began to strip off his clothes. He was about to become a different person.

Back-to-back behind Jalaly's house stood a similar one facing the next street. According to tax records the neighboring house was owned by Jim Rowland, an unsophisticated man who affected the biker lifestyle. No one knew where he came from or what if anything he did for a living. People tended to leave him alone.

Rowland was, in fact, the third face of Jalaly. Now the spy pulled on leather motorcycle pants. He donned a black leather jacket with chrome studs and buckles. On its back was a picture of an American eagle grasping a flag in its claws. He pulled on a pair of heavy oiled leather boots.

The transformation was astonishing. Fitted with lifts the boots added nearly three inches to his height and the pants and jacket were padded to add the appearance of bulk to his body. To complete the change he inserted colored lenses into his eyes, turning them from blue to brown. He used theatrical spirit gum to cement a large, unruly false beard and eyebrows onto his face. He applied a latex device to change the shape of his nose. As a final touch he slipped on a somewhat greasy wig complete with ponytail and capped with a checkered do-rag. Checking himself in a full-length mirror, Jalaly nodded in satisfaction.

Gone was the professorial Pearson. Standing in front of the mirror was "Jim Rowland," a tough biker whose carefully faked record held no criminal offenses or anything else that could bring him under scrutiny.

He continued down the tunnel and emerged in the basement of "Rowland's" house. The contrast with the first house was startling. "Pearson's" lawn was carefully tended, bushes pruned, grass mowed. "Rowland's" yard was overgrown. An old car was sitting up on blocks in the back yard. In the front a pink flamingo leaned drunkenly in a bed of dead flowers. "Pearson's" house had been painted two years ago. "Rowland's" had needed paint for nearly a decade and the roof was in need of attention.

The only thing the two supposed neighbors had in common was their insistence on privacy. "Rowland's" very appearance was enough to keep people at bay. Neighborhood children were quietly but sternly warned away from the house where "that biker" lived—and Jalaly liked it that way. "Rowland" never made any overt threats, but managed to project an invisible wall that seemed to warn: "Danger: Keep Out."

Both houses had drapes or shades on the windows and were equipped with timers to operate electric lights when the houses were unoccupied. No one ever noticed that "Pearson" and "Rowland" were never at home at the same time. And why should they? The two men were so different that it couldn't be imagined they were one and the same.

In the world of secrecy the two faces of Jalaly are called "legends." Each had been carefully cultivated. If anyone should check "Rowland's" background they would find a certificate recording his birth in California's Central Valley. There were documents indicating that he completed a high school diploma. Papers on file with the Department of Defense asserted he had served in the Army during the First Gulf War. Each month he received a check for disability resulting from fictional wounds. The money was real; the person was not.

Using a carefully cultivated and distinctive handwriting, as "Rowland" Jalaly wrote checks for a mortgage payment, water and electricity bills, and even satellite TV. "Rowland" had no computer, but he sometimes rented action-adventure DVDs at the nearby supermarket. The disks were never actually viewed and always returned promptly.

The effort to create stark contrast between the two legends went so far as to include affectations in personal habits. About once a week "Rowland" bought a case of Coors beer. Over time each can was opened and its contents poured down the kitchen sink. The empties were placed out for the weekly garbage pick-up. Although

Jalaly did not smoke, as "Rowland" he bought a carton of Marlboros every two weeks. Burned butts went in the trash beside the empty beer cans. The mailbox in front of the "Rowland" house received monthly subscriptions to *Playboy*, *Guns and Ammo*, and *Cycle World* magazines, all of which were stored in the garage without having been read.

In the other house Jalaly paid his bills as Harvey Pearson, a respected scientist with personal wealth in excess of three million dollars. He managed the investment account that had been set up clandestinely after the supposed deaths of his "parents." He had a computer and high-speed Internet connection. An examination of his hard drive would show nothing untoward. His mailbox received the weekly magazines *Time*, *Science* and *The Economist*, as well as regular monthly issues of *Scientific American*, *Fortune* and *National Geographic*. These magazines were read.

The two legends were meticulously tended at all times, following the principle that perception becomes reality. Such was the secret life of Sanjar Jalaly.

Ready at last to leave for his rendezvous "Rowland" stepped into the garage and opened the door. Besides a battered riding lawn mower and a workbench with a scattering of tools there was an old red Ford F-250 pickup with jacked up undercarriage and oversized tires. Just inside the overhead door was parked a classic 1983 Harley Davidson Glider Fatboy, a full-sized motorcycle in light tan paint and brown leather saddlebags. He rolled the bike around into the driveway and clicked a remote to close the door. Mounting the bike he stepped on the starter let the engine warm for a moment and drove sedately away.

A half-hour later Jalaly turned into a remote forest service road about fifteen miles south of Los Alamos. Late afternoon light was casting long shadows from the Ponderosa pines as he threaded the bike down the lane and past a stout gate marked "Private. No Trespassing." He went about a mile further on, stopped and turned off the engine to listen and look.

Hearing and seeing nothing he started his motor and returned to the gate. There he once again shut down the bike and observed the quiet forest for several minutes. Finally he dismounted, opened a robust combination lock and slid the gate open far enough to roll the bike through. He carefully closed and locked the gate behind him before restarting the engine and continuing down a narrow lane that wound among the trees.

It was another quarter mile to the building, partly concealed by a thick grove of aspens. It was a lonely and private place, perhaps a seasonal vacation getaway cabin. In fact it was exactly that, but like everything else associated with Jalaly it was a false front. The supposed owners of the property did not exist. It was there to provide him with a bolt-hole in case of trouble and to serve as a private meeting place, as it was today.

He parked the bike near the front porch and shut it down. Standing and stretching he looked around the area, always wary. He stepped up onto the porch and knocked, three quick taps, a pause, a single tap, a pause and two more fast taps. For a moment there was no response. He was about to knock again when a floorboard squeaked behind him. He turned. A tall man had come around the corner of the cabin and stepped onto the porch.

"The door is unlocked. Go inside."

Jalaly nodded deferentially and opened the door. The other man followed him inside and shut the door.

"Salaam, Sanjar."

"Salaam, Uncle."

They stood facing one another for a moment, gazing into each other's eyes. Then each placed his right hand over his heart and bowed to the other in the gracious Persian form of greeting. Jalaly did not know the real identity of this man he called "Uncle." For many years his controller had hovered in the periphery of Sanjar's life, first in Tehran, later in Cambridge, and finally here in New Mexico.

As was the Persian custom, for a few minutes they engaged in polite conversation concerning the weather, the state of the world,

Wall Street news, and the alignment of the planets. At last Jalaly felt the time was right to address the business at hand.

"What is the cause of this pleasure of meeting?"

His handler gazed at him for a moment before nodding. He led the way to two battered easy chairs and gestured to Jalaly to sit down. He lowered his long body into the facing chair.

"There is a reason for everything in the eyes of Allah. Your time of secret jihad is nearly at an end."

Jalaly brightened. "I will return at last to the Motherland?" His eyes sparkled in anticipation. This had been his fondest dream, to return to his roots, to be honored as a protector of Islam, to join the brotherhood of nuclear scientists in the completion of the Great Project.

There was a pause. The handler locked his unblinking eyes with Jalaly's for a long moment before speaking.

"Soon, Sanjar, soon. The time draws near. There is but one more task and your jihad will be complete."

Jalaly tried to hide his disappointment. "And what task is that, Uncle?"

The handler reached into his jacket and pulled out some documents.

"There is a secret project at the lab. It is very interesting to our scientists. It concerns the construction of a new kind of computer, a very powerful computer. Have you heard of this?"

"Ah, yes, I think so." Jalaly squinted in thought. "There've been rumors. It's not in my department and I think it's a secret. But little birds do sing, do they not? I've heard it is a quantum computer that could be a thousand times faster than even the super computer we use at the high explosive lab."

"Yes, that is what we have also heard." The handler passed over the documents. "Read these and destroy them. Your final task is to learn the secrets of this fabulous computer and deliver them to our scientists. In their hands it can be a powerful sword to do the work of Allah."

Jalaly bowed his head in acceptance.

"What must I do?"

The handler stood up and retrieved a small sack from a table just inside the door. Unzipping it he pulled out two thin traveler's prayer rugs. He laid them on the floor, carefully aligning them to some small marks made on a previous occasion.

"Sanjar, it is time. Let us pray. Allah shall guide us."

Jalaly nodded and rose from his chair. Kneeling he placed his forehead on the rug. "Allahu akbar," he intoned, beginning the Muslim prayer for the second time that day. *God is great...*

Chapter 10

Quantum Center
Wednesday, 7:45 a.m. MST

The scene at the Quantum Center was more chaotic than ever. The big day had come and the place was packed with busy scientists and technicians. Tommy Dawson was seated at the center console of the raised control platform with Stuart Reeves at his side. There were five workstations at the control center, each with several panels of dials, switches and banks of colorful LED lights. Above each keyboard were five 40-inch flat panel monitors arranged in an open horseshoe shape.

Reeves and his team of hardware and software engineers were running over some final diagnostics on the interface between the center and the nearby "Qubit Farm" where the quantum computer itself was located. Chris Fisher sat nearby watching and making an occasional note.

Standing beside Chris, one of the technicians was giving him a running commentary. He explained that the core of the Qubit Farm was truly tiny, smaller than could be seen by the human eye. The entire device was the product of nanotechnology, engineering on the atomic scale.

"Due to the Uncertainty Principle no data can be extracted while a quantum computer is performing a calculation," he reminded Chris. "As you know, taking a measurement changes a state, thus making it unknown. That would cause the computation to stop. Only when a computation is complete can the result be read."

Chris watched as Reeves tapped some keys causing lights to flash across the control panel. Schematics appeared on the overhead screens and changed as he manipulated a trackball. He clicked on a menu bar and an electronic desktop appeared, an adaptation of a scientific math app.

"All right, that's the main input screen," the technician explained to Chris. "We'll use this to enter the parameters for a computation, then it goes through the server system to the quantum computer."

Chris nodded and turned to look around the room. He eyed the bank of servers along one wall of the lab, several hundred of them linked together to create a parallel computing array. He knew that even that powerful system was capable only of acting as an interface to process the input and output from the quantum computer itself. The wall of servers was lit up, churning out ozone and heat.

Dawson turned in his chair. Spotting Chris he waved.

"Hello Fisher." Dawson struggled out of his chair and came over. "So what do you think of my little toy?"

Dawson was dressed for the occasion in a brightly colored Hawaiian shirt and electric blue slacks. He was wearing his "Quantum Cowboy" baseball cap. On his bare feet were a worn pair of sandals and he had either failed to comb his hair that morning or had mussed it up later because it stood in unruly spikes. There were gravy strains on his shirt where it curved over his protruding belly.

"I hardly know what to think," Chris admitted. "This is the most amazing thing I've ever seen. If it works, they're going to have to give you another Nobel Prize."

"Bah," Dawson exclaimed. "Those things aren't worth the gold they're stamped from. Just a lot of publicity and hot air if you ask me, all to honor the inventor of dynamite."

"Yeah, but there's a couple million bucks involved."

"Money! What good is that?" Dawson pushed through the crowd of technicians to lean over Reeves' shoulder, pointing at something on the overhead screen to the right. "You got a problem there," he announced.

Glancing up Reeves noted the flashing icon and nodded. "I see it. We're running the final diagnostics now, and it should take care of it in a minute."

Muttering something under his breath Dawson stepped back to survey all five of the large monitors. A number of programs were running, creating patterns of movement, cascading lights and progress bars scrolling across the panels.

Remembering his assignment to watch for anomalies, Chris turned back to the lab and scanned the busy scene. During the past two days he'd met many of the team members and chatted with them about the project. When possible he'd even made small talk about their personal lives. So far, nothing had struck him as even remotely suspicious.

Now he noticed a security guard entering the lab, escorting a woman that he had never seen before. As they approached the control center the guard pointed and told her, "There's Doctor Reeves. Right up there." She nodded and he turned and walked away.

The woman stood for a moment looking around. She was carrying a large bag over one shoulder and had a Canon G12 digital camera hanging from a strap around her neck. She was fairly tall, probably early 30s, brown hair cut to a practical length with a short ponytail. She was dressed in an eggshell blouse with frilly lace down the front, tan pants and brown flats. A silver necklace was her only adornment.

Looking up she saw Chris watching her. With a little wave she stepped up to the platform and held out a hand. "Hi, I'm Kate Elliott with *Science Today* magazine."

"Well hello," Chris replied, taking her hand. "Stuart told me you were coming. You and I have something in common—we're both here just to watch. I'm a physicist, here as an observer for the Navy."

Kate smiled and patted her steno pad that was protruding partway out of the over-stuffed messenger bag. "Then we're in the same boat," she said, glancing around the lab, "mere note-takers. This is really something. Stuart told me a little bit about the center, but to see it in fact is just, well, pretty amazing."

"It is," Chris agreed. "Why don't you check in with Stuart and then I'll show you around a bit? I've been here two days, so I know enough to be dangerous." Kate nodded in acceptance and stepped over to where Reeves and his team were working.

"Oh, hi Kate," the engineer said, turning as she touched his shoulder. "Glad you're here. Sorry to say I'm pretty busy at the moment. We're running final tests on the interface, preparing for the first trial run in a few hours. Let me see if I can find someone to show you around."

"It's good to be here Stuart. Thanks, but this fine gentleman has volunteered to shepherd me." She indicated Chris.

"Ah, perfect," responded Stuart. "Lieutenant-Commander Fisher is just the man. Thanks, Chris."

Kate turned back to Chris with new respect. "Both a physicist and a Navy officer? Have you ever commanded a ship?"

"Well, no," Chris admitted with a sheepish grin. "Actually I was a SEAL so I commanded a Special Forces unit in Afghanistan. But that was quite some time ago. These days I'm just a mild-mannered professor."

Kate cocked her head at him then took his arm.

"Well, Mister Mild-Mannered Professor, shall we get on with it?"

As they explored the huge lab Chris explained that while the quantum computer itself was tiny, the control interface was massive, including all the servers, workstations and other equipment in the Quantum Center along with its staff of dozens of physicists, engineers and technicians.

As the two of them wandered among the scattered workstations and machines he pointed out various features and introduced Kate to some of the staff members. Everyone was busy in the build-up to the trial run, so few had time to say more than a few words before turning back to their tasks.

Chris pointed out that the entire experiment—including the lab and the nearby building where the quantum computer itself was kept deeply chilled—was powered off-the-grid by a General Electric nine-megawatt gas turbine generator big enough to serve a modest-sized town. He described how the control interface in the lab was connected to the isolated Qubit Farm by a high-power microwave beam that connected the two buildings through a heavily shielded tunnel.

At the Qubit Farm diamond atoms were trapped in crystal lattices and chilled to near absolute zero, minus 273 degrees Celsius. At those temperatures matter becomes super conducting — electrons can pass through it without resistance. Almost literally frozen in place, the atoms could generate qubits and function as parts of a computer.

The control was accomplished by a complex matrix of nano-sized lasers, firing precise bursts of photons to change the state of atoms at the astounding rate of trillions of actions per second. Quantum entanglement would create further connections between atoms to make the computing process work even faster.

"At least, that's the hope," Chris concluded.

"Can we take a look at the Qubit Farm?"

Chris shook his head. "No, I'm, afraid not, and there really isn't anything to see. The quantum computer core is extremely sensitive and must be protected from even the smallest extraneous signals. Anything that might disrupt the process during a computation, such as a radio wave, sudden temperature fluctuation, even a stray cosmic ray could cause decoherence, that is, a breakdown of the quantum state. The core's surrounded by several feet of lead and the building itself has walls more than ten feet thick made of various layers of shielding material. All of the controls are here, not over there."

"So everything that can be observed, all the real action, takes place here?"

"That's right."

Kate nodded and turned to explore another section of the lab. She stopped occasionally to ask a question or take a picture with her digital camera. Eventually they discovered a break area with a few tables and chairs. Chris stepped to a coffee urn and gave her a questioning look.

"Yes, please, black with just a bit of sugar. And, oh! I'm having one of these pastries. I got up too early to get anything to eat, and it took forever to get through that security office." Kate grabbed a napkin and picked up an apple-cheese Danish as Chris drew two cups of coffee. They sat down at one of the tables.

Within moments the Danish had disappeared and Kate was wiping her mouth with a napkin. Chris watched with mild amusement as she reached in her bag for a mirror and touched up her lipstick with a pale color.

"Well, the moment of truth is coming soon," he said. "Stuart tells me they think there's a fair chance this first experiment won't work, and from my experience I have to agree. There always are bugs and glitches that can't be foreseen. Remember what happened when they started up the Large Hadron Collider at CERN for the first time."

"I sure do—I was there," Kate said with a grimace. "Talk about a bust."

She gave Chris an appraising glance. "I'm curious, what exactly is your own specialty?"

"I'm a high-energy physicist at Stanford. I have a teaching post, but also do research at the SLAC National Accelerator Laboratory that Stanford operates for the Department of Energy. It's pretty advanced research."

"SLAC, what does that mean?" Kate was scribbling in her notebook.

"The original name of the facility was the Stanford Linear Accelerator Center and it was renamed just in 2008 but we kept the original acronym as part of the new name. The center was founded nearly 50 years ago and it's made some groundbreaking discoveries, including basic work on quarks."

"What are you doing there now?"

"We're building a high-energy X-ray beam that will actually be able to let us see electrons. Real cutting edge stuff—but nothing compared to this, assuming that it works as they think it will."

Kate scribbled for a moment then looked up. "If quantum computing does work, what do you think it will mean? I understand there are some practical applications in way-out science, the ability to run huge modeling programs for example. Stuart told me it could make our on-line security systems obsolete by being able to break cryptographic codes, and that will require some changes for sure. But in the long run, will it really be all that big a deal?"

Chris pondered for a moment, sipping from his coffee cup. He set it down and shrugged.

"There've been a lot of inventions that didn't seem all that important at the time," he said at last. "For example, when the transistor was perfected at Bell Labs back in 1947, it was seen merely as an alternative to the vacuum tubes that were currently being used in electronic circuits. Who would ever have thought that simple invention, the transistor, could have led to the computers we have today?

"I can think of a few other world-changing breakthroughs," he continued, "and few of them were immediately recognized for the eventual impact they'd have. For example, when Wilbur and Orville flew their little joke of an airplane, could anyone have conceived that within only a few decades jumbo jets would be flying millions of people around the world?"

"Yes I can see your point," Kate mused. "I suppose most of the real advantages would become apparent only over time, once the possibilities become reality."

"That's right. Today could turn out to be a truly significant date in history—or just another day. It all depends on whether this works at all—and if it does, how much change it will eventually bring. I have a feeling it could be pretty important."

He glanced at his coffee cup and began to pick it up, then paused with a distant look in his eyes.

"It could give us the key to understanding a lot more about how things work—and who knows what we could learn from that deeper knowledge?"

Several hours had passed and the moment was approaching when the quantum computing system would be turned on for the first time. Dawson sat at the main control panel with Reeves at his right. Section managers occupied the other three positions, one monitoring the Qubit Farm, another responsible for the hardware interface and the third for software systems. All wore headsets with microphone stalks.

Around them on the raised platform and the adjoining space below were clustered several dozen scientists and engineers. Others remained at their remote workstations to manage subroutines and operate special equipment serving minor but still critical purposes in the overall system. There was an atmosphere of quiet expectation, suppressed excitement.

Reeves spun his chair around to face the observers. He was smiling, but it was a taut expression that reflected a nervous sense of uncertainty. The culmination of several years of planning and work was about to be tested.

"All right everyone," he announced. His amplified voice echoed from speakers around the room and every face turned toward him. "We're about to start the test run. This is the final checklist. Interface hardware, report."

"Interface hardware optimal. Systems go," came the reply.

"Software systems, report."

"Software systems optimal. Systems go."

Kate was standing next to Chris at floor level near the front of the gathered crowd. She turned to whisper in his ear, "It's like the countdown for a space launch."

"Qubit systems, report," ordered Reeves.

"Qubit systems optimal. All systems go," responded a third engineer.

Reeves turned back to the controls and tapped his keyboard. A surprisingly long number appeared on the large screen overhead. He looked at the display for a moment before turning back toward the room.

"We've chosen this for the initial test. It's a 167-digit digital key that was first factored in1997 by Dr. Samuel Wagstaff at Purdue University. Factoring is one of the hardest computations, especially for large numbers with more than 100 digits. In this case, this number is the product of multiplying two prime numbers by each other. A prime is a number that can only be divided by one or by itself. To break the code requires the large number to be factored, that is, to find the two primes.

"Factoring a large number is very difficult indeed, so it makes an ideal test for our system. This number is the product of an 80-digit number multiplied by an 87-digit number. The important thing for our purposes is that it's already been factored using standard computers." He glanced up at the screen again before continuing. "It took Doctor Wagstaff's experiment about one hundred thousand hours of computer operation to find the answer, using thousands of linked computers."

He paused for effect before continuing.

"It's our hope that we can break this code much faster than those standard computers did. Our most conservative estimate is that the quantum system might be able to break this code in a hundred hours or less. That's a thousand times faster than the electronic machines were able to do. We hope it will be even faster, perhaps ten thousand times faster, in which case it would break the code in just ten hours based on Doctor Wagstaff's experience.

"Well, I'll just shut up now and we'll see what happens."

Turning to his keyboard he made some final checks, pressed the ENTER key and turned his chair back to face the crowd. "Now we just have to…"

He stopped in mid-sentence, struck by the expressions on the faces of his associates. They were staring up at the screen behind him with surprise, awe and even shock. "What…?"

He spun around to look at the screen and then his face, too, froze in incomprehension. For several moments he couldn't speak, couldn't even move.

On the screen above appeared the words:

COMPUTATION COMPLETED

There was a long moment in which time seemed to have stopped then a strident voice rose out of the silence.

"I knew it! I knew it!"

It was Tommy Dawson, who now leaped ponderously from his chair and began to dance around in a circle, waving his arms and continuing to shout. The crowd broke into busy chatter.

"What's going on?" Kate asked Chris.

"I have no idea."

Regaining his self control at last, Stuart called for quiet. Even Dawson eventually settled down and all attention returned to the project manager.

"Let's not get excited," he said. "According to the readout here, the calculation was completed in less than four seconds, even including the interface processing. That's far beyond all theoretical possibility—so we're dealing with a system error here. This just can't be right."

"But it is!" chimed in Dawson, dancing to his feet again. "You'll see! You'll see! That's no error. Check the answer. Check it!"

Irritated, Reeves turned back to the keyboard and clicked some keys. Strings of binary code appeared on the screen above him, the true factors of the long number. He hit some more keys and a second set of strings appeared, this time the answer provided by the quantum computer. He aligned the strings for comparison.

They were identical.

At a complete loss he sank back in his chair, staring at the display. For a moment silence again sank over the group, only to be broken once more by the voice of Tommy Dawson. Shifting from one foot to the other he whooped and shouted. "See! See! I knew it! I knew it!

He spun like an overweight dervish, spittle flying from his lips, his eyes flashing with excitement.

"It's the key! The key to the fucking Universe!" he shouted.

His voice rang out so loudly that it echoed down the vast room and back, "...ing Universe... Universe... verse..." then died away into the silence as the crowd of dumfounded scientists stared at the monitor and the numbers it displayed.

Chapter 11

Los Alamos, New Mexico
Wednesday, 1:45 p.m. MST

It was several hours later and the quantum experiment had been shut down pending analysis of the first results. Several more tests had been carried out and in each case the quantum computer performed at an incredibly fast rate, yielding correct answers in a matter of a few seconds. Everyone was stunned—except for Tommy Dawson, who continued to chortle and shout as if some great truth known only to him had been vindicated.

Before the staff began to disperse, Dawson addressed the group.

"I predicted this," he told them. "I didn't dare publish until there was proof. Now I've got the proof. We've seen that factoring a number this large is nothing for this computer. You'll see that it can do much more—more than any of you can imagine."

He promised to explain his ideas the next morning, calling for the key staff members to convene in the staff auditorium at the New Mexico Institute.

"There's no use speculating," he added. "None of you could possibly understand this."

At that statement a mumble of resentment spread through the room. Top-level physicists and engineers are not accustomed to being treated like ignorant children. They began to disperse, many shaking their heads or speaking animatedly among themselves. In the end, the amazing fact of the day's trial overcame any resentment.

Kate turned to Chris with a wan smile. "You know, I think I could use a drink. How about you?"

He nodded and turned to lead the way outside. They rode together in his Jeep, looking for a quiet refuge. They found an intimate restaurant and lounge and claimed a comfortable corner booth in the bar. The waitress set a bowl of peanuts on the table and asked for their order. She was wearing an apron decorated with atomic symbols and a ball cap bearing the name of the bar.

"You know, something to eat would be great," Kate said. The waitress handed her a bar-food menu. "First I'd like something tall and cold. How about you Chris?" He nodded in assent and Kate sent the waitress to bring them a pair of Fat Tire ales. When the drinks arrived they had decided to split a plate of nachos and a basket of hot wings.

"It's a good time to pig out," Kate announced, shifting around to settle into the padded seat. "I still don't know what happened there today, but it was sure interesting. I've never seen so many people with their jaws hanging open. It was like one of those, what do they call them—'defining moments in history.' Didn't you think so?"

Chris glanced up from his PDA where he had been making some notes.

"Yes, it was a 'eureka' moment," he responded. "I can't think of anything even remotely similar. It was like seeing an amazing magic act and having no idea how it was done."

"Well, you're a smart physicist and you've even worked with Doctor Dawson, so you must have some idea what he's on to?"

"Kate, I don't. I've been racking my brain but I just don't have a clue of what the explanation could be. It's obvious that Tommy's developed a new theory that explains the fantastic results we saw. I have to expect that his theory will change our entire view of physics."

"Why do you say that?"

Chris pondered for a moment, then pulled out a pen and began to doodle on a bar napkin as he spoke.

"Well, for one thing what we saw today doesn't fit with the commonly accepted model of quantum mechanics. Which means that model is going to be pitched out. And those other ideas that are still only hypotheses—string theory, supersymmetry, ideas such as that—they're going to be pitched out too. Hell, it might even make Einstein's relativity theories redundant."

"You think it's that big?"

"It could be. In fact, after what we saw today it's a virtual certainty. According to everything we knew or had theorized, those calculations we saw today were flat-out impossible. The fact that the quantum computer worked so much better than expected opens a door to completely new science. It could be, as Tommy said, 'the key to the Universe,' whatever he meant by that."

They sat in silence for a moment, sipping from the cold bottles. Their food arrived and Chris motioned at the bottles to order a second round. He picked up a wing and held it up for inspection before taking a bite. Kate fumbled with a nacho, twisting it to wrap melted cheese around a slice of jalapeno. They ate in silence for a moment before Chris spoke again.

"Well, we should find out something tomorrow when Tommy explains what it is he's cooked up. I have the feeling he orchestrated this entire experiment for the real purpose of demonstrating his theory."

"Yes, he was the only one who wasn't surprised by the result," Kate mused. "He was expecting it, wasn't he?"

"Oh yeah, he definitely was. In fact, I think he was waiting for it, waiting with anticipation."

Kate took a sip of the ale.

"One thing I have to wonder about," she mused. "It was rather rude of Doctor Dawson to imply that nobody else would be able to understand his theory. Why do you think he said that?"

Chris chuckled.

"Well, I can explain it on a couple of levels. First, as you probably know Tommy's autistic, not seriously affected but enough to make him, well, different from the rest of us. He has an underdeveloped sense of the social graces. We might see it as a lack of manners, but he can't help it."

"Oh, yes, I knew about his autism and of course autistic people sometimes tend to grate on others. What's the other level?"

"Well, as you also probably know autistic people are sometimes savants, that is they have minds that can function in seemingly impossible ways. For example, they might be musically talented, able to play a complicated piece of music after hearing it only once. Or they might have mathematical abilities—do you remember the old movie *Rainman*? That told the story of a mathematical savant."

"So you're saying that Doctor Dawson has special abilities?"

"I think he's a savant in some way, besides being a world-class genius."

"And his special skill is…?"

"Well, it's actually hard to say but I think it has something to do with visualization. It's not exactly like anything I've ever seen or heard of before. It's like he can 'see' things that other scientists can only express in mathematics. He can 'view' the answers to complex formulae, as if they were solid objects. I've never been able to understand it…"

"Are you saying that most scientists, when they try to understand a mathematical theorem or something like that, that they can't really do any more than trust what the numbers are saying?"

"Yes, that's kind of what I mean."

"But Doctor Dawson actually 'sees' the answer, like an object? As if the solution was…something tangible?"

Chris remained silent. Kate looked away for a moment, thinking.

"You know," she mused, "I remember reading that Einstein used to tell people that he thought in images instead of words. Is that something like what you're trying to describe?"

"Yeah, could be. Let me tell you about something that happened when I was one of Tommy's graduate assistants."

Chris bit into another hot wing and followed it with a draught from the bottle of ale before continuing.

"One day some of us in Tommy's lab were chatting about the idea of hyperspace, the concept that forms may exist in more than three dimensions. These things can be described in mathematics. For example, we can write a formula to describe n-cubes, cubes in more than three dimensions, having more than six sides and eight corners in different dimensions. They're called hypercubes and they're merely mathematical constructs. No one could actually visualize one.

"So Tommy was sitting nearby and he got up and came over to listen to what we were saying. He seemed to be fascinated by the fact that we kept remarking that we couldn't see geometric shapes in hyperspace. Finally he spoke up, saying 'You're wrong about that.'

"I remember that we all looked up at him with expressions kind of like some of those we saw at the center today, expressions of surprise and disbelief. 'What do you mean?' someone asked him.

"And that seemed to surprise him, because he didn't know how to answer. He looked embarrassed, and just turned away muttering. I heard him say to himself, 'Well, I can see them.' I had a strange feeling that I'd received insight into a very unique mind."

Kate sat quietly for a moment after Chris finished his story. "So you think this man can 'see' in more than three dimensions? Is that what you're saying?"

"Well, more or less. Yes. He has a savant ability that's unique and which gives him analytic powers that few if any others can match. There's another example of that that I think makes the point, something I just learned yesterday. As they were developing the software for the quantum computer project, Tommy kept referring to strings of code as if they were objects. Nobody could understand that, so finally he wrote a program that translates program code into images. Stuart Reeves demonstrated it to me and it's amazing. They have big plotters that print out diagrams that are like blueprints except they don't represent physical things like the design of a machine or a building. They're visual representations of the software code itself.

"When it was first demonstrated, Stuart told me that Tommy said 'Now you have some idea of how my mind works.' In other words, he can visualize mathematical concepts as if they were physical objects."

Kate sat back in the leatherette booth and stared at Chris.

"That…why that's just incredible," she said at last. "And you think this savant ability lets him figure out things that others just cannot?"

"That's right," Chris said. "And I can tell you one thing, and that's that when he makes his presentation tomorrow morning it's going to be like the greatest brain teaser exam you ever imagined."

"Actually I never imagined anything like that," she said with a laugh. "You're going to have to translate for me, because it sounds like he's going to be speaking a language from some other planet."

Chris chuckled along with her then drained the last of his ale.

"I'll do my best, but I suspect I'll have a hard time myself. Tommy can get so far ahead of the rest of us that sometimes it's hard to see him off there in the distance."

Chapter 12

Los Alamos, New Mexico
Wednesday, 9:45 p.m. MST

It was well after dark when Chris dropped Kate back at the Center to pick up her rental car. Following their conversation at the bar he had taken her for a driving tour of the area and they ended up having dinner at the Mexican restaurant where he had met Reeves and Bonner. This time they sat in the main dining room and enjoyed Margaritas with salsa and guacamole dip followed by plates of chicken enchiladas and flan.

Chris was relaxed as he drove back toward the inn, looking forward to a good night's sleep after a stressful day. He and Kate had developed a comfortable relationship, sharing stories about their interesting lives. As she prepared to leave his Jeep she surprised him by leaning over to give him a quick, affectionate kiss on the cheek.

Fumbling with his key Chris unlocked the door to his suite and was prepared to swing it open when he paused. Something was out of place, but he didn't know what. He glanced around and saw nothing suspicious outside the room. Cautiously he opened the door a crack, leaning close to listen and smell, seeking any clue to his instinctive reaction.

He opened the door further and reached around to turn on the light in the entryway. *Hmm, getting a bit paranoid am I?* He shoved the door open and stepped inside, then froze as a voice greeted him.

"Late night out?" inquired Jeff Jones, the Naval spook. Jones had been waiting in the dark, relaxing on the couch.

"Jesus Christ," Chris exclaimed. "What the hell are you doing, trying to give me a heart attack?"

Jones reached over to turn on the table lamp next to where he was sitting.

"Nah, it would take more than me to scare a tough guy like you," he remarked. "I needed to have a little face time with you and thought I'd make myself comfortable while I waited. Didn't expect you to be out so late, though. Finding anything interesting?"

Chris shut and locked the door. He took off his jacket, hung it in the closet and walked over to the easy chair in the corner. He sat down, leaned back and closed his eyes.

"Yeah, it's been very interesting," he sighed. "I think this thing is about to go ballistic."

"So I hear," Jones agreed. "I know the trial run went better than anyone expected. A lot better from what I was told. Is that your assessment?"

"Oh, absolutely. It wasn't just better—it was beyond all theoretical possibility. That quantum computer can do more in one minute than all the computers in the world put together could do in an hour."

Jones stared at Chris for a moment as if suspecting he might be pulling his leg.

"No shit?"

"No shit." Chris thought for an instant. "Well, maybe not quite that good, but it's in the ball park."

The two men sat for a moment contemplating what that might mean, then both started to speak at the same time.

"What..."

"How..."

They stopped and Chris waved to Jones to continue. The agent sat back on the couch and crossed his legs.

"I was just wondering, how could it be that fast? Do you have any idea?"

"No," Chris said flatly. "No idea. Dawson's going to explain some of it to us in the morning, but knowing him none of us will be able to understand it even after he tells us three times and draws pictures. Would you like something to drink? Beer? Mineral water?"

"Yeah, give me a cola. I could use a shot of caffeine."

Chris stepped to the mini-bar and grabbed a Coke and a bottle of water for himself. They popped the tops, toasted each other and drank.

"Okay," Jones said at last. "This is going to push things to the next level. If we thought there was interest in this before, when word gets out about the success today the spook system will be going nuts. And, shit, word's already out. Our SIGINT units have picked up some chatter about it on the 'net, and NSA has deciphered some messages."

"Then you think there are already agents in place here, in the project?"

"Oh we know they're here. We were pretty sure of that before, and this just makes it certain."

Chris digested this for a moment before speaking.

"I've been talking to as many of the people as I can, discussing technical stuff like you told me to and trying to draw them out about personal matters if I could. Didn't see anything that raised my antennae though."

"Well, you probably wouldn't," mused Jones. "After all, the pros haven't been able to track down these guys. You're doing okay. Just keep on asking questions and keeping your ears cocked. Now that things are getting more interesting chances are someone might get careless."

"Just what are we talking about? For example, what countries, what organizations?"

"Well, based on some signals intel we suspect Iran. There could always be a Russian connection—the KGB never did go away you know, just changed its name and kept on doing the things they always did so well. And, crap, the Chinks might be in it too."

"What about terrorists, al-Qaeda?"

"Nah, they aren't sophisticated enough to get within a thousand miles of this. Those guys aren't even in the picture."

"That's a relief." Chris sipped his water and thought for a moment before continuing. "What are the chances they might have sabotage in mind, instead of just stealing secrets?"

Jones sighed and set the empty cola can down on the end table beside him.

"It's something to think about," he said noncommittally. "I think they'd want both—first to learn the secrets, then to sabotage the quantum device to deprive us of it. And we're concerned about something else…" He stopped as if hesitant to continue.

"And that is?" Chris coaxed.

"Well, it's the fact that there's only one man who knows what's going on. We're frankly worried that Tommy Dawson might be a target. Whether to kidnap him, or to silence him…"

"Christ!"

"Yeah, you see the picture."

"But surely you can give him protection?"

"We can, up to a point. But we have to be careful. He's too freaky to trust. You know the guy, can you imagine him cooperating with a security detail?"

"No, not very well," Chris said. "He'd probably kick their asses out, or call 911."

"Okay, Bonner's got him under surveillance with an eye out for his safety. I wanted you to be aware of that. Be prepared for the possibility that some guys might want to play hardball. That reminds me, when you got the briefing packet did you find something you didn't understand?"

"Yeah, a key with the number 42 on the fob," Chris replied. He groped in his pocket and pulled it out. "Is this what you mean?"

"That's it. It's your get-out-of-jail card. If the shit starts to hit the fan, that key will give you access to an emergency kit, stuff you can use. Don't go near it unless you have to, but that key fits a luggage locker at the bus station. That's all I'm going to say about it."

Chris examined the key with new interest then shrugged and slipped it back in his pocket. He stood up and walked over to the rear window, pulled the drape aside and looked out.

"Do you think they're suspicious of me?" he asked quietly.

"We don't think so. Your cover is solid and we haven't picked up any chatter about you. That could change any time."

"I want to ask you something else. How far can I go with the security director, Paul Bonner? He told me to report anything to him. What's his background and is he someone we can implicitly trust?"

"Yes, absolutely. Paul's ex-FBI, served his 25 years and made SAC, special agent in charge at the Phoenix office. He's solid as a rock. Law degree from Penn State. His father was a federal judge."

"What resources does he have?"

Jones mused for a moment. "He's actually a bit undermanned, budget cuts and all that. Plus his security team mainly just provides uniformed patrol duties. It's a long time since this was the most secret location on Earth. But, that said, the FBI is available to provide intelligence and forensic assistance when Paul calls for it."

"Are there FBI agents on the job now?"

"Paul has called for some assistance, yes. But as you've learned, they don't have the ability to mix and mingle with the super geniuses here. That's why we called on you."

"As you are the living proof, Navy Intelligence is involved. What about other intelligence services, like the CIA?"

"My boy, the CIA is always involved, always. That's their whole purpose, being involved. And messing stuff up."

"What do you mean?"

"Well, they're famous for it, Bay of Pigs and all that. Every time the Company gets its fingers into something, you can be sure it'll go haywire. It must be in their charter."

Chris returned to the easy chair and sat back down. He leaned back with his hands clasped behind his head and looked up at the ceiling. "All right. What else do you have to tell me?"

"That's it for now. As Bonner told you, you can report anything suspicious to him. Paul and I are in contact, so he can get a message to me if you need to. Remember the emergency stash if you get in trouble."

"Okay, thanks. Now if we're finished, I'm beat and we've got an early session tomorrow to hear Tommy explain his new theory."

Both men stood up and Jones shook Chris's hand. "Sleep well. I'll be in touch when I need to."

He stepped toward the door, then paused and turned back. "Stay cool," he said and flashed a peace sign before opening the door and disappearing into the night.

Chapter 13

Quantum Center
Wednesday, 10:45 p.m. MST

Shadows were deepening as evening turned to night. In the Quantum Center a single security guard was on duty. He strolled casually from one end of the lab to the other, stopping occasionally to lounge in one of the many swivel chairs scattered at workstations around the lab. Sometimes he toyed with the computer keyboards but all were password protected and he could only observe various screen saver patterns. Some of them were quite clever, although many reflected scientific jokes that were beyond his understanding.

His name was Jesse Carroll and he was an ex-cop from Denver. Sometimes he wished he had remained in the DPD, although his failure to rise above the rank of patrolman after seven years of service had been a constant disappointment. He had chafed with well-concealed jealousy as he watched others he considered less capable rise through the ranks. He'd always kept a clean record, putting in his hours during day after day of boring, mind-numbing routine.

After three years on the LANL security force his expectation that he would achieve greater success than in Denver had faded. The world seemed aligned against him. Bad luck, lousy karma? He didn't know why but things always seemed to be stacked against him. Hell, the sergeant he reported to was five years younger than he, nothing but a damned greenhorn, and there he was strutting around with his stripes and his shit-eating grin.

Carroll was dressed in a tan uniform with black shoulder tabs and pocket flaps. He wore a standard patrolman's utility belt containing a 9mm Glock pistol, handcuffs, a can of Mace, a high-intensity tactical flashlight, portable radio and a black nylon baton with side handle. Sometimes the tools of his trade felt like the weight of the world, pulling him down toward a pit of despondency.

Bored, he sat down in the rest area and began to read the sports section of a newspaper that someone had left behind. He noticed with disgust that the Colorado Rockies had lost another game and were out of the running for the season. *Idiots!*

Carroll checked the time and stood up to resume his patrol.

The hours passed without incident. At 1:45 a.m. Carroll heard the entry door open. Figuring his relief must be early he meandered toward the entrance, pausing to discard an empty soda can from which he had been drinking.

The lab was mostly in darkness with only a few recessed spots to light certain areas. Others were filled with shadow. A low, continuous hum filled the vast space, the sound of server cooling fans and ventilators. He'd become used to the faint odors of machinery and electronics that pervaded the lab, but it was a little spooky.

"Yo," he said, looking around the lab. "Where'ja get to?" There was no immediate answer, although he thought he saw a figure move in the deep shadows to the right side of the door. What the fuck? "Hey, over here!" he said, shouting now.

There was a slight rustling sound and a dark-haired man appeared from behind some storage cabinets just a few feet from Carroll. The patrolman jumped and pulled his pistol halfway out of its holster. "Whoa, you ain't Pete!"

The stranger held out both hands, their palms forward in a sign of reassurance. He spoke quietly.

"Pete will be along soon," he said.

"Raise 'em," Carroll commanded, jerking the Glock the rest of the way out of its holster and holding it down at his side. "What're you doin' here?"

"Hey, put that away," said the stranger, stepping out of the shadow. "I'm here by authority. Here's my ID." He reached inside his jacket and Carroll jerked the pistol upward, then dropped it again as he saw the hand holding out a badge wallet. He stepped forward to take it, then pulled a mini-Maglite from his shirt pocket and snapped it on. He examined the credentials closely then looked up curiously.

"Huh, the CIA. I'm not surprised, what with all that's goin' on." He handed back the badge case and snapped off the flashlight. "So, whatcha doin' here?"

"Same as you, just checking on things," the agent told him. He began to walk further into the room, looking from left to right as he moved toward the raised control center at the far end of the room. Carroll ambled along beside him like an obedient dog.

The agent stepped up to the central control panel and sat down. Carroll stood uneasily a few feet behind him. The visitor swiveled the chair around. "I'm going to check this system to make sure it's secure," he said. "That's why I'm here."

"Oh..." said Carroll doubtfully. He thought for a moment, then added: "That's okay then, I guess." He sidled off and sat back comfortably in another chair to watch. "You a computer expert then?"

"Oh, yes," the man said. He was busy at the keyboard and after a moment the screensaver faded and data appeared on the overhead screen. He typed some more and the data began to scroll. He reached in a pocket and pulled out a compact Firewire hard drive about the size of a pack of cigarettes. He plugged it into a port and began downloading data.

"Hey, you s'posed to do that?" Carroll asked, suddenly sitting forward in his chair.

The agent turned around again and looked directly at Carroll, his eyes narrowed.

"Officer Carroll, you saw my credentials. Please let me get on with my assignment. You have quite done your job here; now let me do mine." He turned back to the workstation then spun back again as if with an afterthought. "Why don't you get us a couple of coffees? You look like you could use one. Black for me would be fine."

"Yeah, uh, okay." Carroll got up and began to amble toward the rest area. He looked at his watch. It was six minutes after 2 a.m. *What the heck? Pete's late.* He thought about that as he found two cups and filled them with coffee. *He's never been late before. Better take a lookey-see...*

Setting the coffees down on a table Carroll walked toward the door at the far end of the lab. Opening the door he peered out into the night. To the right he saw his own patrol vehicle, one of the standard white Chevy Tahoes used by the LANL security force. There was another vehicle parked beyond it and he could see from the low profile light-set on the roof that it was a second security car.

Quietly he walked around the front of his own Tahoe and approached the second vehicle. He placed a hand on the hood, an old cop trick. Its warmth told him the car had been driven recently. He tried to see into the vehicle but the windows were dark.

"Pete, that you?" There was no reply. He walked around toward the driver's side door. As he rounded the front fender he jumped back with a yell. A human body lay alongside the truck in front of him. *Holy crap!*

Responding to his EMT training Carroll knelt and felt for a pulse on the man's neck. Nothing. He turned the body over and stared into the lifeless face of Pete Chavez, the man who was scheduled to replace him. There was a dark line around his neck.

Just then a voice came from behind him, the soft, monotonic sound of the agent. "Patrolman Carroll, please put your hands up on the side of the car...now!" Slowly Carroll complied, turning his head to look. The agent was standing a few feet away, a gun in his hand. It was pointed right at him.

"What have you done?" the agent demanded. Confused, the patrolman shook his head.

"I didn't do nothin'!" he said. "It's Pete, Pete Chavez. He's dead. I just found him. I wondered why he was late, so I come out to look... Oh Jesus, this's awful."

The agent looked around suspiciously and stepped nearer, keeping his face in shadow. "Stand up, keep your hands on the car and move away from the body. Let me see." Keeping his eyes on Carroll he felt for a pulse just as the policeman had done. After a moment he stood up.

"He's dead alright. Seems to have been gone for a while. You can put your hands down now." Carroll did so and was relieved to see the agent return the pistol to a shoulder holster. "Did you see anything? Anybody?"

"Why, no, I jus' spotted Pete's car and then saw him lyin' there." Carroll pointed vaguely at the body then looked around guiltily as if the killer might be standing nearby, waiting to be arrested.

"Okay, we're in danger here," said the agent quietly. "You've got to help me. Let's go back inside so that I can finish...checking the computer system." The pause was too brief for Carroll to pick up on it.

"No, I've got to report this," the patrolman said, reaching for his portable radio.

"No, not yet," the agent declared. There was more force in his voice now. "It's too dangerous. Let's wait until we know we're safe. We need to secure the lab. Come with me," and he turned and walked back into the building.

Carroll looked at the radio in his hand, shrugged and let it slide back into its holster. *Well, he's CIA. Gotta know what he's doin'.* He followed the agent back into the lab, having to trot to catch up. They jogged back to the control center where the stranger looked up at the time-remaining bar that was tracking the download to his hard drive. It was nearly done. In a moment it reached the end and flicked off. Quickly he pulled the connector cable out of the port and shoved the drive into his pocket.

"What're we gonna do now?" Carroll asked, looking around nervously. "I really gotta report in..." He reached for the radio, turning away from the agent as he prepared to call.

Suddenly he felt something slip over his head, something thin and cold. It tightened around his neck and in an instant his breath was cut off. He felt a brief moment of surprise, then a spasm of horror. A cold steel wire knifed into his neck. In seconds his life was over.

The stranger carefully coiled the makeshift garrote, a length of thin piano wire with small wooden handles at each end. Placing the device in his pocket he lifted the corpse and carried it over to the swivel chair where Stuart Reeves had sat earlier that day. He arranged the body as if it were still alive, left hand stretched out on the keyboard, right hand curled over the track ball, head cocked upward as if looking at the screen above. Satisfied he stood back to admire his handiwork.

"Good night officer Carroll," he said, his thin voice as dead as the man he had just killed. He turned and walked swiftly away.

Chapter 14

*New Mexico Institute
Thursday, 6:45 a.m. MST*

The little auditorium was filling up long before the 7 o'clock time that Dr. Tommy Dawson had set for his briefing. Kate and Chris grabbed front row seats clutching cups of coffee and bagels from the refreshment table set up in the back. Kate busied herself getting organized, pulling out her mini video camera and checking the memory card on her camera.

Chris looked around the room. All the key players in the project were there. Some were already seated, others were still milling around the coffee bar or standing in small groups talking among themselves. He noticed a couple of men that seemed to be staying apart from the others and focused his attention on them.

One was Rahman Singh, a computer scientist that he had met. With the cooperation of Paul Bonner Chris had glanced over Singh's security jacket along with a number of others. A third-generation American of East Indian origin, he recalled, going down a mental checklist. Graduated from M.I.T., got his doctorate at the University of Illinois, home of the National Center for Supercomputing Applications. Tenured full professor at Tulane. Married to a medical doctor from an old-line New England family, three children.

There was absolutely no reason to suspect Singh of anything, and Chris shook his head. *I'm falling into the trap of racial profiling here,* he told himself with disgust. *But still...* As he watched he noticed Singh look around carefully, as if checking to see if anyone was watching him. Chris averted his eyes but used his peripheral vision to continue observing.

Singh reached into an inside pocket and drew something partway out, looked at it, then placed it back out of sight. Again Chris shook his head, trying to allay his suspicions. *Jones has me on edge,* he thought. *He'll have me seeing spies everywhere.*

Chris glanced away and sought the second individual who had seemed to be remaining aloof from others, but now the man had engaged two others in conversation and was laughing at some remark. *I've got to stop imagining so much,* Chris told himself.

He turned to Kate who had finished setting up her equipment. She had her Mac open on her lap and had clipped the mini video recorder to the top, facing the stage. Her note pad was folded open to a fresh page and the digital camera was close at hand. She turned to Chris and smiled sheepishly as he eyed her array of equipment.

"Well, have you had any further thoughts since yesterday?" she asked.

"Not really," he replied. "Whatever it is, it's going to be pretty complicated. We'll see if Tommy can explain things in ways we can understand. He can actually be pretty good at that, now that I think back. When he's in that mode he's like a teacher talking down to a class of morons, and that's pretty hard on big egos like the ones in this room. But he can get the points across."

"Well, as you can see I'm ready to take it all down. Maybe we can make sense of it later if he can't illuminate the subject for us."

"Great. I'll be glad to try to interpret for you afterward—if I can understand any of it myself."

Chris shifted in his theater-style chair and looked around the room again. Nearly every seat was filled now and there were others standing along the back wall, coffee cups in hand. He checked his watch. "It's nearly show time," he told Kate.

In a moment the house lights dimmed. Dawson stepped from behind a curtain and strode up to the lectern. He was wearing a T-shirt with the logo of the Albuquerque Isotopes baseball team, jeans with a western belt and a pair of suede wrangler boots. He picked up a wireless mike and a laser pointer and turned to his audience. For a moment he just stared around the room, a look of curiosity on his face. Then he laughed out loud and turned toward the projection screen behind him. He lifted the mike.

"All right boys and girls it's time for today's lesson. Are you ready to have your minds blown? I hope so, because I'm here to do it."

A murmur rippled through the room.

"He really can be grating can't he?" Kate whispered to Chris.

"Oh yeah, he really can."

"First slide!" Dawson commanded and a simple diagram appeared on the screen. It showed nothing but a dot, a line, a square, and a two-dimensional representation of a cube. Dawson swept the laser pointer's red arrow across the screen.

"This is your basic geometry according to Euclid and all those Greek buddies of his, back around 300 B.C.," he said. "Here we have what we call a point," he wiggled the red arrow over the dot. "A point is a non-dimensional object. It has no dimensions at all."

He moved the pointer to the line.

"If you move a point through space, you create what we call a line. That's a one dimensional thing." He ran the arrow across the line to demonstrate.

"Chris, this is too simple," Kate whispered.

"It's okay," he replied. "I think he's got to start from here to get to wherever it is he's going."

Dawson had continued on to the square. He drew the red arrow up and down the left side of the square saying, "Now you may notice that the side of this next figure, which we call a square, is actually a line. Just as we moved the point in order to create a line, we can move the line sideways to create a square." He moved the pointer across the square to the other side. "Now we have a two dimensional thing. We can't really call it an object, because that implies something with more than two dimensions."

He moved the pointer on to the representation of the cube.

"Now here we have exceeded the ability of our two-dimensional screen to portray the next geometrical shape, which is supposed to be a cube. We can create a cube by moving a square through space, just as we used a line to sweep out a square. Bear with me here and imagine that this cube actually exists in three dimensions. If it really did, it would be the first object that we've seen here.

"These three dimensions are what we've used to measure the Universe, ever since the Babylonians and Egyptians figured out how to cut stone and build temples. Next slide!"

He walked over to the lectern and poured himself a glass of water. Sipping from it he scanned the crowd suspiciously, like a mall guard watching for shoplifters. He set the glass down and turned to the slide, which was a picture of Albert Einstein.

"You all know who this guy is. He came along about a hundred years ago and upset everything by claiming that there are actually four dimensions, not just three. As you probably know, he demonstrated that time could be considered as a dimension, just like the three dimensions of space. He created the concept of what he called space-time, a four dimensional matrix that he thought encompasses the Universe, explains gravity, and a lot of other things. Big stuff! Made quite a stir."

He whirled the red arrow around Einstein's face for a moment then commanded, "Next slide!"

Another portrait appeared on the screen.

"This is a chap whose name was Max Planck. Imagine having a name like that—it reminds me of a very large wooden board." He paused for a beat and there were a few chuckles among the audience. "Anyway, he and some other fellows figured out the basics of quantum mechanics. And it turned out the rules that worked in the larger Universe, as devised by Newton and Einstein, didn't work at the very small scales of atoms and particles."

He whirled around and faced the crowd. "But you all know that, don't you?" He pointed his finger, picking out a physicist in the second row. "You know that, right?" The man nodded uneasily. "And you," he said, moving his finger to another face, "you know that?"

Dawson did a little dance on the stage. The red arrow of the laser pointer jiggled wildly across the ceiling.

"Well, what you know is wrong!" he shouted. "Part right, but entirely insufficient. And now you're going to learn the rest, the part you don't know."

There was a stir as his audience broke into whispers, many shifting uncomfortably in their seats. Dawson strode back to the lectern and took another drink of water.

"Next slide!" he demanded, turning back to the screen.

It seemed that a mistake had been made. The diagram that appeared was the same as the first one, the one that showed a point, a line, a square and a cube. Dawson contemplated the screen for a moment then turned back to the audience.

"Error!" he shouted. "That's what you're thinking—error! Old Tommy fucked up, right? Is that what you're thinking?"

There were murmurs but no one spoke up.

"Well, no I did not, because I'm about to show you something that's going to blow your mind. This is the moment, boys and girls, when everything you know about quantum mechanics and relativity—how the Universe works—is going to go right into the old trashcan. Right there, into the burn bags, dumpsters or shredders—whatever you use to get rid of old, useless stuff."

He aimed the laser pointer at the diagram on the screen, sweeping it from left to right to indicate the four images.

"You think this looks familiar, don't you?" he demanded. There were some grunts of agreement from the crowd. "And it is familiar, it looks just like the first slide I showed you, the one that explained the three dimensions of space. But it's not the same, not at all. And why do you think that is?"

He whirled around again, scanning the theater with suspicion. Satisfied that no answer was forthcoming he turned toward the screen.

"What's he doing?" Kate whispered in Chris's ear. "Has he gone nuts?"

Chris grimaced and gave a little shrug. "Let's see where he's going with this," he whispered back.

Dawson pointed the laser arrow at the dot on the left side of the diagram and raised the handheld mike to his lips.

"This is a point," he said. "But it is not a geometry point like we investigated in the first diagram. Imagine that this is a point in time, not space. Yes, time, that so-called fourth dimension that according to Einstein made up part of his oh-so-famous space-time continuum. But see, since it is a point it cannot have a dimension any more than a point in space can."

He glanced around and noted that his audience was struggling to keep up with him. There were more expressions of incomprehension than enlightenment..

"Just bear with me here, I'm trying to make this as easy as possible. Come on, children—imagine with me that there can be a state in which time can have no dimension. That state existed before the Universe came into being. Until that instant, the moment of the Big Bang, time had not yet begun. Like space it sprang out of nothingness and began to expand.

"And see what happens! As this point in time moves, it traces out a line, just as a geometric point does." He swept the pointer arrow up and down the vertical line. "Time ceases to be non-dimensional and becomes one-dimensional. That's how we've always viewed it, as a one-dimensional line, the so-called Arrow of Time. That's how old Albert imagined it.

"But think about this: What if we could sweep that line of time sideways, just as we did with the geometric line." He pointed at the square. "We would form in two dimensions of time something you've never imagined before, a *time-square* if you will."

He swirled the arrow around the square.

"And then, if you were to move that square in yet another direction in time, you'd create a *time-cube*, wouldn't you?" He pointed to the drawing of a cube.

"And there's the answer, boys and girls. We do not live in a four-dimensional Universe, as we've believed. No, not at all, for you see time can exist in more than one dimension!" He paused to glare around the room, which had grown eerily silent.

"We live in a far more complex Universe than any of you have ever imagined—a six-dimension space-time matrix—a Universe that has three dimensions of space"—he raised his voice to be heard over the clamor now beginning to rise from the audience—"and three dimensions of time!"

At that point the auditorium broke up into chaos. People began to stand up, questions were shouted, strident voices rang from around the room and heated arguments began to break out.

In the center of the stage Dawson stood calmly above the melee, his legs planted squarely, arms crossed with the mike and laser pointer clasped in one fist like the scepters of a king. Placidly, a faint smile on his lips, he gazed down at the agitated mass of people like some mythical patriarch, a bringer of wisdom far beyond the realm of the world as it had been known and understood.

Just then there was a crash in the back of the room. The door slammed open and Paul Bonner strode in, his face grim. Ignoring the chaos he walked up to the stage and turned to the crowd, shouting for attention. He waved for Stuart Reeves to join him then gestured to Dawson to come closer. He took the mike and faced the room, which had grown quiet once more.

"Please, everyone, stay calm. There's been an attack on the Quantum Center. During the night two of our security officers were killed. The bodies were discovered less than thirty minutes ago. I need to ask Doctor Reeves to come with me right away to aid in the investigation. Please go about your regular duties and stay away from the center until further notice."

Grasping Reeves by the arm he turned and led him quickly away.

Dawson greeted this news with irritation, part concern, part disappointment at the interruption to his harangue. He shook his head like a vexed bull then raised the mike and declared, "Well that seems to be all for now. Class dismissed!"

Chapter 15

Quantum Center
Thursday, 10:15 a.m. MST

It was several hours later and the Quantum Center was nearly empty. The control center was festooned with yellow tape left by the crime scene technicians. Stuart Reeves gathered his key supervisors around the large work area used to examine plotter diagrams. Chris and Kate had been invited to observe and were standing a few feet back from the rest. There were now four security officers stationed around the room, uneasy expressions on their stern faces.

"We've determined that the intruder, the killer, attempted to download the results of our trial run earlier today," Reeves announced to groans of dismay. "But it's all right," he continued, "They didn't get what they were after." The groans turned to sounds of relief.

"Somehow they were able to break through the outer password firewall, but we had a backup in place," he explained. "There was a fake set of data behind the front password, and that's what they downloaded. They didn't see the real code behind the second firewall. They're going to be pretty surprised when they learn they were outfoxed."

The status report continued for several minutes while Reeves explained what they knew or could guess about what had transpired during the night. After answering several questions he turned the discussion to what Tommy Dawson had revealed to them a few hours before. A heated discussion began but in moments the group fell into silence as they contemplated the possibilities of a universe with multiple dimensions of time.

Reeves wrote a few notes on a yellow pad then looked around at his associates. Their expressions were blank. They could have been refugees from some terrible disaster.

"We can agree that there's a lot more to this than Tommy's told us," Reeves summarized. "None of us can understand his theory because he hasn't shown us the basis behind it—but just to assume he's right, that there are three dimensions of time—leads to some interesting possibilities.

"As several of you have suggested, if qubits can be created in six dimensions, including three of time, it might explain the way the quantum computer can operate as fast as it does. In effect, the time we observe, the few seconds that it seems to take the computer to perform a calculation, might be irrelevant to the time dimensions in which the operations actually take place."

"So...you think it might actually take a lot longer, but it's happening in some other dimension?" one of the engineers asked. "That sounds like something out of science fiction." Several others grunted in agreement.

"Well, remember that many things we now take for granted were originally suggested by science fiction writers," Reeves remarked with a wan smile. "Just because we don't know something doesn't mean it isn't possible. If that weren't true, we'd still be living in caves trying to figure out how to invent fire."

"Yes, that's right," chimed in an eminent physicist. "But I just can't see how time could exist in more than one dimension without our having noticed it before."

A new voice rang out. "That's because you've all had blinders on!"

Tommy Dawson was striding toward the cluster of scientists. He still wore the Albuquerque Isotopes T-shirt, now enhanced by a large coffee stain. He was wearing his "Quantum Cowboy" cap.

"It's been right there in front of your noses all the time," he declared.

As he came nearer the group spread apart to form a semi-circle around him. Dawson stepped up to the table, turned to lean against it and raised an index finger in the air. "Are you ready for the next lesson?" he asked. A dozen heads nodded.

"Okay, let's talk about space again, those three dimensions that you all agree exist. What makes them different?" He looked around inquiringly.

"Well, they measure different vectors…" offered an engineer.

"Ah, yes—but what's different about dimension A, which might measure in direction X, and dimension B which might measure in direction Y? There isn't any difference except in their orientation. If you were to rotate an object so that dimension A took the place previously occupied by dimension B, and vice versa, would you be able to see any difference?"

"Umm, I guess not," the engineer admitted.

"No, you wouldn't because the three dimensions of space are really the same except in their orientation. Is that clear?"

Several heads nodded. Kate turned to Chris with a perplexed look. "What's he getting at now?" she whispered. He shrugged and shook his head.

"So you agree that space consists of three dimensions that are actually identical to each other? That have exactly the same qualities?" There was some hesitation, but again several people nodded.

"Now let's look at time, which is the real issue here. You say that there couldn't be three dimensions of time." He suddenly pointed his finger at a nuclear engineer. "Why do you think that?" he demanded.

"Er, well, because we, uh, measure it as a single quantity," the man stammered. "You can use a clock to measure time and it always goes in only one direction. We don't see time going sideways or backwards, after all."

"How do you know that? Just because you can't 'see' it or measure it, does that mean it doesn't exist? Come on, does it?"

Reeves chimed in. "We were just discussing the fact that because we don't know something doesn't make it impossible."

Dawson whirled toward him. "Exactly! That's precisely correct. But what seems so strange to me is that we have, all of us, seen different versions of time right in front of us, from the day we were born."

There was a moment of confused silence. Dawson smirked and took a moment to boost up and sit on the table edge. He placed his palms on his knees and beamed at his "class" like a Zen master holding court.

"Well, can you see it?" He spoke softly then his voice rose to a shout. "Can you! It's right there, right in front of you! Just look!" His beaming smile turned into a glare.

No one spoke and for a long time Dawson just stared curiously at the group. Finally he sighed.

"All right, I'll point out the fucking obvious. You!" he pointed at a computer scientist, "What do you call time when it applies to something that happened last year?"

"Why, I... I'm not sure what you mean..."

"You call it the past, you idiot," Dawson roared. "That's one kind of time, *past time*. Now," he pointed at the nuclear engineer, "how do you refer to the time for something that hasn't happened yet? Never mind, I'll tell you. You call it the future. That's another kind of time, *future time*."

"Look," the first man protested, "I hardly think insults are in order here..."

Dawson turned his gaze back to the computer scientist and grinned. "Okay, I apologize. I'll give you a chance to prove yourself innocent of idiocy. What's the third kind of time that we're all familiar with, that we see every day?"

"Why, er, I guess you mean the present," the computer man said.

"Yes! Give that man a Havana cigar!" boomed Dawson. "You can take off your dunce cap and come back out of the corner. Yes, exactly, the third kind of time is the present. *Present time*. Now, what is interesting about these three kinds of time?" He looked around inquiringly.

"Why, they're all different!" said a mechanical engineer, a surprised look on his face. A murmur went through the circle of scientists and engineers.

"Exactly, they're all different. Compared with the three dimensions of space, which are actually just the same except in their orientation, these three kinds of time are quite different from each other. Let me tell you just how different they are."

He hopped down from the table and began to pace in a little circle. The ring of listeners widened to give him room.

"First, let's think about past time. That's the kind of time most familiar to us, because it exists in a linear way, stretching back from present time. We visit it in our memories. What is past is fixed. It cannot change. Everything that's happened since the Big Bang, which is the moment that time and space began, is stretched out through the Universe like an expanding web. Each event that has ever occurred, from the flipping of an atom to the collision of galaxies, is frozen in the record of past time. It's an archive of everything that has ever happened."

Spotting some bottles of mineral water on a nearby table he paused for a moment, and made a beeline to grab one. Opening it he took a gulp, then wiped his mouth on his sleeve and picked up his narrative.

"Because light does not travel instantaneously, we can actually observe events that happened millions and even billions of years ago. When our telescopes detect a supernova in a distant galaxy, we're seeing an event that occurred far in the distant past, before we humans existed, perhaps even before the dinosaurs. The farther out into space we look, the farther back we see into past time. Look out far enough and we see events that took place even before our Earth condensed from a molecular cloud."

He raised an index finger in the air. "That's point one. We can think of past time as a backup of every event that has taken place during the entire life of the Universe. Now, what about future time? It's not the same as past time, is it?" He pointed at the computer scientist. "Tell me how it's different," he demanded.

"Why, I guess it's different because it hasn't happened yet," the man replied. "But...doesn't that mean that it doesn't exist? How can any kind of time exist until it has actually happened, that is, uh, well, until it's in the past?"

"Aha, no!" Dawson crowed. Future time hasn't 'happened' yet'—but that doesn't mean it doesn't exist. How can the past be created if there is no future? The two are linked by present time, and past time is not created out of nothing. In quantum terms, future time exists in the same way that a nuclear particle does under the concept of the Uncertainty Principal. It exists as an unrealized potential."

He took another swig from the bottle of water, dribbling some down the front of his T-shirt.

"We've seen that past time is fixed, frozen. Future time, on the other hand, is in a state of flux, uncertainty. You might try to visualize it as a cloud of virtual potentials, all the things that might happen. We're quite familiar with that concept. Think of actuarial tables that insurance underwriters use to guess whether you're going to live for another fifty years or another five minutes? Both of those are possibilities.

"And what happens as future time grows nearer, as we approach a given location in the future? Why, the potential begins to narrow doesn't it? Let's say that you bought that insurance policy fifty years ago. Now half a century has passed so a great many of the uncertainties about your lifespan have been eliminated. We know that you didn't drop dead five minutes after signing the contract. We know you made it through fifty years of possibilities that no longer exist. They've dropped out of the cloud of potentials."

He turned suddenly and pointed at Chris. "And what does that mean? Come on Fisher, you were always a smart one."

Chris thought for a moment before replying.

"I think it means that future time can change, and since it can change it does exist."

"Bingo!" Dawson stepped forward and gave Chris high fives. "You've got it dead right. Yes, future time does exist and it is a cloud of ever-changing potential that comes into focus as it draws nearer to the present.

"And now for the 64 billion electron volt question: What is present time? Anybody have any ideas about that?" He stared around at the gathered experts, then drained his bottle of water and tossed it in the general direction of a waste can. It clattered across the concrete floor, accentuating the growing silence.

At last a hand went up. It was Kate's and everyone turned with surprise.

She hesitated, looking around at the staring faces of preëminent scientists and engineers. "I probably shouldn't even say anything..." Dawson waved her on.

"Well, it seems to me that the present is, well, whatever it is that turns the future into the past. Something like a loom, weaving fixed and absolute past time from the uncertainties of future time."

For a moment even Dawson seemed taken aback. He looked at her with astonishment then whirled around to face the others.

"From the mouth of a babe," he cried. He turned back to Kate. "My congratulations, ma'am. Let me introduce myself. I am Tommy, the time genius. And you are...?"

"Why, I'm Kate Elliott. I'm a science writer with..."

"*Science Today*," Dawson finished for her. "I've read some of your articles. Nice job on the super hadron project at CERN." He turned back to the gathered scientists and engineers.

"Yes, present time is exactly as she described it, like a filter, or as she said a loom, that processes future time to create past time. Actually, I like to visualize the present as a wave front moving forward into future time and condensing a trail of past time behind it. According to my theory, the time dimensions are expanding just as we've observed the expansion of the Universe in space."

"Wait a minute," one of the engineers interrupted. "I know the Universe is expanding, but you're saying time is expanding too?"

"Of course," Dawson replied with impatience. "Present time is the result of the expanding time element of six dimensional space-time."

"But," the engineer persisted, "the expansion takes place at the edge of the Universe, and yet we see present time everywhere."

Dawson looked down his nose at the questioner before replying disdainfully.

"You seem to think that space is expanding like a balloon, and the only change takes place on the surface of the balloon. That is false. Space is expanding everywhere. So is time. Present time exists everywhere."

Dawson turned away dismissively before continuing.

"Another thing about present time: It really has no length at all. And what other observation about the three kinds of time does that suggest?" He glanced around then shrugged. "Well, I'll save you a few neurons and tell you. Past time is long, as long as the Universe is old. Future time is even longer, perhaps not infinite but extending all the way to the end of the Universe, should that even happen. And present time? It's infinitely short. That is to say—and this is deeply ironic—present time cannot be measured in any unit of time that we know. No matter how small a unit we might use—microseconds, nanoseconds, femtoseconds—it will not suffice to measure present time. It constantly moves on into future time leaving past time behind it. From our perspective, present time is merely an interface between past time and future time."

"Jesus," breathed the eminent physicist. "You're right, there *are* three kinds of time, and they *are* very different from each other, far more different than the three dimensions of space. But...how can we recognize these different kinds of time as dimensions?"

Tommy Dawson smiled and his eyes twinkled.

"It was proven by the results of the trial run yesterday," he asserted. "I predicted, privately of course, that the computer would work far faster than could be explained by the old laws of quantum mechanics. I based that prediction on my theory of six-dimensional space-time, something that I've been working on for years. I haven't published my theory because I've been waiting for proof—the proof we saw yesterday. You can be certain that it's all written down, and my journals will be yours to examine just as soon as I can organize them."

He turned to Reeves and held out his hand. "Thanks for listening to me. I hope you've all learned something. I know I can be a bit, well, I've been called rude and I apologize for that."

In an almost courtly manner quite unlike his usual demeanor he proceeded to shake hands with each man and woman in the group, ending with Kate. He grasped her right hand in both of his and awarded her a broad grin.

"Ms. Elliott, thank you for your interest in this work of mine. I don't often grant interviews, but my door will always be open to you."

And with that he strode away as the meeting broke up into buzzing speculation.

"Jesus," breathed the eminent physicist.

"You can say that again," an engineer remarked.

"Jesus," repeated the physicist, "Christ Almighty."

Chapter 16

Los Alamos, New Mexico
Thursday, 11:45 a.m. MST

Chris and Kate split up outside the Quantum Center, agreeing to meet later for lunch. Kate said she was going to her hotel room to freshen up and call her office. While he didn't mention it to Kate, Chris wanted to check in with Paul Bonner.

As he walked into Bonner's headquarters he spotted the security chief in an intense discussion with a small group of his senior staff. They were gathered around a large whiteboard on which a list of clues and speculations about the night's killings had been written. There were question marks next to most of the items.

Bonner glanced up as Chris entered the room and raised a finger to indicate he would be with him in a minute. While he waited Chris gazed at the whiteboard, absorbing the information. It was clear that evidence was lacking. The mode of death for both men was strangulation by a thin cord or wire. No fingerprints or other forensic evidence had been found that could link to any suspects. There was a note indicating that the killer or killers had apparently had inside knowledge, including the ability to access the building and to break the outer password firewall on the computer.

After a moment the meeting broke up and Bonner waved Chris into his office. He closed the door and walked over to a small conference table in one corner. He pointed to a chair and sank into another with a tired sigh. For a moment the men sat looking at one another.

"Looks like you don't have much to go on," Chris offered.

Bonner sighed again and leaned back in the chair. Staring at the ceiling he twisted his neck first left, then right. Chris heard the soft pop of stressed tendons.

"No, it's like one of those locked door mysteries you read about in novels," Bonner replied. "We don't know how the killer got into the lab; we don't know what he did there except to try to download the results of the experiment; and we don't know how he managed to overcome the guards."

"I saw a note about strangulation," Chris said. "That sounds like he may have used a garrote."

"That's our presumption. It appears the men were strangled quite violently by a thin strand, probably a metal wire. Their necks were cut all the way around. They died almost instantly. In the case of Officer Carroll the jugular vein was severed so he actually bled out." He shuddered. "It was nasty, real nasty."

Chris said nothing. After a moment Bonner sat forward in the chair.

"Look, I know you didn't plan to get involved in anything like this. Jones told you there might be some chance of danger, but believe me we had no idea that it would come to murder." He paused and Chris could read the pain in the man's eyes. "And to think that two of my men..." Bonner grimaced then reached up to wipe his forehead with a sweaty palm.

"I understand," Chris said quietly. "I'm glad to keep working with you but I feel like I haven't been much use."

"Well, join the club," Bonner grunted. "Look, there isn't much I can tell you. The lab's going to be in lockdown for at least a couple of days now. The Feds are bringing in some more support for us. Until we find the killer all of us have to be on guard."

Chris drew circles on the tabletop with an index finger. "I'm particularly concerned about Tommy," he said. "He's admitted he has documentation of his new theory, and anyone who wants to know his secrets could get them either by kidnapping him, or stealing his papers and lab notebooks. He could be in personal danger."

"Yeah, I know. It worries me too. We've got a plainclothes team watching him around the clock, on the sly. As you know he's pretty independent and not very cooperative sometimes."

Chris laughed. "You got that right." He thought for a moment. "I probably know Tommy as well as anyone here. I think he trusts me. Would you like me to have a chat with him?"

"Well, it wouldn't hurt." Bonner rubbed his forehead again. "Ideally I'd like to get him clear out of the picture, in a safe house somewhere. Think you could convince him to take a little vacation?"

"I can try. Why don't I go see him right now? Do you know where I can find him?"

"Oh I hope to heck I can! If my surveillance team's lost track of him I'll have their asses stuffed and mounted on plaques for my trophy wall." He grinned and picked up a phone. "Check whereabouts of Tango Delta," he told the person at the other end. Covering the mouthpiece he turned to Chris. "That's our rather lame code name for Tommy," he explained. Chris nodded, understanding the military alphabet code for Dawson's initials.

In a moment Bonner turned back to the phone, listened for a moment then hung up the receiver.

"He's en route to his house at the New Mexico Institute. Should be there in about three minutes."

"Great. Can you tell me how to get there?"

"I'll do better than that, I'll take you there myself. Come on."

Bonner grabbed his Stetson from a hat rack and pushed in onto his head. Chris followed him through the office and into the parking lot. Bonner clicked the remote on his distinctive black Tahoe, parked in the number one position right outside the door. "Hop in."

"Actually, I think I'd rather follow you," Chris told him. "I have another appointment and that'll save time later. Also, you won't have to hang around and wait to bring me back."

"Sure," Bonner said and opened the door. "Where you parked? Oh, there in the guest spot—at least that's one mystery this genius can solve." He shook his head glumly and slid into the seat as Chris jogged to his Jeep and prepared to follow.

Less than ten minutes later Bonner pulled through an ornate gate onto the grounds of the New Mexico Institute. Chris followed the Tahoe around several winding curves, past the main buildings and into the residential area. Bonner pulled over and signaled for Chris to come forward. He buzzed down the side window and leaned out, pointing.

"That's his house, the third one over from that big pine. I'm thinking he shouldn't see me, so I'm going to leave you here. Thanks, and good luck. Let me know how it works."

"Okay, sounds like a plan," Chris responded. "Where are your security guys?"

"Oh, they're around. I hope you don't see them. Let me know if you do and I'll kick their butts." He chuckled, put the truck into gear and made a wide U-turn, giving Chris a two-fingered salute as he spun his wheels in the gravel.

Chris watched the black Tahoe disappear around a curve then turned to examine the house. It was one of the larger units in the Institute's residential area, as suited Tommy Dawson's seniority. All were essentially the same Santa Fe style, adobe structures modeled on ancient pueblo designs. Expansive lawns and landscape plantings surrounded the scattered buildings. Beyond was the edge of a thick forest.

Chris parked in front of the house and got out of his Jeep. He noted Tommy's green Range Rover parked to one side. There was no sign of anyone else around. He walked up to the door and lifted the heavy metal knocker, using it to tap three times. There was a muffled shout from within.

"Well, whoever you are, come on in!"

Chris grimaced and shook his head. *I could be a team of terrorists armed with machineguns.* Sighing he opened the door and stepped inside. Dawson was standing in the middle of his living room, leaning over the keyboard of a Sun workstation. He glanced up with a little look of surprise then waved Chris forward.

"Fisher my boy! What a pleasure! Welcome."

"Thank you Doctor Dawson, it's good to be here. I wanted to drop by to see you in person, about what happened last night."

Dawson's face darkened and he turned away for a moment.

"Terrible thing, terrible." He fumbled at a box of tissues, extracted one and used it to wipe his eyes then blew his nose. "I don't understand such things," he said, looking searchingly at Chris. "Here, here, sit down. Make yourself comfortable." He pushed a swivel chair in Chris's direction and perched himself on the edge of the massive partner's desk that dominated the center of the room.

Chris looked around. The space was more like a computer science laboratory than a living room. There were several office chairs and workstations, but no couches, easy chairs, end tables or other examples of everyday furniture. Chris could see to the left into the kitchen. A dining area was in the far corner, a table with three chairs. The remains of a breakfast and a laptop computer were on the table along with piles of papers and journals.

"Until this week I haven't seen you since you left Cal-Tech," Chris began.

"Yes, oh yes, but I kept track of you, I did," Dawson said with a warm smile. "Great work! That thesis of yours, one of the best I've seen. And, those fatuous dolts that put you through the Spanish Inquisition..." he broke off at the sight of Chris's questioning look and added, "You know, your oral exams? Anyway, they were so impressed it's a wonder they didn't fail you just out of pure cussed jealousy." He laughed and leaned over to slap Chris on the knee.

"Well, that's very kind of you to say...," Chris murmured, looking slightly embarrassed. "I had no idea you'd followed my career. Thank you. I owe a lot to my favorite mentor, you know."

"Oh? And who would that be?"

"Why, you of course."

"Oh, well thank you. It was my pleasure. I just wish we'd been able to have more time together there at Pasadena. I missed being able to see your doctoral work through to the end." Dawson hopped down from the desk and walked into the kitchen, shouting over his shoulder, "Coffee? Fresh brewed, not like that shit at the lab." He returned in a moment with two steaming cups and handed one to Chris. "Still take it black I presume?"

The two men sipped in silence for a moment then Chris set his cup down and looked Dawson in the eye.

"Doctor, I don't presume to understand your new theory but, well, I wonder if even you realize how important it is?"

Dawson settled in one of the desk chairs and swung it around to the left, then the right, and finally all the way around like a listless child.

"Well, I do understand," he said at last with a tone of smugness. "In fact, there's more to it than you can even imagine. As I said yesterday, it's the key to understanding the entire Universe…"

"Yes, I have no doubt," interrupted Chris. "But do you realize that your discovery could be something that others would want to get?"

Dawson smiled. "Why, of course they'll be interested in my discovery. This is going to be the greatest thing since Galileo…"

"No, no, you still don't understand," Chris interrupted. Dawson looked up in confusion. "I mean that there are enemies of our country that want to have your secrets for themselves. They've already killed two men last night."

Dawson absorbed that for a moment.

"Oh," he murmured at last. "You think that's what it is? I thought it was just a burglary or something…"

"No, Doctor Dawson…" Chris began.

"For God's sake, call me Tommy. And quit pussy footing around. Spit it out boy!"

Chris began to talk and Dawson listened impassively. Long minutes passed as Chris explained the dangers, the fears that the secrets of Dawson's theory might be stolen and put to use by enemy states, the possibility that Dawson himself might be in danger because of his special knowledge.

"You can see this is serious," he concluded. "I think you need to drop out of sight for a while, go to a place where you'll be safe."

Dawson pondered this for a moment then nodded slowly.

"I see your point," he mused. "Where would you suggest? I could check into a hotel... There's a Motel 6 not far away."

"No, not that. We need to get you into a secure place where you can be protected around the clock."

"Well, I'd need to have my papers, my computers or at least one and a few hard drives..."

"All that can be arranged," Chris assured him. "Let me make a phone call and we can get started right now." Dawson nodded toward a phone and Chris dialed Bonner's cell number.

"Paul, Chris Fisher. Tommy's ready to cooperate... What? Oh, sure. Thank you." He set the phone back down and looked at his watch with an ironic smile. "Help is on the way," he said watching the second hand.

Thirty-two seconds later there was the sound of two vehicles pulling up in front of the house. Four seconds after that there was a knock on the door and a moment after that a face appeared in the rear window. After eight more seconds four plainclothes agents were crowded around an astonished Dawson. He turned to Chris with a look of incredulity.

"How'd you do that?" he demanded. "What are you, David Copperfield or something?"

Chris chuckled and shook his head. "They've been watching out for you all along," he explained. "They'll help you get your things together and escort you to a safe place. I've got to go now, but I'll be in touch when I can. Thank you Tommy."

"Well, thank you too, I guess," Dawson mumbled uncertainly. He turned to a stack of documents and started to shuffle through them. "See you later then."

Chapter 17

Los Alamos, New Mexico
Thursday, 2:20 p.m. MST

Chris and Kate met for a late lunch at a family cafe. They sat at a square table in the middle of the room. It was past two o'clock and the place was nearly empty. Chris worked on a bacon cheeseburger with slaw and a side of onion rings while Kate nibbled on a salad with grilled chicken and honey mustard dressing. Finally she laid down her fork and looked up with a plaintive expression.

"Well, do you understand anything Dawson's said?"

Chris laughed and took a drink from his iced tea. "Well, not really. Remember that it was quite a long time before very many scientists understood Einstein's theory. Later every physicist claimed they could, but you still have to wonder about some of them."

"You mean they're unable to understand it?"

"Well, it's because the theory of relativity describes a model that isn't intuitive. It has to be 'understood' through the language of mathematics, and with a bit of, well, I guess you'd call it 'faith' to make it fit together."

"Oh, faith is it? What are you, a closet priest?"

"Not that kind of faith," Chris replied. "At least, not exactly. I mean it's not something you can just look at and say, 'Oh, I see, that's it,' like looking at a photograph. You need to learn different ways of thinking to get your mind around it. And it's my guess that if understanding relativity was difficult, getting Tommy's new theory is going to be twenty times harder."

"I can certainly believe that," Kate mused. "We're like the members of some primitive tribe who've never even seen a book of matches, suddenly confronted by a nuclear aircraft carrier."

"Heh, yeah, something like that." Chris picked up the last onion ring and took a bite. The cafe used heavy ironware dishes and the onion rings had been served on a thick six-inch diameter plate. He pushed it aside and took another sip of tea. The last customer left money on the table and left.

"Chris, I'm worried about Doctor Dawson. After last night I'm afraid something might happen to him. Do you think he's safe?"

Chris smiled. "Oh, yes, he's being looked after." He noted the relief in her eyes. "There's something going on that we don't understand, but he'll be okay."

He relaxed and let his attention wander around the room. The waitress had disappeared into the kitchen leaving them alone. The cafe had large plate glass windows opening onto the street, with red leatherette booths along the front and right side wall. The entrance was on the left, facing a cash register and reception counter. The rest of the room was taken up with ten square tables including the one at which they were sitting.

Idly Chris watched as a black Ford Expedition pulled up to the curb. The windows were blacked out so he couldn't see inside. He watched as two men stepped out of the SUV. They were large, broad shouldered and tough looking, both over six feet and in their late 20's or early 30's. They wore sunglasses and dark jackets cut full but not quite full enough to conceal the bulges in their left armpits. *Federal agents?*

Chris watched curiously as the two men stepped onto the sidewalk. They looked up and down the street then turned their attention to the cafe. As they stepped toward the door some animal instinct stirred in Chris's mind. He shifted his chair around to be able to stand up quickly. He caught Kate's eye and cocked his head slightly toward the men as they came into the cafe.

The first man through the door had blond hair in a brush cut. He was wearing a blue shirt and tie beneath a navy blazer, unbuttoned, with gray Dockers and heavy black shoes. He glanced around the room, seeming to avoid looking directly at Chris and Kate.

The second man was right on his heels, dressed in similar clothes but with a black sports coat and tan pants. Oddly he was wearing a pair of brown leather Wellington boots with thick soles. He had a hatchet face with a boxer's nose and dark hair with premature gray streaks. He too took a quick look around then stepped away from the doorway as if taking up position as a sentinel. The hairs on the back of Chris's neck rose as he sensed palpable danger.

The first man stepped further into the room, moving between the tables as if seeking something, like a wolf trying to look innocent while prowling among sheep. He continued to avoid looking their way. In a moment he was between their table and the door to the kitchen, about five feet away from Chris. He looked around as if choosing a place to sit. The second man stayed close by the door, about fifteen feet from where Chris was sitting.

Suddenly the situation changed. Both men looked directly at them and pistols appeared in their hands.

"Okay, keep your mouths shut and we won't hurt you," the first man told them in a low voice. "You're comin' with us. Don't ask questions."

Chris gazed steadily at the first man. He'd seen guys like this before. In SEAL training they were the ones who started out playing the tough guy act and invariably ended up washing out. These were common thugs, not federal agents.

He glanced at Kate and was surprised to see that she was remaining calm. *What do they want with us? Oh, I get it! They're working for someone who wants to pick our brains because we've been observers on the project. They think we know stuff—and, we do!*

He shifted further in his chair, now turned sideways to face the nearest man. The second man by the door was behind him now but Chris had his position mapped in his mind. He looked up at the blond and raised one eyebrow.

"I think you've made a mistake," he said mildly.

"There's no mistake. Just come along quietly." The man made a prodding motion with the pistol.

"No, you really have made a mistake," Chris said levelly and glanced at Kate. She moved her arm and knocked over her iced tea glass, causing a clatter as the ice cubes and liquid spilled across the table toward the thug. Startled, the man turned partly away from Chris—and it was time to act.

It had been years since Chris had relied on his primal instincts and defensive arts, but it was as if they had always been there. Time seemed to slow down as he made his play.

Beginning to stand he reached out with his right hand and grasped the heavy six-inch plate by its rim. He whirled around to the right, flinging the plate backhanded like a frisbee directly at the man near the door, flying fast and true right at his face.

He continued to whirl clockwise as he rose. Just as his right hand was releasing the plate, his left grasped the back of the chair he had been sitting on. By the time he was standing he had the chair in both hands and was lifting it straight up from the floor. Completing the 360-degree pirouette and taking a quick step forward he whipped the chair up behind him and brought it down like an axe blow on the blond thug. Two of the chair's legs broke on the man's shoulders and the solid seat sheared into his right arm. There was a crack of breaking bone and the pistol clattered onto the floor.

Continuing to spin Chris faced back in time to see the second man slumping to the floor. The plate, thrown like a blunt and heavy Ninja throwing star, had struck him on the bridge of his nose just across his eyes, a potentially killing blow. The man's sunglasses had shattered and shards of dark plastic were embedded in his face.

Whirling back to the first man, Chris stepped forward and kicked the pistol out of reach under the nearest booth. The man was still standing but dazed from the blow. His right arm hung uselessly. Chris delivered a karate chop to his neck then kicked him in the left kidney as he spun around and fell to the floor.

Total elapsed time was approximately two and a half seconds.

"Let's go!" he ordered and Kate grabbed her messenger bag and jumped up. They ran out of the cafe, leaving the two men unconscious, one possibly dead. Chris pointed at his Jeep, parked half a block down the street. "This way! Forget your car. Come on, move!"

They ran and Chris clicked the remote door opener as they approached. They jumped in. Chris fumbled with the key for a second then the engine roared and he spun the wheels as he tore away from the curb and down the street. For several minutes he dodged and jinked up side streets, down alleys and around traffic.

Kate rode in silence, half turned in her seat to look behind them.

"I don't think anyone's following us," she said at last. She looked at Chris with curiosity. "Just where did you learn, well, whatever it was you did back there?"

"Former Navy SEAL, remember?" He clenched the wheel, using the side mirrors and glancing from side to side to watch for threats. *It's like Afghanistan all over again.*

He slowed down and pulled into a parking lot beside a church, partly hidden by a row of bushes. Leaving the engine running he put the transmission into neutral and set the parking brake. He opened the door, jumped out and dropped to the ground to inspect the underside of the Jeep. Springing up he circled the vehicle, dropping several more times to look. At last he got back in the driver's seat and took a deep breath.

"What was that all about?" Kate asked.

"Looking for bugs," he said.

"Oh!"

She remained silent for a moment before speaking again.

"What did those men want with us? What's going on?"

Chris looked over at her then looked away to continue his surveillance.

"I think it's because of Tommy Dawson's time theory," he told her. "They think we know something about it because we've both been observing the experiment. In fact, we probably have a better overall perspective on the project than almost anyone else except maybe for Tommy, Stuart and a couple of others."

"Oh my God! You mean they were trying to kidnap us for information?"

"Exactly."

"What're we going to do?"

He smiled grimly and put the Jeep back into gear.

"We're going to get out of Dodge," he said, popping the emergency brake. "But first we need to collect a few things."

Kate remained silent as he drove back to a main street and continued toward the center of town. He pulled up beside the bus station. Leaving the engine running he set the brake. Reaching into his pocket he fished out the locker key and held it up.

"This fits a locker, number 42. Go get what's in it and bring it back here. Throw it in the back seat, jump in and we'll be out of here like a scalded cat."

She grabbed the key without hesitation, jumped out of the Jeep and ran inside. Chris counted the seconds in his head, recalling how he used to do that when his patrols were entering combat. *Thirty-eight, thirty-nine, forty, forty-one...* Kate was back. She opened the rear door and swung a large duffel bag onto the seat with a grunt, slammed that door, opened the front door and jumped in. Before she could even reach to close the passenger door Chris had popped the brake and dropped the Jeep into gear. With a screech of tires on asphalt it jumped away from the curb.

"You didn't tell me how heavy it would be," Kate accused. "It practically pulled my arm out of its socket."

"Sorry. I didn't know.

"Well, okay, just don't ever do it again." She smiled grimly and Chris chuckled.

"What do we do next?" she asked after a while.

Chris thought for a moment, trying to concentrate. *Good question. Gotta find someplace safe and take a look at that bag, see what Jones left for me.* He drove steadily toward the east end of town, looking for a place to get the Jeep out of sight. A secluded parking lot behind a row of apartments served the purpose.

They stepped out of the Jeep. Chris opened the rear passenger side door and unzipped the duffel bag. *Ah! Fun toys!* He spread it open so that Kate could take a look.

"What is it, an arsenal?" She reached down to touch a military style rifle, wrapped in a thin layer of bubble wrap. There were six loaded magazines beside it, and other objects could be seen below.

"That's an M-4 carbine with a folding stock," Chris told her. "It's a more compact version of the M-16 military rifle. These are 30-round magazines. And here," he lifted up the next layer of bubble wrap, "We have a pair of Beretta nine millimeter pistols, not my favorite but they'll do. And here's a bundle of cash, ten grand in new hundreds. Oh, and look here: cell phones, six of them. Must be the prepaid kind, use-and-pitch, untraceable.

"That reminds me," Chris said with a start. "We need to go into stealth mode. Let me have your cell phone and anything else that broadcasts a signal. We need to remove the batteries."

In a few moments they had turned off and disabled their cell phones, Kate's BlackBerry, and the wireless modem in her Mac laptop.

"From now on we've got to think and act as if we're in constant danger," Chris told her. "We're going to become invisible. But first, we need to call the cavalry." He picked up one of the pre-paid cell phones, punched the speaker button and dialed a number from memory. After three rings a deep male voice answered.

"Perry."

"Leonard, this is Chris Fisher."

"Sir, good to hear your voice."

"Likewise. Leonard, we have a situation. How many of the old team are you still in contact with?"

"Well, I think about a dozen sir."

"Can you mobilize them, as many as you can, for an action mission?

"When?"

"In three days."

"Where?"

"New Mexico."

"What kind of action?

"Be prepared for full combat op."

"That might take some doing, sir. You realize none of us are still active Navy. I've been retired for two years..."

"I understand. I've been recalled to duty and bumped to Lieutenant-Commander. I need backup I can trust."

There was a long silence, then: "Where do you want us?"

"We'll rendezvous in Albuquerque at 10 a.m. three days from now. Leave a message for me at the Avis desk at the airport. Use code names. I'm Mr. Frederick. You're Paisley. Have a hidey-hole set up somewhere nearby. Rent the best SUVs you can get, but nothing flashy, one for each two men. Use different rental agencies. I'll fill you in then, but you can tell the guys it's a national security issue, and it's big. I'll guarantee expenses."

"Aye, sir."

"And, Chief, keep it on the QT. This is dangerous stuff. No cell phone chatter, radio silence like in Afghanistan. Get prepaid phones for everyone. Remember the old rules."

"Of course sir. See you Sunday morning."

Chris clicked off the phone and turned to Kate. "That was my old second in command, Leonard Perry, former chief petty officer, one of the best Navy SEALs that ever was. I'm calling up some of my old unit."

"So I see," Kate said. "I won't even ask why a self-described 'mild-mannered professor' is suddenly a military superman."

"No, don't ask, not now at least. We need to find a way to hole up somewhere safe for the next three days. But first I need to make another call."

In a moment Chris was reporting to Paul Bonner about what had happened at the cafe.

"Jesus!" the security director swore. His voice sounded tinny on the cheap phone's tiny speaker. "Come on in and we'll put you and Kate on ice along with Tommy."

Chris looked inquiringly at Kate. She shook her head impatiently.

"No, we're going to stay in the open but out of the way. I've got some ideas. I'll be in touch soon. Meanwhile, don't try to contact us. As far as you know we aren't even around. Got it?"

"Got it. Anything I can do to help?"

"Not right now. You need to protect Reeves and his key guys, though. We've disabled our personal cell phones so there's no way to reach us. Talk to you soon." He cut the connection, turned off the cell phone and tossed it into a thick bush.

"Now we just need to figure out where to hole up. We need to get away from here, someplace where there are few people around. We need to avoid hotels, motels, restaurants and any public area. We need somewhere isolated, where we'll never be found."

"I know just the place," Kate said. "Remember the archaeological site I told you about? It's in the middle of nowhere, up in the San Mateo Mountains. The dig is shut down for the season now. We can camp there and I'm sure we'll never see another soul."

"You're up for roughing it?"

"Absolutely, I'm an old Girl Scout. But we need to get some equipment. We need more than just guns to survive in the desert."

"You're right. There's a WalMart in Espanola, about a half hour east of here. I checked it out online the other day. They'll have everything we need."

"Well, let's go then," Kate declared.

Chapter 18

Santa Fe National Forest
Thursday, 10:35 p.m.

Rahman Singh was a worried man. He clutched the wheel of his car tightly, his palms slick with sweat. He was steering down a winding dirt road in the Jemez Mountains, about fifteen miles south of Los Alamos. The surface was rutted and studded with rocks. The car bounced and the undercarriage sometimes grated on stones or hard packed earth.

Singh glanced at the clock on the dashboard. It was nearly 11 p.m., late in a very long day. He had followed his instructions, secretly recording Tommy Dawson's early-morning talk, an event that ended with the startling revelation of the killings that had taken place in the Quantum Center.

Singh had spent the day sequestered in his room, immersed in personal conflict and grief. He suspected that he was at least in part responsible for the tragedy.

Singh was not a religious man but on this day he prayed to the gods his devout grandmother had taught him to know. He tried to address his appeal to Vishnu, the four-armed god who maintained peace and balance in the world. But try as he might to focus on Vishnu, his mind conjured up another deity—Kali. Each time her image entered his thoughts he shrank with fear, huddled in a fetal position on the couch of his hotel room.

Kali, an avatar of the ancient Divine Mother, was a ferocious goddess whose role was to protect the Hindu pantheon. Singh remembered that she could also reward protection from the First Chakra, the instinctive fear of death that stalks the unenlightened.

But he could not free his mind of the fact that Kali was also the Hindu goddess of Time.

The words of Tommy Dawson had transfixed Singh and now that revelation was overlain by the intruding image of Kali and all that she embodied. Time. And Hindus believed in three separate worlds, those of the earth, the astral plane and the celestial sphere. Three dimensions of time?

Now as he drove down the rutted track his thoughts swirled, returning again and again to the day's events. *Murder!* He could not believe that his actions could have led to that! Were his hands stained with the blood of those two innocent men? *No! There must be a mistake, something terribly wrong.*

His attention wavered and one of his car's wheels caught in a deep rut, briefly throwing it out of control. Shaking, he pulled to a stop and turned off the engine. Wiping his sweaty face with a handkerchief he slumped in the seat for a moment, overwhelmed with grief and, yes, fear of the visceral kind that Kali was said to engender in the unenlightened.

It had started innocently enough, or so it seemed. Three days earlier a stranger had approached Singh. He seemed vaguely familiar, someone he might have seen around the LANL. He had identified himself as a government agent. He showed the computer scientist what seemed to be authentic credentials and sought his help.

There were spies threatening the quantum project, the man had told him. He and others from the government were seeking to identify and thwart those spies. Singh was asked to provide information that could be used in that effort. The approach was subtle, appealing to his patriotic duty. The computer scientist was convinced and at first it seemed completely innocent. The agent had merely asked for some general information about the quantum project, nothing that seemed out of line, nothing that could help any enemy of America.

Such a fool! He led me like a goat to the sacrifice.

Again late in the afternoon of the previous day Singh had returned to his hotel to find the agent waiting in the parking garage. The man had suddenly appeared from behind a concrete pillar. He stepped up to Singh's car as he parked. He gestured for Singh to unlock the doors, slipped into the passenger seat and directed the scientist to back out and drive away.

They rode in silence for a few minutes, the agent using hand gestures to indicate turns. They approached a dark street and Singh was guided to a secluded parking lot.

"What is it?" Singh had asked, turning to the stranger with a questioning look. "Is something wrong?"

For a long time the man had said nothing, merely looking around as if idly observing the scene. When he spoke his voice was cold and eerily calm.

"Doctor Singh, we're entering a very dangerous time." He'd paused, staring meaningfully into Singh's brown eyes. "Danger often calls for bold actions."

"What do you mean? Danger from what?"

The agent looked around again, then his gaze returned to Singh's face.

"The spies are at work. It's imperative that we gain the information we need to protect this project. You must give me the entry code to the lab and the user name and password to the main control center. As a member of the cybernetic engineering team, I know you have that information."

Singh was overwhelmed with surprise and shock.

"You know I can't give you that!" he blurted.

"Yes, you can—and you will." The voice assumed a threatening note.

"No, please, I mean it. I've sworn under the Official Secrets Act not to reveal any sensitive information…"

Glaring, the agent held up a hand and Singh stopped talking. He was seized with sudden confusion and fear.

"We can forget about what you've sworn. How much do you love your family? They are safe in New Orleans? There in your two story house with the little dog and the two goldfish? Your lovely American wife Caroline? Little Bobby? And the twins, let's see, they're in second grade now aren't they? Carla and Charlene. Cute as buttons, aren't they?"

A cold chill ran through Singh's spine. *What's this? Is he threatening my family?*

"How dare you!" he had half-shouted, turning in the seat toward the agent. He found himself staring into the mouth of an automatic pistol.

"You'll do as I tell you or your family will pay the price," the man had told him in a matter of fact way. "It's not a question of choice, you see."

Singh fell back into his seat. A rush of adrenalin had poured through his blood stream and he could feel a pounding behind his temples. Fear and anger blended in a mind-stopping mix of emotion.

"All right," he had said, his voice low, defeated. "I'll give you the pass codes, but you must promise my family will be safe."

"Of course. And there's one other thing."

"What?"

"There's a briefing tomorrow morning. I want you to record it. Here is a machine. You turn it on by switching this button upward." The agent had demonstrated. "It's very sensitive. Keep it out of sight. You can tell when it's recording by this little green light." He held it up for Singh to see then turned the switch back and the light went out. "Don't attempt to play back the recording. You will return the machine to me tomorrow night at eleven, at a remote place. Here are the instructions." He handed Singh a folded piece of paper. "Do everything as I've said and your family will be safe."

Back in his hotel room that afternoon Singh had picked up the phone and called his home in New Orleans. A recorded message began to play. It was a man's voice speaking in flat tones. A cold voice.

"You've reached the home of the Singh family," the recording had said. "No one is here right now. We don't know when they'll be back. This machine is not equipped to take messages." There was a click and the connection was cut.

Singh had fallen onto the hotel bed clutching his head in both hands. Tears seeped from between his fingers and his body shook violently for long minutes.

Now, having made the recording of Dawson's presentation as ordered, Singh was driving to his rendezvous with the agent, deep in the Jemez forest. He had not slept for more than 36 hours and his frazzled nerves were on edge. Before leaving for the rendezvous he had drunk three cups of coffee, adding extra packets of sugar to each for an energy boost. *God, I'm a nervous wreck! What's going to happen to me? To my family?*

His car jolted over the rough road. It was now ten minutes after eleven. *I'm late! Oh God! He'll be angry that I'm late.*

He tried to drive faster but the car began to swerve and buck so he had to slow down to a crawl again. He leaned forward across the wheel to look for the marker for which he had been told to watch.

There! At last he saw it, a blaze orange strip of plastic tied to a tree branch, just beside a narrow track that led off into the forest. Slowing he turned in, drove about a hundred yards and stopped. He turned off the engine and sat still, listening to the ticking of the cooling engine and the wind sighing in the darkness of the forest. As he had been told he left the headlights on.

"You're late!" Startled Singh jumped in the seat then turned to see the supposed agent.

"I'm sorry! The road..."

"No excuses! Get out."

Singh opened the door, released his safety harness and got out of the car. The agent stepped back a few feet and held out a hand.

"Where is it? Give it here."

Silently Singh handed over the recorder. The man examined it briefly then slipped it into his pocket.

"Is it okay now?" Singh asked plaintively. "My family...?"

The stranger grinned at him. It was not a friendly expression, nor a humorous one.

"Sure, your family's fine," he intoned.

They were the last words Rahman Singh would ever hear. The agent pulled out a pistol and shot him in the forehead at point blank range.

"They're just fine," the Iranian controller repeated as the computer scientist dropped to the ground.

Chapter 19

San Mateo Mountains
Friday, 5:20 p.m. MST

It was late the next day. Chris Fisher and Kate Elliott had spent most of the previous night winding their way through the back roads between Los Alamos and the San Mateo Mountains northeast of Grants, New Mexico. They had slept for a few hours in the Jeep, hidden in the trees off of a remote logging road in the Cibola National Forest.

Now they were settled near the ruins of the ancient pueblo, the place where archaeologists were uncovering evidence of the Anasazi people who had built the immense city in Chaco Canyon to the north. Here had been an outlying trading center of that long-lost civilization. For centuries it lay undiscovered beneath drifted sand and overgrowth of trees and brush. Now it was emerging once more before the trowels and brushes of the scientists.

Arriving around midday the pair had erected a tent, unrolled sleeping bags inside it and slept for more than five hours. Now as the western sky erupted into a spectacular Southwestern sunset they sat in folding chairs around a campfire. An iron coffee pot was simmering at the edge of the embers.

The previous night they'd unpacked the duffel bag and organized its contents. Chris now carried a Beretta pistol in a shoulder holster and on his belt was a combat knife, twin to the one left behind in his motel room. Locked and loaded, the M-4 carbine stood against a nearby stump. He had decided not to give Kate the second Beretta until she had some basic training, which they planned for the next morning.

The kit also contained a bullet-proof vest; some blocks of C-4 plastic explosive with detonators; a machete-like axe; goggles; a pair of tactical radios; night vision binoculars; a utility belt; two canteens; a well broken-in pair of size 12 combat boots; and other essentials including a stock of MRE's, the infamous meals-ready-to-eat.

They had just finished consuming some of the military rations. Kate rose and busied herself policing the site, putting the trash into a garbage bag. She grabbed two bottles of spring water from the back of the Jeep and handed one to Chris.

"Coffee should be ready soon," she said, sitting back down and admiring the fading afterglow in the sky. "Look, the stars are starting to come out! This sky is just amazing."

Chris took a sip of water as he eyed the dome of the night that was spreading from the East. There was no Moon and stars began to appear as the darkness spread.

"Look, there's a bright one," Kate cried, pointing toward the West. "Do you know which star that is?"

"It's not a star, it's Venus. Well, they call it the Evening Star when it's in the Western sky, but it's a planet." Chris stood up to admire the sight. "And, look, above it is another planet, Jupiter I think."

The fire sizzled as sap boiled from juniper logs. Pungent smoke accented the crisp mountain air. Chris placed three more logs onto the bed of coals. The coffee pot was adding a wonderful scent to the air. Chris picked it up.

"Ready for some java?" Kate nodded and he poured boiled cowboy coffee into steel cups. She accepted one gratefully, holding it in both hands to warm them. Now that the Sun had set the temperature was dropping fast. She got up and retrieved a down vest from the tent.

Now the sky had darkened enough for the luminous band of the Milky Way to appear. It wound across the zenith to plunge down to the Southwest horizon. Far to the east the distant lights of Albuquerque made a tiny, pale glow, no match for the glory of the Universe spread out above them.

Sitting back down, Kate curled in the folding chair, gazing at the sky.

"Chris, what Tommy said, about how his theory would explain everything? Do you think it could be true? That we can ever come to know…that!" She pointed up at the stars, her hand outlined by the flickering red light of the fire. "After all, it's so enormous, so complex…and we're so small and insignificant."

Chris glanced at her but said nothing. His eyes returned to the spectacle above their heads. *I wonder too.* He shifted in the chair and glanced around. Something was making him nervous, tickling his instincts. *Probably a coyote.* He settled back in the chair and sipped from his coffee cup.

"Well," he replied at last, "you're right about that. But I can but hope that human intelligence will someday reveal the truth in its entirety."

"I suppose so," Kate murmured. "But it seems that every time we make a scientific discovery, find the answer to something, we glimpse more questions beyond that. The Greeks once believed that everything was made of just four elements, earth, air, fire and water. Then Democritus came up with the idea of the atom, and after that came protons and electrons, and then quarks, and now even tinier particles. There's dark energy, dark matter, the Higgs boson, string theory—there seems to be no end to the puzzles."

Chris nodded. "You're right, but if Tommy's theory is correct—that there are six dimensions instead of only four, and that three of them are dimensions of time—perhaps that could explain all those puzzles…"

"You mean that everything we think we know is wrong? Will we have to start all over again, with earth, air, fire and water?"

"I don't think we'll have to go quite that far back," Chris replied with a grin. "But a lot of physicists are going to be really upset when they have to chuck out all their precious theories."

Kate laughed, a musical two-toned sound that split the darkness around them. Just then Chris stiffened, reacting to a scraping sound from nearby. *That's no coyote.* In one fluid motion he pulled out the pistol, rose from the chair and stepped away from the fire.

"Who's there?" he demanded.

There was another sound then footsteps and a figure emerged into the outer glow of firelight.

"Whoa, take it easy," said a deep-toned voice. The figure stepped closer and Chris could see that the intruder's hands were held out in front of him in a sign of peace. He lowered the pistol but kept it ready.

"What are you doing here?" he demanded.

The man stepped closer and came into clearer view. He was not tall, about five foot seven Chris guessed, and had the complexion and features of a Native American. His hair was dark, glossy and woven into two braids that dangled on his shoulders.

"I must ask you the same question," the stranger said. "This is a sacred place. I have placed it under my protection."

"Chris, I remember this man," Kate said urgently. "He was part of the crew here at the dig. I never spoke with him, but they said he was a medicine man or shaman or something, and that he'd volunteered to help."

"That's right," the man told them. "Now I've stayed to protect the spirits of this place until the winter snows come. And mainly to keep away pot hunters."

He came nearer and squatted beside the fire opposite Chris, who had put away the pistol and resumed his seat. "Is there any coffee left?" Kate ran to the Jeep for another cup and poured it full of the steaming brew.

"I remember you, too," the shaman said to Kate. "You're with that science journal. You came to write a story about this place. I remember you for the Thunderbird you wear. I call you Thunderbird Woman."

Kate glanced down self-consciously and her left hand touched the pendant on its silver chain. "Oh, yes," she said. "But what a strange thing to remember…"

"No, not strange. My cousin the Zuni silversmith made that piece. His name is Robert Red Horse."

"Why, that's right!" Kate said. "I bought it from him at the Zuni pueblo. Are you Zuni then?"

"All the tribes are my brothers and sisters," the shaman said vaguely. He picked up a stick. "We are all one people here beneath Father Sky"—he pointed the stick up at the starry night—"and here on the bosom of Mother Earth." He pointed to the ground between his feet. "Why did you choose to become Thunderbird Woman?"

"Why, I don't really know!" Kate exclaimed. "In fact, I didn't know I had—but you're right, I feel something special about the symbol."

"It is the spirit speaking to you." The shaman gazed steadily at Kate, his luminous brown eyes sparkling in the dancing light of the fire.

Chris sat back in his chair and watched the conversation with a touch of amusement. *Thunderbird Woman, huh. Might explain a lot.*

"You are a distant sister of the People, are you not?" the shaman asked Kate.

"Why, yes. My great grandmother was Cherokee."

"I thought so. You have the spirit within you. And you," he turned to Chris, "are Anglo?"

"Why, European I guess, just plain American. My grandfather was named Josef Fischer, spelled in the German way, and he was from Dusseldorf. He was a sailor who fell in love with an Irish girl. He jumped ship in Dublin to marry my grandmother, Molly McGowan. Her kin were not pleased so they ran away to America."

Chris smiled as another thought occurred to him. "I did have one ancestor on my mother's side of the family who had some slight connection with your People. He was a cavalryman with General Custer…"

The shaman let out a guffaw. "Then you are a brother by blood," he cackled. "Did he die bravely?"

"Well, no," Chris replied sheepishly. "He actually deserted a week before the Little Bighorn. Lived to be 92."

"Ah," the shaman said approvingly. "Then your clan has a strong sense of survival, does it not?"

"Well, yes, I suppose you could say that."

"For your ancestor to have left General Custer to his fate was a brave act of intelligent self-preservation. The stupid die, the smart survive.

"Like your great grandfather," the shaman continued, "I am Joseph. Joseph Eaglefeather is my name. I am what you call a mestizo, a man of mixed blood. Just as you," he turned to Kate, "are one-eighth Indian, I am one-eighth Anglo. My great grandfather was a soldier, too, an Irelander like your ancestor from Dublin," he nodded at Chris. "But we can forgive such things because every man and every woman are of the tribes of Man. We all are brothers and sisters, except that some are near and some are far away."

Drinking from his cup he looked around the camp.

"You seem to be prepared to stay a while," he observed. "Again, may I ask why you're here? What brings you to this sacred place?"

"Why, it was my idea," Kate blurted out. "I don't know why... I just thought it would be safe here."

Eaglefeather gazed at her for a moment.

"Thunderbird brought you here," he said solemnly. "Your spirit guide is looking after you. But, from what danger do you hide?"

Kate looked at Chris. He shrugged and turned to the shaman.

"We're involved in a scientific experiment at Los Alamos..." he began.

"Ah, the place where White sorcerers brought down fire from Father Sky."

"Well, I guess...the atomic research laboratory. We're working with a great scientist, a genius who was able to imagine a new theory of the Universe, of space and time."

"Ah, yes, space and time. Our ancestors knew much of these things," Eaglefeather said with an airy tone.

Slightly taken aback by the comment Chris paused for a moment before continuing.

"Yes, I suppose they did," he replied dismissively. "Anyway, there are men who want to steal this secret. They killed two policemen night before last, and yesterday they tried to kidnap the two of us."

Chris paused then held out his hand. Eaglefeather grasped it.

"I'm Chris Fisher, and this is…"

"Yes, I know, Thunderbird Woman, or as she is known to the wider world, Kate Elliott, senior editor of *Science Today* magazine."

Chris sat up straight with a start.

"You seem to know a lot," he said.

"Oh, there is no secret to it. I merely asked one of the archaeologists who she was. I sensed the spirit within her and was curious. One learns much by asking."

"Oh." Chris relaxed. "Then you know nothing about me?"

Eaglefeather gave him a sideways glance.

"That is not true. I know that your name is Chris Fisher, great grandson of Josef. I know that you are of the tribes of Germany and Ireland. You have told me these things yourself. Also I know from the evidence of my eyes you that you are an honorable man and I see that you are not only brave but very smart."

"That's probably a pretty good idea of who I am," Chris admitted. "To be more precise I'm an atomic physicist like those who, as you put it, brought down the fire from Father Sky."

"We call such people as you magicians." Eaglefeather stared into the fire for a moment. "I, too, am a magician, but of a different style. I study and practice the ancient ways, not the uncharted paths of the future."

"Why, you speak of the future as it if exists as a special time," Kate remarked, looking questioningly at Chris. "That sounds familiar to us, like something we learned from the man Chris spoke of a moment ago, the man with a new theory of time and space."

Eaglefeather rose and stretched, then squatted again. He folded his legs beneath him like a Yoga master and took three slow, deep breaths.

"I, too, can tell you something about time," he said at last.

Chris and Kate said nothing. The shaman's eyes seemed to focus somewhere far away.

"In the cultures of the People we believe that time does not travel forward forever as you scientists think. Our people believe that time travels in circles, that it comes around again and again, like a wheel. Time is connected to the Earth below and it connects us to the Universe above. The ancient astrologers of Babylon and wise men of even earlier times believed this. They recognized the cycle of the ecliptic, the turning of the heavens. The Egyptians built their pyramids in resonance with the turning of the stars, and unknown people thousands of years ago built great clocks such as Stonehenge to measure the passage of time."

"How do you know this?" Kate asked with surprise.

Eaglefeather's eyes came back from the distant place and turned to her.

"Why, in my Master's thesis at Arizona State I compared the ancient myths of America and the Old World," he told her, trying not to smirk. "In a popular version it was published by the University of New Mexico Press as "Braids in Time—How Ancient Myths Intertwine."

"Oh," Kate said, feeling a bit foolish. Chris turned away to hide an amused smile.

"Did you think me an uneducated heathen?" Eaglefeather asked with a wry grin.

"No, no, of course not. It just surprised me that you had knowledge beyond that of the Native American world."

"You don't need to apologize. Let me continue." The shaman's eyes again seemed to focus somewhere far away.

"The ancients had gods for everything, you know. It was their way of explaining the mysteries of the world around them. One of the greatest and most powerful was the god of Time. Chronos was the name the Greeks gave to him, but he appeared under many names and in many guises throughout the early ages of civilization and, I am sure, long before that in the era when men painted pictures on the walls of caves.

"Chronos was once envisioned as a great serpent, wrapping his coils around the primal world-egg before the universe was born. It was said that he mated with the serpent goddess Anake, whose name means 'inevitability.' If you will pardon the obvious crude joke, it corresponds to your theory of the Big Bang, for their union created the Universe."

Eaglefeather paused to poke at the fire with his stick. A cloud of sparks rose into the night sky, mingling with stars.

"In later Greek and Roman times Chronos was imagined as a Man-god, turning the wheel of the Zodiac to make the seasons change. Earlier myths spoke of him as having three heads, one of a man, one of a lion and one of a bull..." He stopped as Kate gasped.

"Three heads? You said that Chronos, Time, had three heads?"

"Why yes, that's right," Eaglefeather responded. "Does that surprise you?"

Chris and Kate were both dumbstruck. At last she spoke.

"The man we told you about, the man with a new theory, says that there are three kinds of time, three dimensions of time. It just startled me when you said that the ancients thought of Chronos with three heads. It seemed such a coincidence..."

Eaglefeather smiled and shook his head.

"In the affairs of Man and the vastness of time there are no such things as coincidences," he told her solemnly. "The ancients were honored for their wisdom. There was a reason for that."

"Jesus," Chris muttered under his breath.

Eaglefeather yawned then stood up.

"I can see that I have given you enough for one lesson," he said. "It grows late and I must return to my hogan for the night. Please come tomorrow as my guests for breakfast. I am living just a half-mile up that track." He pointed to the Northwest. "Come at seven and I will prepare croissants and espresso. Or waffles if you prefer."

"You have an espresso machine in your hogan?" Kate asked with surprise.

"One might truly wonder at what today's hogan may contain," Eaglefeather told her, adding: "Mine is of the Winnebago style."

"Oh, the tribe from Wisconsin?" asked Kate.

"No, the Winnebagos of Iowa, builders of fine motor homes. My hogan on wheels boasts a four kilowatt generator, air conditioner and many other examples of the insidious manner in which the White man has corrupted our pure Indian ways."

And flashing a broad grin Eaglefeather turned and sauntered into the darkness. Kate and Chris looked at each other and broke out laughing.

Later they rinsed out the dishes, poured the remaining coffee on the fire and shoveled dirt to smother the last embers. Working by the light of an electric camp lantern they put away the folding chairs and other items in the Jeep and prepared to spend the night in the tent.

"It's getting really cold," Kate said. With a shiver she zipped the down vest tighter around her neck.

"This is usual weather for the high desert," Chris told her. "Once the Sun goes down, temperatures drop by 30 degrees or more. It's the thin air that does it. I hope those sleeping bags you got us are warm enough."

Kate looked at him appraisingly, as if weighing something in her mind.

"What?"

"Oh, nothing. Well, just an idle thought that occurred to me…"

"Yes?"

"Well, when I picked out those sleeping bags, I happened to notice they could be zipped together…"

"Oh?" Chris's lips curled in the beginnings of a smile.

"Yes, so they could make a single large bag, you know?"

"Yes, I know." His smile grew wider.

"So, if it gets really cold we could, umm, keep each other warm…"

Chris broke into a full grin. He stepped forward and took Kate in his arms.

"I think that's a wonderful idea."

"You do?"

"Oh, yes. In fact, I was wondering about that same thing myself."

"You were?"

They both began to laugh then their mirth was extinguished in a long, sensuous kiss.

"Let's see about those sleeping bags," Chris said when they broke for a breath. "I'm feeling warmer already."

Quickly they finished cleaning up the campsite. Chris locked the Jeep. He released the holster and pistol and placed them inside the tent along with the M-4 and a three-cell tactical flashlight. Kate was busy fumbling with zippers.

"Damn," she swore. "It said these things would zip together!" She flipped one of the bags over to try another combination. "Oh, there, now I think I've got it!" There was a long sound of zippers meshing. She laughed triumphantly.

Chris helped her spread out what was now a double sleeping bag. He reached up to close the tent flap and snapped off the camp lantern. Darkness descended on the interior of the tent.

There was silence except for some muffled movements, and then Kate's voice: "Aren't you going to take off those combat boots?" she asked.

"Oh, yeah, sure," mumbled Chris, sitting up to loosen the straps and cords. "I was just about to do that…" And again they broke into laughter.

Chapter 20

San Mateo Mountains
Saturday, 6:35 a.m. MST

Chris and Kate woke early, just as the Sun was beginning to peek over a distant mountain. They dressed quickly and headed in the direction Eaglefeather had indicated. They were curious about their new acquaintance and looking forward to the promised breakfast.

The shaman was waiting for them, dressed in jeans, boots and a beaded deerskin jacket over a denim shirt. As they approached he raised his right hand in greeting, reminding Chris of an antique cigar store Indian.

"Well, my guests are just in time." He ushered them to the door of his Winnebago motor home. It was a bus-like vehicle more than 30 feet long with two slide-out sections to make more space inside when parked.

"Some hogan," Chris remarked.

"Truly, no one can claim we Indians have forgotten how to live in tune with Nature," Eaglefeather said with a straight face. He swung open the door and waved them inside.

The main living space was about 15 feet long and 12 feet wide with the slide-out extended. A couch and easy chair were at one side and in the opposite corner was an efficient kitchen. The promised espresso machine was waiting on the counter. On the opposite side were a table and three chairs, set for breakfast with a flask of orange juice, butter, jam and a jug of milk.

"Welcome," the shaman said, steering them to the couch. "We can start with coffee, then I can prepare croissants or waffles. What is your pleasure?"

"Oh, I love waffles," Kate said.

"Then waffles it shall be, with bacon. And I have some genuine Indian-style syrup made from the sap of the maple tree. Would you like your espressos straight or foamed with milk?"

"Foamed please," said Kate. Chris nodded agreement.

Cranking up the generator Eaglefeather got busy with the coffee machine and soon handed them tall cups of cappuccino. They sipped the sweet beverages as he beat flour, milk, eggs and sugar to make batter. The waffle iron was warming and bacon was frying in a pan.

Chris looked around the comfortable space. Noticing a rack of books he got up to take a closer look. There were volumes on history and Native American lore, others on geology, botany, and even a few novels. He spotted Eaglefeather's own book and lifted it out.

"That's my opus," the shaman told him, pouring batter into the waffle iron. "It's much easier reading than my thesis. The editors at the university press were very kind to help me use my words correctly."

"I think your words are fine," Chris replied, flipping through the pages. "This is quite impressive. I'll be sure to buy a copy when I get back to civilization."

Eaglefeather set down the bowl of batter and put his hands on his hips. He scowled.

"You think we are beyond the pale of civilization?" He swept his right arm around the luxurious motor coach with an inquisitive look on his face. Then he smiled and picked up his bowl.

"You must learn that civilization may not exclusively be found in cities, or in universities for that matter. Civilization exists wherever men of wisdom and good faith are found. It comes from inside the brother or sister," he touched his chest, "not from without." He turned back to check the bacon.

"Yes, I see what you mean," Chris said. "I knew that, too, it's just that we tend to forget..."

Soon crisp bacon and brown waffles were served with butter and syrup. Orange juice and glasses of milk were poured. Cups were refilled with frothy cappuccino. The three new friends sat down to eat.

Afterwards Chris and Kate sat on the couch and Eaglefeather lounged in the easy chair. "It's good to have company," he said. "The life of a shaman can be a lonely one and I enjoy the excuse to make a good meal. Most days I merely eat a bowl of oatmeal and a piece of toast."

"Sounds like my Irish grandma's idea of a breakfast," Chris remarked.

"Yes, it's another sad example of the Indian's submersion in the ways of the Anglos," Eaglefeather sighed. "In the old days we would be satisfied with a handful of grubs or some pemmican made from dried berries and bear grease..."

"Yuck!" said Kate. "That doesn't sound so good."

"Alas, no. The insidious ways of the white man are all too tempting, leading the Indian from his pure pathways." Eaglefeather grinned. "And what lies in store for you two now?" he asked. "You can't hide forever from this danger. Winter is coming and your tent will soon be buried in the snow."

Chris leaned forward on the couch.

"No, we're leaving tomorrow. I've called on some old friends to help me. They'll arrive in Albuquerque tomorrow morning."

"Who are these friends? One cannot ask just anyone to join in facing danger."

"Oh, they aren't 'just anyone,' not at all," Chris declared. "They're members of my old SEAL special forces unit. We fought together in Afghanistan."

"Ah, then you are a soldier and not merely a magician of physics," the shaman remarked. "That does not surprise me. I had wondered how you were able to escape from the men who intended to kidnap you. Not 'just anyone' could do that, I think?"

Kate chimed in.

"They were like sheep to the slaughter," she declared. "I've never seen anything like what happened, not even in the movies. There were two of them and they had guns pointed at us…" And she described the scene in the cafe that left the two men unconscious or possibly dead after the sudden seconds of fury.

Eaglefeather gazed at Chris with new respect.

"Truly you must have some blood of the People after all," he told him solemnly. "What occurred has the flavor of Indian prowess, in particular the counting of coup."

"What do you mean?" asked Chris.

"In combat the Indian who counts coup on his enemy is much honored. Your strike with the dinner plate was done in the way a brave Sioux warrior might have counted coup, because it was a humiliation to the enemy and took much courage. And the way you struck the second man with your hand and your foot, again you counted coup on him."

"Okay," agreed Chris doubtfully. "I didn't think of it that way. I just wanted to win."

"Of course, as any warrior does. What will you do when your friends arrive?"

"I don't know yet, but at least I'll have men around me that I can trust, to cover my back."

"Yes, and you have a powerful friend already." Eaglefeather glanced meaningfully at Kate. "She is beside you, and brings the spirit of Thunderbird to protect you both."

Kate chuckled and fingered the necklace. "I don't think Chris needs much help."

"Oh, no, never discount the power of the spirits. You've already saved him twice."

They both looked at the shaman with questioning eyes. He nodded soberly and informed Chris: "First she acted to distract the kidnappers by spilling her tea. If she had not done that, would you have prevailed over them?"

"Perhaps not. And the second time?"

"Why, by allowing Thunderbird to direct you here, to this sacred place of safety, this place where I am here to advise you with native wisdom. And, perhaps most important, to serve you breakfast and thus save you from those dreadful MREs I saw in your camp." He chuckled and got up to clear the dishes from the table.

Chapter 21

San Mateo Mountains
Saturday, 2:10 p.m. MST

That afternoon Chris began to train Kate in the use of the Beretta pistol, beginning by familiarizing her with the gun.

"This is a model M9A1 Beretta 9mm semi-automatic pistol," he told her. He released the magazine and cleared the action to make sure it was unloaded before handing it to her. "This is an improved version of the model originally developed for the United States Marine Corps. SEALs can use a variety of sidearms. The Beretta was never my favorite, but it's a fine weapon with an excellent record for reliability."

"It's kind of heavy," Kate said, taking the pistol in both hands as if it were something fragile.

"It weighs about two pounds unloaded," Chris told her, taking it back and popping in the magazine. "It holds fifteen rounds. Weight is a factor with any handgun, but the heavier the gun, the less recoil. If you were to fire an ultra-light 45 caliber pistol, you'd have some idea of what it's like to be kicked by a mule."

He demonstrated how to field strip the automatic. As he did so he explained each part and showed how the pistol operated. He removed the cartridges from the magazine and put them in his pocket.

"Always remember the basic rule of firearm safety, which is never to point a gun at anything unless you intend to shoot it. That applies even if you think it's unloaded, because when you always follow that rule you can't go wrong."

"Makes sense," she said. "Show me again how the safety works. You'd always want to have the safety on, right?"

"Usually, until you are ready to shoot. Most of the time it's best to keep the chamber empty when carrying the pistol. You can always chamber a round quickly by simply pulling back the slide." He demonstrated the action.

"See how the action locked open when I pulled it back? That's because the magazine is empty. When you fire the last round it does the same thing, automatically prepared to chamber a new round when you slap in a fresh magazine."

He explained several more features, letting Kate handle and operate the unloaded weapon. Next he let her practice loading and unloading the magazine, making sure she knew how to seat the cartridges properly.

Finally he showed her how to hold and aim the pistol, explaining the sight picture and the importance of squeezing rather than jerking the trigger. At last he put a loaded magazine into the Beretta and handed it to her, the barrel pointed at the sky.

"Now let's see if you can hit anything."

They walked a short way into the desert and Chris tossed three empty soda cans about fifteen feet in front of them.

"Okay, see that Coke can on the right? You're going to shoot it, so what do you do first?"

"First I make sure nobody is in front of me, or as you said, in front of the firing line. So step back please." Chris grunted approvingly and took a step back. "Next I load a shell into the chamber." She pulled back on the slide and let it slap back. "Now I point the pistol at the target, and it's okay to do that because I intend to shoot it. Then I release the safety and...squeeze the trigger." *Bam!* "Oh!"

At the crash of the shot the pistol jumped and so did Kate. A spurt of dust showed that her shot had gone several feet beyond the Coke can.

"That's okay," Chris said. "You'll get the feel of it. Hold it a bit more firmly in your hand and relax your arm to let it take the recoil."

Bam! Kate fired a second round. This time she held steady and the bullet struck closer to the target.

"This is harder than it looks," she said, pointing the barrel up and turning toward Chris. "Tell me again about the sight picture."

"Sure, just put the front sight in between the slot in the rear sight so that all three points are level with each other. Then place the spot you want to hit right on top of the front sight, as if it were a ball and you were balancing it there."

"Right." Kate aimed again. *Bam!* The can leapt into the air and she gave a squeal of delight. "I did it! There's one bad guy down! Mister Coke."

Chris laughed. "Good shooting. You'll be another Annie Oakley before you know it. Now try again, and this time fire three times as quickly as you can. See if you can hit the target each time."

Bam! Bam! Bam! The can leapt into the air after the first shot, but the next two went astray.

"The lesson here is that it's hard to hold your aim between multiple shots. It takes practice to time each shot, getting the sight picture back between each one. Try it again."

Bam! Bam! Bam! This time the first and third shots hit the can.

"You're getting it now," Chris said. "Now try shooting all three cans in sequence, starting with the one on the left and moving right."

Bam! Bam! Bam! Again, two out of three were hits.

"I'm really getting it aren't I," Kate said with satisfaction. She pretended to blow smoke from the barrel like some movie gunfighter.

"Yeah, you've got an eye for it. You've got three more cartridges so why don't you try the same pattern once more."

Bam! Bam! Bam! This time all three cans were hit and Kate let out a whoop.

"Okay, now you need to change magazines and then we'll try shooting from a bit farther away."

They practiced for another half hour as Chris gave her more pointers and safety tips. At last he concluded the training session, telling her to remember that the gun was for defensive purposes, to be used only for protection in case of deadly threat.

"And there's another fact to keep in mind. You're not licensed to carry a concealed weapon. In this state you can keep a gun inside your vehicle, even hidden, but if you carry it hidden on your person you'll be violating state law. As a serving military officer I have an exemption, but if you're caught carrying the weapon concealed on your person it could result in a criminal charge. So keep it in the car unless absolutely necessary, and if you need to use it, take it out of the vehicle in the open."

Back at their camp they started a new campfire and prepared to make a pot of coffee. The Sun was already sinking toward the Western horizon, casting long shadows. They would have to get an early start the next day to meet Chris's team in Albuquerque.

"What will happen then?" Kate asked. "And what role do you have in mind for me?"

"I don't know yet what we'll do, but as far as you're concerned I just want to keep you safe. I should have insisted on letting Paul put you into a safe house, but the situation just seemed too dangerous to take chances."

Kate looked away. She got up and went to the Jeep to put away the Beretta, first making sure the magazine was properly inserted and the chamber empty. Chris was adding logs to the fire. She pulled out the two folding camp chairs and carried them over beside it.

"I'm glad I stayed with you," she said as they sank into the chairs, holding their hands out to warm in the flickering firelight.

"Yeah, so am I," Chris replied. "But I'm concerned about you. I'd tell you to get clear out of here, go back to your office in Boston—but as a dedicated journalist I know what you'd say to that."

"Hah, yeah," Kate responded. "You got that right, buster. No way I'm gonna miss out on how *this* story ends."

Chapter 22

Safe House, Los Alamos
Saturday, 8:45 p.m. MST

Dawson and Reeves had been moved into a guest house right on the grounds of the LANL, protected by the lab's high security perimeter. Instead of outsourcing the protective mission to the FBI, Bonner had chosen this spot right in the heart of the action and assigned a team of his own plainclothes officers to low-profile surveillance. The lab's security had already been penetrated once, so they were taking no chances. Six men were posted around the house, divided into teams of two. Uniformed patrols in squad cars manned a wider perimeter.

Reeves' wife Valerie and their daughter had been evacuated by helicopter to Denver International and put on board a United flight to Minneapolis. Two federal agents met them there and escorted them to the home of Valerie's parents in nearby Eau Claire, Wisconsin, far out of harm's way.

Dating from the early days of the lab, the guest house had four bedrooms, a large family room, dining area and a rather old-fashioned kitchen. Dawson had staked claim on the dining table and proceeded to cover it with a surprising amount of equipment and documents. A 30-inch flat-screen monitor with keyboard and trackball held the center space surrounded by stacks of papers and Blu-Ray data disks. On the floor beside the table was a powerful tower computer daisy-chained to a massive twelve-terabyte RAID hard drive assembly. A large format laser printer was balanced somewhat precariously on an end table procured from the living area.

It was now early evening. The two had dined on Chinese carryout brought by a LANL patrolman. Dawson was back at his computer, oblivious to the surroundings. He had spent most of the last night and nearly all day working on his notes and reports, aiming to complete organizing them so that he could share the details of his new theory.

Reeves, on the other hand, felt like a prisoner. At frequent intervals he used the landline phone to check with his administrative assistant or other staff members. Still, he chafed at being confined. At times he caught himself pacing nervously, hands behind his back and head down, staring at the shabby carpet. It was a welcome relief when he heard the doorbell ring to announce a visitor. It was Bonner, escorted to the door by one of the surveillance officers.

"Paul, come in. We're going crazy here."

"Speak for yourself," Dawson called out, and returned to his work without acknowledging the security chief.

"Well, okay, Tommy's not going crazy—just me I guess," Reeves admitted with a wry grin. "What's been happening? Here, come in, sit down." He steered Bonner to an easy chair and settled on the nearby couch, then jumped up again. "I bet you'd like a cup of coffee. Let me get it."

In a moment the two men were sipping from mismatched cups, one featuring a slogan from a long-ago jazz festival, the other a picture of Pluto the Dog. Bonner stretched his legs in front of him and glanced around the room before beginning his report.

"It's been an interesting day. It started this morning at the hospital where they took those two goons who tried to kidnap Fisher and Elliott. I guess you didn't hear the details of that, but Fisher did a pretty good job on them. One's still in a coma. He got hit across the nose with the edge of a plate. They think he's going to live, but may have permanent brain damage."

Reeves whistled under his breath. "Hit with a plate? How's that work?"

"Thrown like a frisbee from halfway across the room, had to be traveling 60 or 70 miles an hour when it nailed him right across the eyes. Speaking of which, he'll regret not choosing the extra-cost shatterproof lenses for his sunglasses, because the broken plastic did a lot of damage to his face and one of his eyes is jelly.

"Anyway, the other guy got off easier—just a broken arm, a ruptured kidney and some pretty gruesome bruises. He didn't seem too happy about it."

"Did he say what happened?"

"Actually didn't seem to remember much. He kept muttering about how they should have been warned, something like that. I gather they didn't know about Fisher's martial arts background."

"Did he say who they're working for?"

"Nope. He made like a clam when I tried to interrogate him. A couple of FBI guys are coming in to work on him. We're in touch with the attorney general about getting him a reduced charge on the attempted kidnapping in return for his testimony. My bet is he'll see the light."

"What about the rest of my supervisory guys? Are they safe?"

"Yeah, sure. We put 24-hour security teams on their residences and locked down the neighborhoods with squad patrols. Nobody can get close to them without running the gauntlet. I'm maxing out my overtime budget."

"Don't worry, I'll approve it."

"But there's another development. One of the visiting staff people, a computer science guy from Tulane…"

"Oh, yes, Rahman Singh. What about him?"

"He seems to have disappeared. And, funny thing, we can't get hold of his family in New Orleans. The FBI sent an agent out to his house and it looked like nobody was home."

"What do you make of that? Singh seemed like an okay kind of guy."

"We're not sure what to think. I do know that he was one of the people that Fisher asked me about. He looked through Singh's personnel jacket in fact. But he vetted a lot of other files, too, and he didn't seem to have anything in particular against Singh."

"You say Singh disappeared. When?"

"According to the hotel clerk he walked out of the lobby about 10 p.m. the night before last. He wasn't carrying anything and all his clothes and stuff were in his room, including his laptop and PDA. He just drove away."

"Another mystery then?"

"Yes. We're continuing to look into it. But there's something I wanted to ask you about. We've been wondering how the intruder got the entry code and computer passwords to the quantum center. Could Singh have possibly had that information?"

"My God! Yes, absolutely. He was one of the men responsible for computer security. He could have had those codes. And, he didn't know about the fake firewall on the main system—we kept that little secret between just me and Tommy—so that could fit."

"I was afraid of that," Bonner mused. "We've got an APB out for him with the State Police. Maybe he'll turn up. Wait, I've got a call coming in."

He keyed his portable radio to accept the call. He listened for a moment then clicked off with a curt "Thanks." He looked at Reeves. "That Navy intelligence guy I told you about, Jeff Jones, he's coming in out of the cold. He's on his way over. He might have some new intel for us."

"Great, I've been wanting to talk with him. I'll start a fresh pot." Reeves headed for the kitchen. A moment later the doorbell rang and Jones entered, introduced himself to Reeves and sank down on the far end of the couch. He gratefully accepted a cup of coffee.

Bonner told him the news about Rahman Singh. Jones nodded gloomily.

"I'm afraid I have some bad news about that," he said. "We asked the Fibbies to get a warrant to enter Singh's house in New Orleans. His wife and kids were there, in the basement…"

"Oh, Jesus! Don't tell me… Are they all right?" Reeves blurted.

"They're okay, but in the hospital for exposure and dehydration. They'd been gagged and restrained with plastic wire ties for more than 48 hours."

There was silence for a minute. Bonner muttered under his breath.

"It looks like Singh was acting under duress when he gave away the codes," Jones said. "Obviously his family was threatened to make him go along, so he was a victim rather than part of the operation."

"What kind of monsters are we dealing with?" Reeves rubbed his face with the palms of his hands.

Jones shook his head.

"We still don't have a clue on who's behind this. Naval Intelligence has picked up some sat phone chatter but it's heavily encrypted. Even NSA can't crack it."

There was a sound from the dining room. Dawson had pushed back his swivel chair and turned toward them. He smiled.

"NSA are wimps," he declared. "We can crack it in a minute." He turned back to his work.

"Oh, for crying out loud!" Reeves slapped his forehead. "Of course, the quantum computer. Can we get those transmissions?"

Jones jumped to his feet, pulled out his secure cell phone and hit a direct dial code.

"I'll get the intercepts e-mailed directly to the lab," he said as he waited for an answer on the other end. "What's the email address?" Someone picked up the call and Jones quickly gave instructions.

"Tommy, we need to go to the lab. Come on, let's go."

"Oh crap, I should have kept my mouth shut," the physicist grumped. "I was just getting on a roll here… Oh, okay, let me back up my files and get my cap."

Chapter 23

*Quantum Center,
Saturday, 10:30 p.m. MST*

Ten minutes later the four men pulled up at the Quantum Center. Two squad cars had accompanied Bonner's black Tahoe. The patrolmen jumped out and formed a protective circle as Reeves punched in the pass code to the door. Inside, two more uniformed officers waited.

"How long will it take you to get the system up and running?" Jones asked Reeves.

"Everything's in a sleep mode, so it shouldn't be too big a job. As long as nothing goes wrong, we should be okay."

"What can go wrong?"

"Well, just about everything," the engineer admitted. "Every sub-system in this whole center is dependent upon everything else. When we're up to speed we generally have several techies and programmers on hand to monitor things. Let's just hope we can make a clean start-up."

He headed for the control center at the far end of the room. The crime scene tape had been removed. Reeves hesitated for a moment when he noticed that a new chair was in place at the control panel. He suppressed a shudder, realizing that the old one must have been soaked in patrolman Jesse Carroll's blood. Shaking his head sadly he sat down. Dawson took his place at the central station and Reeves sat at his right. Bonner and Jones stood back to watch.

Within a few minutes the system was coming to life. Dawson was running some diagnostics while Reeves walked back and forth checking the three unmanned control stations.

"The Qubit Farm is checking out okay," he reported. "That's the best news. And the server array is on-line, too. It has automatic redundancy, so even if one of the boxes blows up the system will still work."

Jones and Bonner said nothing, merely watching with curiosity as the screens scrolled text and schematics. Cued by the heat generated as hundreds of computers came to life, air conditioning ventilators clicked on and the gentle sound of moving air filled the vast space.

Finally satisfied that the system was ready, Reeves called up the email program and downloaded the encrypted messages that had been captured by Naval Intelligence.

"It looks like there are three separate messages here," he said, turning to Jones. "Do you have any information about where they originated?"

"Apparently right here in Los Alamos, but they were sent through a satellite phone that had been blocked so that we couldn't trace it."

"And do you know where the messages went?"

"No. They could have been received anywhere."

Reeves loaded the first message into the workstation and started a program to identify the encryption code. As they waited Jones looked around with interest.

"Fisher told me this thing can do more computing in a minute than all the computers in the world could do in an hour," Jones said. "That can't be right, can it?"

Dawson swiveled his chair around quickly and caught Jones in a gimlet-eyed stare. Then he shrugged and turned back to his keyboard, muttering to himself.

Jones caught Reeves' eye and the engineer gave a little nod. "Chris wasn't exaggerating," he said softly. "You'll have to excuse Doctor Dawson. He gets a little touchy sometimes.

"Now," he continued, speaking louder, "let's see if we can factor this bugger. The encryption code is 214 digits long, a monster." Tommy grunted with disdain. "I'm entering the number into the quantum processor now." He punched the enter key, crossed his arms and sat back in his chair.

"So, what happens next?" Jones asked.

"Nothing."

"What do you mean 'nothing'?"

"I mean it's already done," stated Stuart, pointing to the screen above him. "It factored that number in about five seconds, and there are the keys."

"Jesus! That quick? Fisher really was serious wasn't he?"

Dawson let out a disdainful snort. "You don't know the half of it," he said. "I haven't even told anyone the best part yet."

The three men looked at him curiously for a moment. Jones turned to Reeves and raised one eyebrow in question. The engineer shrugged and shook his head.

"Okay, come on, let's run that message through the translator and see if we can read it," prodded Bonner. Reeves turned back to the keyboard and punched up some software.

"It's running now," he said. "This will take a bit longer because now we're on standard computer time, not quantum time." He pointed to a scrolling bar that was counting down the seconds as the translator operated. In a moment it finished and popped off, revealing a document.

"The phone call was transmitted as a text message, not voice. It seems to read okay, which is a relief. I was afraid it might have been double encrypted, in which case we would have to start all over again with a second set of keys."

Bonner and Jones stepped closer and stared up at the brief paragraph of text.

Quantum machine successful. Much faster than expected. Major threat to security. More later.— Eminence Grise.

Bonner drew a sharp breath. "It must be him, the guy who broke into the lab and killed my men." He clenched his fists. "What do you make of this?" he asked Jones.

"Well, it's a direct report on the successful test. Notice that the sender signed himself 'Eminence Grise.' I think that's French..."

"What the hell does that mean?" Bonner interjected. "Are these guys working for the God-damned French! Our friends the French?"

"No, I don't think so. Let me think, that phrase rings a bell..."

"It means 'Gray Eminence' in English," Dawson announced. "It refers to Francois de Tremblay, an influential monk who worked with Cardinal Richelieu in the 1630s."

The other three men stared at him. After a moment of silence Dawson turned his chair around and glared at them.

"So now you know what it's like to have an actual memory," he asserted. "I read the *Encyclopedia Britannica* when I was seven you know. Never forget anything..." His voice trailed off. "Well," he murmured, "sometimes I forget to go to bed..." He turned back to his workstation with an embarrassed look.

"But, what does that have to do with this?" Jones asked plaintively.

"Nothing, nothing at all," Dawson told him, glancing over his shoulder with contempt. "It's just a phrase used to identify a powerful person working in secret. In this case, a good pseudonym for a spymaster." He spun his chair back. "Let's get going on the next translation shall we? Perhaps we'll learn some answers. The system's working fine."

"Okay, sure," Reeves muttered, starting to set up the decryption. In a few minutes the second message was displayed on the screen.

Quantum center penetrated by others but no data obtained. Two guards dead. Dawson source of new quantum theory. Red Falcon en route. More later.— Eminence Grise.

Again the men stared at the message. Jones broke the silence.

"Who's this Red Falcon?" he wondered. "And what does it mean 'by others'?"

Bonner grunted and sat down in a chair. "Crap, I think it means there's more than one game in play. Let's factor the third message, Stuart."

The engineer went to work again and soon the third and final message was being processed. "Look, here it comes."

Red Falcon, assignment is go. Falcon Stoop set for tomorrow.—Eminence Grise.

"Oh Jesus!" Bonner exclaimed, shaking his head like an angry bull. "What does that mean? What's a 'falcon stoop'?" Dawson rolled his chair back to view the new message. He read it in silence then went back to his keyboard.

"It's simple enough," he murmured. "Couldn't be any more obvious really."

"What?" Bonner nearly shouted. "This could be a matter of life or death. What does it mean?"

Dawson looked up, wondering about the outburst.

"You mean you still haven't figured it out?" He looked around at the three men. "No, I suppose not," he sighed. "It's not even code words, just English. In relation to someone called 'Red Falcon,'" it means an attack. When a falcon drops onto its prey at up to 200 miles an hour the dive is called a 'stoop'." He swooped one outstretched palm downward in the motion of a diving bird.

"And the target? Why that's me of course." Dawson smirked. "Pretty lucky for me I won't be at home, huh?"

"Jesus," breathed Bonner. "It's just what we feared—they want to kill you to destroy the quantum project. And you're right, it's perfectly clear…once you explain it."

"'Falcon Stoop,' pretty exciting stuff," murmured Dawson. "And to think, all I did was come up with the most incredible theory since Democritus dreamed up the atom…" He giggled, took off his

"Quantum Cowboy" cap and pointed at its symbol of an atom. "It's quite a compliment really, isn't it? Very flattering."

"Uh, yeah, I guess," Bonner said, shaking his head. "Hope to Christ nobody ever takes a mind to flatter me that way!"

"Oh, I shouldn't worry about that if I were you," said Dawson gazing appraisingly at the security chief. Bonner grimaced and swore under his breath.

Chapter 24

San Mateo Mountains, New Mexico
Sunday, 4:40 a.m. MST

Chris and Kate were up before five a.m. and coffee was ready in twenty minutes. Chris suggested they get breakfast later in Grants. Looking at their supply of MREs, Kate curled her upper lip and readily agreed.

As they filled their cups with coffee they heard the scrape of boots and Eaglefeather appeared out of the pre-dawn darkness.

"*Hola*, White man," he said, making a peace sign with his right hand. "*Ya te hay*, sister Thunderbird. Did you make enough for three?"

Kate poured a cup as the shaman sat down by the fire and crossed his legs yoga-style. He accepted the coffee gratefully and gazed approvingly at his new friends. Wisps of juniper smoke stirred in the cool morning air.

"I heard the sound of your shots yesterday afternoon," he told Kate. "Did you master the firestick?"

Kate laughed and even Chris chuckled at Eaglefeather's droll use of English.

"She did very well," Chris told him proudly. "An excellent student in the science of weaponry."

"Ah! I presumed as much." He beamed at Kate, who blushed. Then he turned to Chris with a somber look.

"From today you will again face danger," he intoned. "You're a fighting man so you know what this means. Kate is no warrior, but I feel the power of the spirits within her." He turned to Kate. "Thunderbird will be with you. His strength is always secret and may not be apparent, but never forget that you can draw the power of his spirit into your own."

He turned back to Chris.

"You are one who is far away from the People. Though you may not recognize them, I know that you have spirits of your own to look over you. But one can never have enough of spirits so I have brought another to be with you in times of need."

He unfolded his hand to reveal a thin silver chain with a dangling pendant. He held it up for Chris to examine, letting the adornment swing in the firelight. The artist had used silver, turquoise, coral and jet to create the image of a dog-like animal, its muzzle raised into the air as if howling at the Moon.

"This is Coyote," Eaglefeather told Chris. "This lovely piece was made by my cousin the Zuni artist, the very same who created Kate's Thunderbird. Your own personal spirits will serve you well in danger. Now you will have another in times of need."

He held the fetish higher. The polished metal and semi-precious stones sparkled in the firelight.

"The spirit of Coyote is often underestimated. Sometimes he's said to be a clown, a trickster. But above all he has wisdom and is very clever. I give you the spirit of Coyote to stand beside you. He will be your spirit guide. But first you must bond with him."

He reached into a leather bag that hung from a thong at his side.

"This is my medicine bag, the hallmark of the shaman," he explained. "Here I keep the magic that shapes my art."

He withdrew two little rolls of doeskin and laid them on his lap.

"Among the People there is a blessing ceremony that accompanies the passage of a sacred spirit to the keeping of another. It is quite long and filled with much ancient wisdom. But you don't know our language or ways so I will give you the *Reader's Digest* version." Chris smiled and nodded. Kate sat back in her folding chair and watched with fascination.

Eaglefeather laid the Coyote charm beside the two rolls and opened the first package. Chris watched his hands intently. Folding the doeskin flat the shaman revealed a little pile of fine powder.

"This is pollen from the Sacred Datura, a plant that grows here in the Southwest deserts. You Anglos call the Datura a hallucinatory drug. We native peoples know it lets us travel into the spirit world. It is used as a medicine and sometimes by evil ones to perform witchcraft." He frowned, adding: "No worries, for I do not deal in evil things."

He took a pinch of the pollen and sprinkled it on the silver pendant, speaking some words in a melodic language. When he was done he shook the grains of pollen into the fire where they made a little sparkling burst of light.

"I have spoken to the spirit," he said. "Now the pollen of the Sacred Datura has lent its strength to Coyote. He can guide you into the spirit world and let you return safely."

He carefully re-folded the doeskin swatch and returned it to the medicine bag.

"And I have another gift for Coyote as I prepare his spirit to look over you." Eaglefeather opened the second deerskin roll to reveal a little nest of tiny pink crystals. "This is rose quartz, a rare and powerful mineral."

He picked up one of the crystals between thumb and forefinger then folded away the rest. It was no larger than a grain of rice. He held the tiny gem up to the firelight, letting its facets shine. For a long moment he held the crystal still.

Feeling a strange sense of relaxation Chris and Kate stared at the flickering reddish light that sparkled from the stone, entranced by its mysterious force. At last Eaglefeather spoke.

"Now you have felt the power of the rose quartz crystal. It focuses the thoughts. It acts as a lens for the mind. I will now add this power of concentration to the spirit of Coyote."

Once again he spoke mysterious, lilting words, passing the crystal over the silver emblem of Coyote. Then he raised his face toward the sky, now beginning to brighten into morning. As his voice rang out with a final phrase he dropped the crystal into the fire. He held the pose for nearly a minute until the first rays of the rising Sun touched his face.

Enchanted, Chris and Kate remained still. At last Eaglefeather lowered his gaze to meet theirs. He stood up and moved behind Chris, passing the silver chain around his neck and fastening the clasp.

"Wear this always and Coyote will be with you," he said. "And now it's time for us to part. May the spirits protect you and fortune always be yours. I say to you now the Navajo words '*hozhoogo naninace doo*.' These words are spoken to say 'goodbye,' but their true meaning is 'May you walk in beauty.' It is a prayer for peace and well-being, which I offer to you."

Eaglefeather shook hands, first with Chris then with Kate, and without saying another word he turned and walked away into the growing dawn light.

They sat for a long time without speaking. Kate leaned forward to pick up the coffee pot and poured them fresh cups. Still they sat, sipping and contemplating Eaglefeather's words.

"May we walk in beauty," Kate breathed after a while—and the spell was broken. Chris lifted the Coyote pendant, examined it closely then tucked it and the chain inside his shirt.

"Let's get going," he asserted. "We've got to get back to the world." He grinned. "All four of us."

Kate laughed and began to pour the remaining coffee on the fire. A cloud of steam rose from the drowning flames, mingling with blue-gray smoke. She picked up the camp shovel and began to cover the embers while Chris folded and put away the folding chairs. In minutes they were driving down the rutted road that wound through the desert toward the town of Grants.

Chapter 25

Safe House, Los Alamos
Sunday, 8:10 a.m. MST

It was just after eight o'clock in the morning and Bonner's cell phone was ringing. He set down his cup of coffee and pulled it out of his pocket.

"Hello."

"Paul, it's Chris Fisher." The security chief sat up straight in his chair.

"Fisher, where are you? Are you okay?"

"Yes and so is Kate Elliott. We've been laying low until I could get organized. What's the situation there?"

The security chief rubbed his forehead with a sweaty hand, then switched the phone to his other ear.

"We've made no progress finding the guy that killed my two officers," he said ruefully. "And the two thugs that tried to kidnap you and Elliott aren't talking. Well, one of them isn't talking. The other one's in a coma."

"The one I whacked with the plate?"

"Yeah. That was some move you made on them. The waitress saw it from the kitchen door and their surveillance camera got the whole thing. We've got a full picture of what went down. Gotta hand it to you."

"I'd kind of like to see that footage myself sometime."

"Oh, you will of course. There's another thing to report. One of the computer scientists working on the project has disappeared. We think he's probably dead."

"Rahman Singh?"

"Yeah," Bonner said with surprise. "How'd you figure that?"

"I've been doing some thinking. I realized that with his computer science specialty, Singh might have been able to get the entry codes to the Quantum Center. I saw him acting strange at Tommy's briefing the other morning, so I just added two and two and it made four."

"Well, you were right. His family was threatened. They were found tied up in the basement of their house in New Orleans."

Chris said nothing for a moment. Bonner could hear the rush of wind and the whine of tires in the background.

"And there's another thing…" Bonner added.

"Yeah, what?"

"Navy Intelligence caught some calls being made from a sat phone, here in Los Alamos. They were heavily encrypted. NSA couldn't even begin to crack them."

"So they're useless?"

"Well, no. Doctor Dawson's quantum computer broke them like they were encoded with something that came in a box of Cracker Jacks. We read the messages. They may have been from the guy that killed my men. But even more important, it gave us a heads-up that they're planning to try to kill Dawson, just as we feared."

"But you have him under wraps, right?"

"Yeah, he's safe enough. But they don't know that."

"What're the details of their plan?"

"We don't know much, but an assassin is apparently en route."

There was a long silence between the two men before Chris spoke.

"You know," he mused, "if they don't know we have Tommy safe, that could give us a chance to make a tactical move on them."

"Yeah?"

"Sure. One thing I learned with the SEALs is that when you have intelligence about the enemy, you always look for a way to use it against him. In this case, we know they're coming after Tommy so it suggests the possibility of setting up an ambush. Do you know when?"

"We think tonight. But I don't have the resources to pull off something like that. It would take a SWAT team to ambush Red Falcon. Oh, by the way, that's his code name. We think he's coming in today but we don't know from where."

"If the strike's set for tonight it would be at Tommy's house. Only thing that would fit. Probably o'dark thirty."

"O'dark thirty?"

"Navy slang for the darkest time of the night," Chris explained.

He thought for a moment. Again Bonner heard the sound of traffic, big trucks moving fast.

"Where are you now?" he asked.

"Traveling east on I-40, toward Albuquerque. Kate's with me."

"Okay. When do you expect to be back here?"

"Later today, but I want to keep a super low profile."

"That's fine with me. Until we catch this guy nobody's going to be safe."

"That's why we need to be pro-active," Chris told him.

"So tell me more about your idea?" Bonner coaxed.

"Well, I might be able to put something together."

"You?" Slightly incredulous note in Bonner's voice.

"Yeah. With some trusted friends of mine."

"Friends! What are you doing Fisher?"

"I'm just covering our collective asses. Well, mostly mine. Look, do you know how I can get hold of Jones? I need to talk with him."

Bonner laughed, the first cheerful note in the conversation.

"You probably won't believe this, but he's sitting right across from me. We're drinking coffee together in the safe house where we're keeping Dawson and Reeves. Here, I'll put him on."

He tried to hand the phone to Jones, but the agent waved it away.

"Oops, he says he doesn't want to talk. Wait, he's signaling me to silence this call. Back in a flash." Bonner punched the mute button.

"Make sure he's using a throwaway phone," Jones told him. "If he is, get the number and tell him I'll call back on my secure phone." Bonner nodded, thumbed the phone back to life and spoke to Chris. He grabbed a pen from his shirt pocket, wrote a number on his wrist then hung up. Thirty seconds later Jones was on the line with Chris using his own encrypted phone.

"Commander Fisher, what the hell are you doing?" he rumbled.

Chris laughed. "Being in command," he said enigmatically. "Your instructions to me were pretty vague and I figured the situation called for independent action. You told me not to trust the CIA, and the FBI seems to be baffled. People are getting killed and those thugs tried to kidnap Kate and me, so I figured it was time to cut loose."

"Okay." Jones sounded noncommittal.

"I'm setting up my own response team," Chris told him. "It needs to be strictly on the hush-hush."

"You've called on some members of your old SEAL unit," Jones guessed.

"Yeah, I'm meeting them this morning. We'll need support."

Jones was silent for a long moment, then:

"What kind of support?" He sounded slightly more encouraging.

"First, I need authorization from the Department of the Navy to re-up retired or reserve personnel on an emergency basis, something like what you did to me but with more finesse and human kindness."

Jones chuckled. "Cute, Fisher. Go ahead."

"I'll call later with a fax number where the authorization can be sent. I need it in a couple of hours. That's point one."

"All right," Jones said. "I can make it so. What next?"

"Equipment. I need full combat rigging for my team."

"How big a team do you have? Your original SEAL platoon had sixteen men in it, but surely you haven't got them all?"

"No, and I won't know how many could make it until we meet up this morning. Let's aim for ten. That's probably more than I'll have but should be on the safe side. Do you have a pen and paper to take down my shopping list?"

"Don't need it. Took a memory course years ago. Part of my spook skills. Go ahead."

"All right, let's start with battle dress. I need combat boots, camo trousers and blouses, Kevlar helmets and flak jackets, three pairs of socks and three shirts for each man. Split them fifty-fifty between large and extra-large; there aren't very many wimpy little SEALs. In fact, better throw in a triple-x large extra tall set in case one particular guy shows up. Also civvy-style duffle bags for each man, large ones."

"Okay, got it."

"I want a sidearm for everyone. The Berettas like the ones you left at the bus station will be fine. Three magazines and 150 rounds for each, Winchester Ranger SXT expanding bullets, the ones they used to call Black Talons. Shoulder holsters for each gun."

"Fine."

"Ten M-4 rifles, each with three 30-round magazines and 200 Remington jacketed hollow points. Each rifle equipped with night vision scopes and flash suppressors."

"Yeah, go on."

"Throw in four Heckler & Koch MP-5s, each with two extra 50-round magazines and laser sights. Same 9mm SXT bullets as for the Berettas."

"Fisher, what are you planning—World War Three?"

"A SEAL is always prepared, sir. You know our motto don't you?"

"Yeah, 'The only easy day was yesterday.' I know it."

"And I'm not done."

Jones sighed. "Okay, go on."

"Hand grenades. A mixed case of flash bangs, fraggers and tear gas. Plus four rifle grenade launchers and a case of munitions for them. And twelve sets of comm gear with headsets, the encrypted kind for battlefield use."

"What else?"

"Ten night vision goggles, sets of camo paint, all the kind of stuff you'd need for a surveillance and night action mission. Ten sets of four throwaway cell phones. Ten sets of gloves and balaclavas in desert camo. Ten combat knives."

"Okay. And?"

"Cash. We need some operating funds. You gave me ten grand and that will give us a start, but I need more. Let's say another five grand for each man, fifty in all."

"Planning on having a party?"

"Something like that, sir."

"Tsk tsk." Jones clicked his tongue against the roof of his mouth in admonition. "Is that it?"

"I think that will do for now, sir."

"You sure you don't want a couple of Aegis missile cruisers? A flattop? A nuclear attack sub?"

"If they'd do me any good here in the desert, I'd be tempted sir."

Jones laughed out loud.

"What about transport?" he asked.

"We're taking care of that. We're coming in stealthy, which means civilian vehicles."

"Good thought. So your men will be in civvies, too?"

"That's right. The gear will be kept under cover until if and when it's needed."

"Listen Fisher, I want to go on record here. I don't like this and it might cost me my career. But I know you're a smart guy and I'm the one that stuck you with this FUBAR, so I'm gonna give you what you're asking for."

"Thank you sir. We're going to need the stuff PDQ. Can do?"

"I think so. There's a Marine Corps unit stationed at Kirtland, the Air Force base in Albuquerque. Give your name at the gate and ask for the Marine quartermaster."

"Got it. How can I contact you to get the fax?"

"Call Bonner and give him the number. Use another throwaway phone. Use a fax you can walk away from and wipe the memory before you do."

"Aye aye sir."

"Good luck Fisher."

"Thank you sir. Would you put Bonner back on please?" Jones handed over the phone. 'Paul, we need to prepare the battlefield. In case anyone's watching Tommy's house, can you set it up to look like everything's normal? I'm thinking if you can get someone who looks like him to make a few trips in his Range Rover, that kind of thing."

Bonner thought for a moment.

"Yeah, I've got a guy that might work for that," he mused. "Sergeant Boyd. He's been eating a few too many donuts. Dress him in one of Tommy's T-shirts and cap and he could pass at a distance."

"That'll work. Sneak him into the house then do the clothing switch and have him drive around a little, but not get too close to anyone. We want it to look as if Tommy's following his usual pattern. I'll get my team on the scene after dark and take it from there. Better touch bases with security at the New Mexico Institute, let 'em know what we're doing."

"Okay, good thinking. I'll get Boyd on it right away."

"Great. See you soon."

Chris turned off the phone. Kate was driving and he was reclined in the passenger seat. He took the memory card out of the phone and bent it several times until it broke in half. He threw the pieces out the window. He removed the battery from the phone and it too went onto the verge, followed about a quarter mile later by the phone itself, which Chris tossed right into the busy roadway where it would quickly be crushed by hundreds of wheels.

"So, we're on?" Kate asked.

"No, *we* are not on." Chris gave her a stern look. "You're not going to be mixed up in this anymore."

Kate said nothing but her knuckles tightened on the steering wheel. Chris looked away and watched the desert landscape passing by. They were approaching the Route 66 Casino, an anomaly in the middle of nowhere. He gazed in wonder at the sight of hundreds of cars, trucks and RVs parked in front of the gaudy hotel and gambling palace. Images of dice and cards zoomed across the face of a giant animated sign, bait to attract new suckers to engage in games of "chance" where the odds always favor the house.

"You're not going to cut me out of this now," Kate told him. Her jaw was clenched, her eyes fixed on the road ahead.

Chris sighed and shifted around in his seat to turn toward her.

"Look, if it comes to trouble you're not equipped for it, mentally or physically. You'd only get somebody hurt, probably you but maybe someone else, too."

They rode in silence for several minutes. A string of enormous tractor-trailer units roared past in sequence, causing the Jeep to rock in their slipstreams. The diesel behemoths were going ninety miles an hour, Chris estimated.

"I don't want to get in the way, or Lord help me, be involved in anything dangerous," Kate said. "I just want to keep in the edge of things, where I can know what's going on."

Chris cleared his throat as if about to say something, then sat back in his seat.

"Okay, look, I won't leave you in Albuquerque. You can stick with me while I put my team together. But when we get back to Los Alamos you're going to the safe house with Tommy and Stuart. That's going to be the control center for whatever happens, so you can keep in touch with everything."

"Sure, that's fine." Kate smiled. "Don't push me clear out, that's all I'm asking."

"I just don't want you to get hurt."

"Of course, darling. Nor do I—any more than I want you to be harmed. Ever since what happened in that cafe, I've frankly been scared to death. But this is the greatest story I'll ever write and I want to be able to tell it all."

"Okay, that's great. We're on the same page then. You'll be an observer, but stay out of harm's way."

They rode along in silence for several minutes. Chris was studying Kate, thinking that she didn't look very frightened.

"So, you were planning to dump me in Albuquerque?" she asked, eyes on the road.

Chris didn't answer.

"That's what I thought," she muttered.

About an hour later they were approaching Albuquerque's commercial airport. The divided highway leading in to the terminal was decorated with sculptures representing oversized Native American pottery, enormous bowls and jugs taller than a man.

"Take the arrivals lane and pull up in front of the rental car section," Chris instructed her. "We're looking for Avis. I'll go in, you drive around the loop. Keep moving until you see me."

Kate nodded and downshifted as they entered the congested area. Cars were double parked as arriving passengers loaded baggage into cars, waited for shuttle buses or signaled for cabs. Kate slowed in the middle lane, pausing for a moment as Chris jumped out. She drove away.

The automatic doors swished open and Chris stepped inside. He moved out of the stream of passengers to let his eyes adjust to the light and perform recon. He took off his sunglasses and slipped them into his shirt pocket. He lounged for a moment, trying to look innocuous, leaning against a pillar and glancing idly around. Seeing nothing out of order he walked to the Avis counter. Two agents were seated behind the counter. A young man was tending to a customer. The other was a woman about 30.

"Hi," he said, stepping up in front of her station and grinning. "Maybe you can help me?"

"Of course sir," she said, returning his smile. "Do you need a car?"

"Oh, thank you, no," Chris said, putting on a slightly befuddled expression. "No, I'm supposed to meet someone and he said he'd leave a message here. My name is Frederick? I'm supposed to meet Mr. Paisley? Did he leave something for me?"

The agent's smile faded for a moment, then she seemed to remember something. She turned to her associate.

"Excuse me Joel, but didn't someone leave an envelope here earlier this morning?"

"Sure, here it is." The young man opened his cash drawer and pulled out an envelope. He looked at the writing on the front. "Mr. Frederick is it? You said it was from…?"

"Paisley."

"That's right. Here you go."

He handed over the sealed envelope. Chris slipped it into an inside pocket. To assure he would soon be forgotten he gave them a vacant grin of thanks before turning away. In moments he was back in the traffic lane watching Kate approach. She slowed to a stop and he jumped in and latched his seatbelt.

"That was quick, just one turn around the pasture," Kate said. "You got the information?"

"Yep," Chris said, patting his coat pocket. "Soon as we're out of here I'll take a look."

They drove around the loop again and back to the west toward I-25, then north two exits where Chris instructed her to pull off and stop in a hotel parking lot. He took out the envelope. It was imprinted with the Holiday Inn emblem. "Please give to Mr. Frederick...from A. Paisley" was written on the front in bold printed letters. Below that was the word "Personal," heavily underlined twice.

He tore the flap open and unfolded the message inside.

It read: "Hi Fred. If you're reading this, welcome. The sales meeting is at the Downtown Hyatt Regency, just off Second Street. Tell the clerk you're with the Red Ball Construction Co. and ask directions to our hospitality suite. The other five are already here—Paisley."

Chris chuckled. The "red ball" was a subtle reference to a standing joke about Navy SEALs, supposedly balancing balls on their noses like real seals in a circus.

"Head for downtown and look for Second Street," he told Kate, folding the letter and putting it back in his pocket. "Hyatt Regency. We've got a sales meeting to attend."

Kate glanced at him.

"Sales meeting?"

"Yep. Top performers only." He grinned happily. "There's gonna be seven of us on this sales team."

Chapter 26

Albuquerque, New Mexico
Monday, 11:15 a.m. MST

Leonard Perry had arranged a hospitality suite and reserved rooms for the arriving men, guaranteeing the reservations with a credit card that belonged to his second cousin. His cover was a sales meeting for the Red Ball Construction Co. and the bills would be paid with cash to avoid a paper trail.

Standing in the hall outside the hospitality suite, Chris smiled. He was looking at the door. There was a rubber chicken hanging on the handle and a hand-written sign had been taped to the door. It read: "ON THE BALL / Welcome Top Red Ball Salesmen." There were crude drawings of liquor bottles around the legend although he knew none of the men would be drinking anything other than coffee or soft drinks.

"Cute." He knocked on the door. In a moment it opened a crack then swung wide revealing the broad shoulders of Leonard Perry.

"Lieutenant! I mean, Lieutenant-Commander, sir," Perry said, keeping his voice down. "You made it. Come in. And you too, miss," he added, spotting Kate. He turned toward the room behind him and spoke sotto voice, "Cool it; there's a lady present."

Chris stepped past Perry into the room. Kate lingered just inside the door, watching to see what would happen. Five men jumped to their feet, came to attention with a click of their heels and threw a salute.

"As you were men." Chris had a big smile on his face. "You're not on active duty yet. This is a low-profile mission, so military courtesy is suspended for the duration. I'm just Chris to you." He turned to Perry and took his hand. "Lenny, it's great to see you! How long has it been?"

"Too long," Perry replied. "I retired a couple years ago. Finished out my service training brats at Coronado." He was referring to Coronado, California, home of the Naval Amphibious Base where SEALs learn their trade as elite Special Forces fighters.

Chris turned back to the others. For a moment he gazed at each one in turn, smiling and nodding in recognition.

"Men, thank you for coming. Your country needs you, I need you. I've got authorization coming to temporarily put you into active duty. There'll be a bonus, I promise. We're going to be getting equipped this afternoon. Right now I want you to meet Kate Elliott." He took Kate's arm and led her forward beside him. "Kate, these are some of the best, toughest, most loyal men ever to serve in the United States Navy."

He strode to the man standing on the left and shook his hand.

"Kate, this is Martin Sharp. We call him 'Shooter,' as in sharpshooter. His name is Sharp, but he's also a top-rated sniper with a lot of notches on his gun. How many kills Shooter?"

Sharp grinned. He was slender for a SEAL, about six feet tall and 170 pounds. His face was tanned and his hair bleached from the Sun. Despite the friendly smile his blue eyes held the look of a raptor. He stood erect but relaxed. In fact, Kate noticed that all these men had a certain comfortable look while at the same time being almost preternaturally alert.

"Fourteen or fifteen," Shooter replied. "There was one Taliban chief we weren't quite sure about, you remember sir?"

"Oh, I sure do," replied Chris. "You knocked him down at twelve hundred yards with that Barrett of yours, just as he took a sip of mint tea. Never had any doubt it was a kill, but they got away with the body."

Sharp smiled and took Kate's hand.

"Glad to meet you ma'am."

"Hi Shooter." She shook his hand and looked around at the others. "You guys are probably wondering what I'm doing here. There's a lot to tell, and Chris will explain everything. But you need to know that I'm a just someone who got caught up in something. Chris saved my life a few days ago, and he's still protecting me. I'll try not to get in anyone's way."

Six pairs of eyes turned to Chris with interest. He pretended to ignore their questioning looks and stepped to the next man in line.

"This is Carlos de la Vega," he told Kate. "We call him Della, and he doesn't seem to mind." He pumped de la Vega's hand then passed it to Kate. "Della's a decorated hero, twice winner of the Navy Cross for valor."

De la Vega raised her hand in his, bowed in a courtly manner and passed her knuckles near his mouth. "My honor, Ms. Elliott," he said.

He was larger than Shooter, about six-two and broad in the shoulders. He had a Mediterranean complexion with piercing brown eyes. His face was square with a high forehead and beetle brows. His mouth had the sensuous look of a Catalonian nobleman.

"My pleasure too, Mr. de la Vega."

He laughed. "No, please, my friends really do call me Della. You must do so as well."

"Of course, Della. And you can call me Kate. All of you," she added, addressing the others.

Chris turned to his right.

"Next we have our little Mouse." He seemed to be referring to the six-foot-seven giant who was standing in front of the couch. Kate guessed his weight at near 300 pounds, nearly all of it muscle. His hands were the size of three-pound canned hams and he seemed to have footballs stuffed up the sleeves of his shirt.

"Mouse?" she repeated.

"Yes, our precious little one," Perry chimed in. "We thought about calling him Paul Bunyan or Godzilla, but figured that might go to his head."

"Hi, Mouse."

"Hi, Kate. Don't mind these guys. We're all friends but you know how it is. If I wanted to I could wrap these rats around my little finger and flick them across the room like spitwads." He laughed and the others joined in.

"I imagine you could," Kate said, taking his hand cautiously. To her surprise his handshake was gentle and tender. "So, I suppose Mouse isn't your real name then?"

The giant chuckled. "No, my Mama called me Floyd, Floyd Garrett. I'd rather be called Mouse though." She nodded.

"Next is our intellectual," Chris said, turning to the next man in line. "This is Jim Chen. We used to call him 'Teach.' What about it Jim, did you finish that college degree?" Chris shook his hand.

"Oh, yes sir, and a master's too. I'm a high school science teacher now. Living the dream."

"Then I guess all those books you always had your nose stuck into did some good, huh?" He turned to Kate. "Jim's grandfather thrice removed came over from China in the 1870s to help build the transcontinental railroad. He eventually brought over his sweetheart and they settled in Colorado. Jim's a third generation soldier. His grandfather was a tanker, a gun loader in Patton's Third Army in World War Two. His father was a Green Beret Captain in 'Nam, winner of the silver star."

Chen was medium build and wore his glossy black hair in a military style. He was dressed in chinos and a professorial tweed jacket with leather arm patches.

"Pleased to meet you, Kate."

"Glad to meet you, too. I hope you may have read some of my articles. I'm an writer and editor at *Science Today*."

"Oh, of course. I thought that name sounded familiar. I'm a fan of yours. I thought that piece about the deep sea thermal vents in the Atlantic Ocean was fascinating."

"Why, thank you. That was an interesting story to do."

Chris stepped to the last man in line.

"And last but not least, this is Joe Tuttle. Goes by the moniker 'Snapper,' as in snapping turtle." Chris chuckled as Tuttle ducked his head as if pulling it inside of a shell. "He used to play with those Ninja Turtles when he was a kid, and with the name Tuttle it was a natural. He's quite a Ninja himself."

"Hi, Kate. Glad to meet you." Snapper shook her hand. He was a blond with blue eyes and had the lean look of a long-distance runner. He grinned, displaying a set of teeth that would make his namesake jealous.

"Thanks for coming," Chris said. "Thanks to all of you. Now it's time to get to work. Let's make ourselves comfortable and I'll bring you up to speed. Do we have any coffee?"

"Yes, sir!" Perry jumped to his feet. "That is, yeah Chris, we have coffee." He sat back down and aimed a thumb at a carafe and cups on a nearby table. "You have to give us time to get used to dropping the military BS," he said. "Help yourself boss."

Chris laughed and walked over to pour a cup. "Kate? Coffee? Anyone else?"

Several hours had passed. Leaving Kate with three of the SEALs at the hotel, Chris had led the way in his Jeep accompanied by Perry and followed by a rented tan Ford Expedition carrying Sharp and Garrett. They stopped at a Kinkos store where Chris obtained the faxed authorization to re-enlist his team. He immediately made duplicates for everyone and gave signed originals to Perry, Sharp and Garrett.

Next stop was the Marine quartermaster at Kirtland where they loaded the vehicles with the items Chris had requested. They took only what they needed, leaving the extras behind. The gunnery sergeant in charge handed a packet to Chris and asked him to sign a receipt for it.

"I don't know what's in this," he said, "but the messenger was accompanied by an armed security officer and it came from the purser's office." He winked. "Could be just about anything, I guess..."

"Yeah, thanks Gunny."

Chris noticed that Sharp was looking around the Marine armory like a kid in a candy shop. "Got an idea Shooter?" he called over to him.

"Yes sir, Lieutenant-Commander," the sniper replied, beckoning Chris over. They had been instructed to follow military etiquette while on the base. "Sir, the Marines use Barretts. Do you think we could get one?"

Barrett Firearms of Murfreesboro, Tennessee is the premiere maker of sniper rifles. Their super accurate weapons with matched optics are the choice of the military's top Special Forces marksmen.

"I hadn't thought of that," Chris said. "Well, of course I didn't know whether you'd be one of the guys to show up. Feeling kind of left out are you?"

"Well, you know how it is." Shooter put his hands in his pockets and looked down at his shoes. "Those M-4s are okay, I guess, but they ain't worth shit beyond a couple hundred yards, pardon the French sir."

Chris laughed. "Let's see what Gunny has to say."

A moment later Shooter was clutching a brand new Model 82A1, Barrett's semi-automatic 50 caliber rifle equipped with a Leupold Mark IV 20x50 scope. He was trying unsuccessfully to contain a wide grin.

Everything was packed in innocent looking civilian duffel-style luggage. Back at the hotel Garrett waved away the bellman and rolled out two of the hotel's trolleys. They piled up the equipment and rolled it to the suite.

Once everything was secured, Chris swore in the other men and signed a copy of the authorization form from the Department of the Navy for each. He broke open the packet of cash and counted out 50 crisp one hundred dollar bills for each man.

"Don't use any charge cards. If you can get a receipt, do so, but don't sweat it. Don't do anything that might raise suspicion. Remember, we're all just ordinary salesmen, so try to look the part. I know it's hard, but slouch a little bit, get some of that stiffness out of your backbones. You're in the Navy now—but we don't want to look like it."

"Gosh, Chris," Perry said playfully. "You got something stuck up your spine too, you know?" Everyone laughed as they tried to act like salesmen. It wasn't easy.

"Hell, I *am* a salesman," Garrett said, 'and even I don't know how to look like one." Everyone turned to him, eyebrows raised.

"Yeah, I sell cars. Best in the dealership," he added.

"What kind of cars, Mouse? You couldn't even get into most of them to give a demo drive," Tuttle quipped.

"Matter of fact, wise ass, I sell those little Smart cars," Garrett replied. "They're bigger inside than they look—and I *am* the demonstration."

"Okay," Chris said with a laugh. "Let's all try to look like Mouse and imagine him squeezing into a Smart car." Everyone chuckled and Tuttle slapped Garrett on the back. "Let's get to work. We've got a plan to make and a mission to carry out."

Chapter 27

Los Alamos, New Mexico
Sunday, 7:15 p.m. MST

Evening was approaching, spreading beams of golden light across the Jemez. Chris and his team had infiltrated Los Alamos and were preparing for the mission to set a trap for the presumed assassin.

Chris and Kate had driven up from Albuquerque in his Jeep. The tan Ford Expedition followed at a leisurely pace carrying Perry and Tuttle. A dark blue Dodge Durango bore Garrett and De la Vega. Sharp and Chen were riding in a white Chevy Suburban. All were dressed in informal civilian clothes, trying to look like salesmen. They coordinated their movements using the encrypted tactical radios.

Each pair had checked into a different motel, paying with cash. They spread out to eat dinner in different restaurants then returned to their rooms to prepare for the night's mission.

Chris and Kate were at the guest house where she was to stay with Dawson and Reeves. One of Bonner's officers had delivered a selection of sandwiches from the canteen. Dawson was still working at his keyboard while the other four men conferred in the living area.

"I hope they didn't bring tuna salad," Dawson called out loudly. "Hate tuna salad. Always felt sorry for Charlie."

"We've got ham and cheese on rye and roast beef on whole wheat with sharp mustard," replied Kate from the kitchen where she was examining the selection. "Big dill pickle slices and little cups of slaw and potato salad. Coffee or soda to drink."

"One of each and a Coke," Dawson demanded, not pausing from his work. "No, cancel that. Make that two ham and cheeses, a beef and two Cokes. What's for dessert?"

There was no answer. Even Kate was learning that it didn't pay to engage Dawson in conversation when it came to the question of food.

Listening to this exchange, Chris smiled and winked at Reeves.

"It's no wonder they put out a contract on him," he quipped.

"I heard that Fisher!" Dawson shouted, concentrating on his monitor, trackball spinning. "Don't think I'll forget it either."

Chris, Bonner and Reeves chuckled and the sound of Kate's two-toned laugh came from the kitchen.

The plan was in place and they were waiting for dark to put it in motion. During the day the over-weight sergeant Peter Boyd had been impersonating Dawson. Wearing some of the physicist's distinctive clothing he'd hung around the house, taken two short trips in the Range Rover, and was now waiting to be extracted.

Glancing at his watch Chris keyed his tactical radio and spoke into the microphone stalk. He was using the general channel so that all six of his men heard.

"This is Falcon Trap One. Check in."

"Falcon Trap Two," came the voice of Lenny Perry, followed by each of the strike team members, responding in order.

"Mission is go," Chris said. "Deploy."

"Aye aye," came the answers from six men.

Chris set the radio to standby. Kate appeared with a tray and he grabbed a ham and cheese sandwich.

At the nearby New Mexico Institute the plan went into motion. A brown UPS truck turned into the Institute grounds and made its way toward the residential area. It stopped twice, first at a science lab, then at a house at the far end. Each time a man in a brown uniform bustled out of the truck carrying packages. Reaching Dawson's house he backed the delivery van up to the portal at the front.

The driver, actually one of Bonner's patrolmen, got out and went to the back of the truck. He slid open the door and grappled with a large box. The porch light was off and no one could see that as he carried the box to the door a second man emerged covertly from the truck and shadowed him to the door, which opened as they approached. The deliveryman seemed to hand the package to someone coming from inside the house. In fact, it was the man who had just come from the truck.

As the driver returned to the truck Sergeant Boyd, no longer wearing Dawson's clothes, slipped out of the door and disappeared into the back of the truck. The driver grabbed a second box and as he carried it to the door a second man slipped out of the truck and into the house, turning to accept the package as if he had come from inside.

The door closed and the truck drove away, meandering through the residential area and making two more stops before leaving the Institute grounds.

Inside the house Perry and Tuttle began to open the boxes. They contained Kevlar helmets, flak jackets and MP-5 9mm submachineguns. One of the boxes contained the head from a clothing store dummy. They quickly put away the boxes and arranged the room to make it look occupied. They organized the pantry just off of the kitchen to create a hiding space.

"He's probably going to have some kind of audio probe," Tuttle said. "We need to do something to create sound ambience for cover."

"Good thought," Perry replied. "Let's turn on the ventilation system for a start." He stepped to the thermostat, set the heat/cool lever to "off," then flicked the fan switch to "on." Air began to rumble through the ducts.

"I'll turn on this overhead fan, too," Tuttle said, pulling on the chain. "Oh, and look at this. If I put some loose papers right under the fan, and hold them in place with this paperweight, they make a nice rustling noise."

"Sweet."

Going into the bedroom the two men loosely rolled a bulky down comforter and placed it under the sheet, bending and forming it to resemble a human body. Tuttle placed a digital micro-recorder under the pillow and turned it on. The sound of gentle snoring began to come from the recorder's speaker.

"Too bad we won't be getting any of those Z's tonight," he remarked. He picked up the dummy head and positioned it on the pillow, facing away from the door. To complete the scene he tucked the sheet around the dummy head. Meanwhile Perry was moving around the room, checking to make sure the windows were closed and locked. He plugged a small nightlight into the wall to backlight the bed, making it hard to see there was no real person in it.

After a final look around Perry spoke into his radio, announcing that he and Tuttle were in place. The other four responded that they, too, were on station.

They had the house under covert surveillance. De la Vega and Chen were positioned among the trees about two hundred yards from the edge of the forest and at 90 degree positions to left and right from the house. Equipped with night vision goggles their task was to watch for an approach through the forest.

Garrett had moved into position near the roadway leading in from the main street. Hidden in a clump of bushes, he could see any traffic coming into the area and also observe the house. He too had night vision goggles.

Sharp was set up in another house across the way. The usual resident was attending a physics conference in Paris and the Institute director had given them permission to use it. It gave the sniper a good position from which he could watch the approaches to Dawson's house from almost every direction. A window was raised about six inches, giving him a field of fire with the 50-caliber Barrett.

"Now we wait," Perry spoke softly into his radio. "Whisper mode from now. When our Falcon gets close, Snapper and I will respond merely by tapping our mikes. One tap yes, two no, three repeat message. You know what to do."

The six men settled into their ambush, prepared to wait silently as the night grew darker.

Chris was edgy. His team was in place and here he was in the safe house. True, it was his plan and he was coordinating it, but he would preferred to be in the thick of the action. He paced in a circle around the living area.

"Hey, Chris, come here, sit down," Reeves told him. "Either the guy's going to show up, or not. If he does, your men can handle him."

Chris sighed and sank into an easy chair. He thumbed the radio and spoke into it.

"This is Trap One. Trap Four, report," he said. Falcon Trap team number four was De la Vega who was positioned in the woods.

"Trap Four. I can see the house. Lights are on. Trap's set."

"Roger. Trap One out."

Chris looked at his watch. It was not even 9:30 yet. Damn, too early to expect anything. He settled back in the chair and looked around the room. Dawson was beavering away on his computer in the adjacent dining area. Kate had gone to her room to get some rest and Jones was lying on the couch staring at the ceiling. Bonner had returned to his office and had promised to stop back later.

"You know, even if we get Red Falcon tonight, that doesn't really solve anything," Chris said to no one in particular. Reeves nodded and Jones turned his head to look at Chris.

"What's your point?" the agent asked.

"We have to catch this Eminent Grease rat. He's calling the shots. He killed three men already. Sure, we might be able to take Red Falcon alive, and he might give us an ID on Grise, but somehow I doubt it. I told my guys not to take any chances. If Red Falcon so much as starts to point a gun at any of them, he's dead."

"That's fine with me," Jones mused. He sat up on the couch. "Last thing I want is for one of your guys to get hurt. But we might be able to learn something from him, dead or not."

"He's not likely to be carrying anything that will help us," Chris said glumly.

"Yeah, but he's bound to have a trail we can follow. Fingerprints, DNA, travel records—Lord knows what we could learn just from his dead body. We'll do a full forensic workup."

"Yeah, we might trace him but that won't help us catch Grise."

"Probably not. That guy's covered his tracks pretty damn well."

The three men stared at each other for a moment.

"We don't even know what he looks like," Chris said. "Hell, he might even be a woman for all we know."

"That's possible," Jones admitted. "In fact, that could explain how he, or possibly she, managed to get close to those two officers on guard duty. A woman would have an advantage."

"Bonner told me his investigators checked the footage from all the security cameras at the main entrances to the lab from the night of the killings. They didn't see anything out of order. Either Grise managed to sneak in somehow, or he or she has authorized access."

"That's right, but we don't know which."

Dawson stopped tapping on his keyboard and in the silence the three men turned their attention to him. He swiveled his chair toward them.

"Would it help if we could see what he looks like?" the physicist asked.

The question was met with silent looks of incredulity. At last Chris spoke.

"Well, there's no way we can do that."

"Hmpf." Dawson turned back to his keyboard and began typing again.

Chris, Jones and Reeves looked at each other.

"Uh, Tommy?"

"Yes, Fisher?" Spoken abruptly and in a dismissive tone. Dawson continued to type.

"Why did you ask us that?"

"Oh, never mind. I was just trying to be helpful."

The three men looked at each other again. Chris shrugged. Jones pointed his forefinger at his temple and twirled it in a circle. Noting this, Reeves shook his head.

"Did you have something in mind, Tommy?" he inquired.

The typing stopped. Dawson sat staring fixedly at his monitor for a moment.

"You know, I sure would like another Coke," he declared, his eyes still on the screen.

Reeves glanced at Chris and shrugged.

"Tommy, you're avoiding the question," he said.

"No I'm not. I'm just saying I'd sure like another Coke. What's wrong with you Reeves?"

Reeves sighed and got up. He walked into the kitchen, opened the refrigerator and grabbed a can of cola. He started to close the door then swung it open again and picked up a second can. He walked into the dining area and set them down in front of Tommy.

"Here, I brought you two of them."

"Oh, great." Tommy popped one of the cans and took a long drink from it. Some of the cola dribbled down his chin.

Reeves pulled one of the dining chairs around and sat down. He beckoned to Chris and Jones and they got up to join him. Ignoring them, Dawson resumed typing.

"Tommy, what did you mean when you asked if it would help if we could see what the killer looks like?"

Tommy looked up with a glare.

"What the hell do you think I meant? It was a simple enough question wasn't it? Except I didn't get a straight answer, so never mind." His attention went back to the monitor. Chris winced.

"Tommy, I apologize," he said. "I didn't mean to blow you off. Do you know something we don't?"

Dawson stopped typing again.

"Let's go into the other room where we can be comfortable," he said, picking up his two cans of cola and lumbering into the living area. He dropped into one of the easy chairs, set the unopened can on the floor beside him and drained the other.

Still seated around the table, Chris, Jones and Reeves glanced at each other again. After a beat Reeves nodded and they got up to return to the living area. Jones and Reeves sat down on the couch and Chris took the second easy chair.

"Tommy, what's this about? Do you have an idea?"

"Do I have an idea!" Tommy roared. "I come up with the most amazing theory in the history of the Universe and you ask me 'do I have an idea'?"

"Well, yes, of course there's your theory…" Chris hesitated. "But what's the connection? We don't understand what you mean."

Dawson looked surprised, then chagrined.

"Well, tan me for a polecat," he said. "I never did get around to telling you the best part of my theory, did I?" He whacked himself on the side of the head with an open palm.

Silence filled the room.

"Well, I didn't, did I?"

"Tommy, we honestly don't know what you're talking about," Reeves told him. "What part is that?"

Dawson chortled.

"Why, the part about being able to view the past."

"What!" Reeves said, sitting up straight.

"You mean…?" Chris began then stopped in confusion.

"Oh good holy Christ," Jones muttered. He buried his face in his hands.

Chapter 28

New Mexico Institute
Monday, 12:45 a.m. MST

It was after midnight and the Falcon Trap team was settled in on high alert. De la Vega had wriggled beneath a fallen tree trunk and was scoping the forest with his night vision goggles. Chen had found a similar hiding place and burrowed under a mat of pine needles. With their camouflage clothing and secure positions it would have been almost impossible for anyone to spot them, even using a thermal vision system.

Down the road from the house Garrett's considerable bulk appeared to have melted into a luxuriant bush. Only his head moved, turning in a regular pattern to observe the area leading into the residences from the main campus.

Across the way a dark window was partway open. Behind it Sharp had arranged a coffee table and sofa cushions to support his Barrett. He was sitting on the floor cross-legged behind the powerful rifle, scoping the open areas around Tommy's house.

Inside, Perry and Tuttle were sitting on dining room chairs, ready to take cover if their prey approached. Hardened by years of training and combat they didn't speak and moved only to check the time or respond to a call from one of the other team members. Chris checked in at 1:30. It was approaching that period of deepest night known to warriors as o'dark thirty. All was quiet.

At 1:42 Della rang in on the general frequency.

"Got a bandit," he whispered.

Everyone's attention was riveted, including Chris's at the safe house.

"What you got?"

There was a moment of silence.

"Someone's coming out of the trees. Pretty big. A man. Rigged for silent running."

A phantom-like figure moved quietly through the pine forest. Narrow beams of moonlight shone through the trees, flashing stroboscopic glimpses of a stocky man dressed in dark gray coveralls and a balaclava. His face was painted with camo colors. He wore thin, dark gloves and unpolished black combat boots.

"How far?" Perry asked.

"He's at my two o'clock, about a hundred meters out. Moving straight toward you. He'll pass near me, too far for Teach to spot him."

"Teach, prepare to close up behind him if you can."

"Aye aye," Chen whispered. "Della, say when it's okay for me to start moving."

There was silence for a long time then Della reported in low whispers.

"Edge of the lawn. He ducked behind bushes. He's on surveillance. You can start to move Teach."

Stepping carefully the assassin reached the edge of the forest and paused behind a dense shrub. He lifted the left cuff of his coveralls to glance at his chronograph watch: 1:57 a.m. He reset the timer hand and shot the cuff to hide the dimly glowing face of the watch.

There was a single click on the comm frequency as the SEAL team went into the next stage of stealth. Chen had tapped once on his microphone stalk.

"Perry, keep the light on in the main room of the house," Chris instructed. "Let's leave him hanging for a while."

Time passed, marked by the gibbous Moon as it drifted toward the West. Fresh mountain air stirred the upper branches of the trees. Presently the constellation Orion began to rise, soon followed by Sirius, the Dog Star, twinkling in the thick air near the horizon. The Moon passed behind the mountains. As its pearly glow faded the stars seemed to grow brighter, the dome of the night divided by the rippling band of the Milky Way.

Sirius rose higher in the sky and began to shine like a beacon with clear white light. Somewhere toward the mountains a single coyote bayed. In a moment a second voice joined the chorus, then a third and yet others until the night was filled with stars and wild music.

Moving slowly, inches at a time, Chen had worked his way deeper into the forest and southward as De la Vega moved forward to meet him. At last the two men were positioned directly behind the known position of the assassin. A half hour passed.

"Perry, turn off the light now." Chris again. "Make it look as if Tommy's going to bed for the night."

Tap.

At 2:32 a.m. the light in the main room of the house went out. Perry and Tuttle stepped cautiously into the kitchen pantry. Perry drew the door closed. They had disabled the latch earlier. The men were crowded side-by-side in the dark.

"Not tryin' to get familiar with me, are you Chief?" Tuttle whispered.

"Shut up."

"Aye aye sir."

In the pitch-black closet both men grinned. Nothing happened for 45 minutes.

"He's moving." De la Vega's whisper.

Tap.

Silence. Then Sharp: "I see him."
Tap.
Chris: "Della, start to move up to the edge of the forest. Teach, move to his flank and close up."
Tap.
Tap.
Through De la Vega's open mike Chris heard a strange sound. "What's that?"
"Coyotes, sir."
"Well, I'll be damned."
"Sir?"
"Never mind. Stay on target."
"Aye aye."
More minutes passed as the assassin worked his way toward the house.
"I've got him." Garrett, a kind of rumbling basso profondo whisper.
Tap.
"I've got the house in sight." De la Vega. "Target moving up to the rear window."
Tap.

The intruder reached the house and crouched beneath the now-darkened window. He reached into a pouch on his utility belt and withdrew a matte-black device. It was half the size of a cigarette pack with two slender wires attached at one end. There was an ear bud at the end of one of the wires. He pressed it into his right ear. The other wire led to a tiny microphone. He pressed the mike onto the glass of the window above his head. Coated with a sticky substance it clung in place, no larger or more noticeable than a housefly.

He listened. There was the hum of a ventilating fan—but also an unfamiliar sound. What is that? Tensing, he pressed the ear bud tighter and blocked his other ear with a finger. Now he could hear gentle swishing and fluttering sounds. After a moment he nodded to himself: a slowly rotating ceiling fan was rustling some loose papers.

Silence. About ten minutes passed.
"Target moving toward north side." De la Vega. "Checking bedroom."
Tap.

Red Falcon put away the listening device and moved carefully to the west side of the house. He drew a matte black Sig Sauer model 226 automatic from the shoulder holster beneath his left arm. Reaching into another pouch on his utility belt he pulled out a metallic cylinder about an inch in diameter and five inches long. Carefully aligning the threads he screwed the silencer onto the barrel of the 9mm pistol. He gently popped out the 15-round magazine, double-checked that it was properly loaded then clicked it back into the weapon. He turned away from the cottage and held the gun against his stomach to muffle the sound as he gently drew back the slide to chamber a fresh round.

More time passed.
"I see him." Tuttle. "He's coming around past the Range Rover." The sniper had the target in his 20X scope. "He's got a pistol."
Tap.
"He's coming around to the front door." Garrett.
Tap.
Silence, then Sharp: "He's at the door. He's doing something to the hinges."
Tap.

Red Falcon crouched for a moment to watch and listen, letting his eyes adjust to the dim light. The coyotes had ended their choral performance and the night was silent. He edged cautiously to the entrance, the Sig held ready at his side.
The door was made of weathered oak, carved in an ornate Mexican folk-art style and fitted with antique hand-forged black

iron hinges. Centered just below eye level was a heavy cast iron knocker in the shape of a lion's head. The assassin produced a small tube, unscrewed its cap and drizzled light lubricating oil on each hinge. He waited quietly for a few minutes, giving the penetrating oil time to soak into the pins.

"On your toes. He's reaching for the doorknob."
Tap.

Unlike the heavy iron hinges the lockset was modern, a pick-resistant Medeco in solid brass. The assassin had ways to deal with such locks, but before beginning he reached out and grasped the doorknob. A thin smile crossed his lips as it turned easily in his hand.

Chris had instructed Perry and Tuttle to leave the front door unlocked. It was well known that Dawson was absent minded, and the physicist had admitted that he never bothered to lock the door. That left the trap wide open and waiting.

Inside the pantry Perry and Tuttle held still, breathing slowly through their mouths. MP-5 machine pistols were slung across their chests. They heard a slight noise as the front door swung open, then the sound of someone moving stealthily into the house.

"He's gone inside." Sharp.
Tap.

Perry shifted his weight carefully to his right leg and put his right hand on the pantry door. The plan was to wait for the sound of shooting, then use the sound as cover to rush the assassin. The two men tensed, ready to pop into action.

Moving quickly now Red Falcon strode to the bedroom door. He turned sideways, placing his back against the right-hand wall. Leaning slightly forward he thrust the Sig with its bulky silencer around the jamb.

He could see the shape of a large body lying in the bed, covered only by a light sheet. His victim's head was visible on the pillow. The perfect setup for a kill. Without hesitation he took two quick steps into the room and fired the pistol. Pfft! Pfft! Pfft! *Three muffled shots went into the body on the bed.*

Perry and Tuttle had heard the sound of quick footsteps receding down the hallway toward the bedroom. Perry eased the door open a few inches.

They heard the muffled noise of the assassin's silenced Sig Sauer, three flat sounds like a paperback book falling on a hardwood floor. Perry swung the door wide and slapped Tuttle gently on the back.

The SEAL stepped quickly out of the kitchen. Nearly soundless he slipped across the main room, glanced down the hall then moved stealthily toward the open door. Perry was right behind him.

"Go!" he whispered.

Tuttle stepped into the bedroom, dodging to one side of the door and kneeling as Perry followed him into the room.

"Freeze!" Tuttle shouted.

"Drop the gun!" Perry commanded. Both SEALs had their MP-5s leveled at the silhouetted shape standing by the bed. The figure began to turn toward them.

"Drop it!" Perry repeated, but the figure continued to turn and now the SEALs could see the ugly shape of a silenced pistol beginning to swing in their direction.

"Shoot!" The machine pistols stuttered, releasing streams of death. They'd assumed the assassin would be wearing body armor and by pre-arrangement Perry aimed high for the neck and head, Tuttle aimed low for the legs and groin.

The pistol clattered on the floor, followed by the body of the late assassin.

"We're clear!" Perry shouted, swinging the barrel of his weapon up toward the ceiling.

"Clear!" Tuttle echoed.

"Falcon Trap team, to me," Perry commanded. "Falcon is down. Repeat, Falcon is down."

Somewhere in the forest a single coyote raised its voice in a final song to the night sky.

The eerie sound echoed from the built-in speaker on Chris's radio. Gathered around him were Reeves, Jones, Bonner, Kate and Dawson. Everyone was silent for a moment after the all-clear message from Perry. Then Dawson looked up and Chris saw there were tears in his eyes.

"That could have been me," the physicist said, his voice unsteady. "They really did want to kill me!"

"But they failed," Chris pointed out.

"Yeah, they failed. Damn them. Damn them for even thinking about it." Dawson wiped his eyes with the sleeve of his T-shirt.

"Well, back to work," he announced, suddenly cheerful. He climbed out of his easy chair and plodded back to the workstation in the dining area. Marveling at the sudden change of mood, five pairs of eyes watched him in silence.

"Look, Chris," Bonner said in a low voice. "You guys get some rest. I need to secure the crime scene at Tommy's house. My forensic team is headed over there now. The SEALs will have melted into the night as planned. Tomorrow when we're not whacked-out tired, we'll get to the bottom of whatever Dawson was talking about."

Chris, Kate and Reeves quickly agreed and Bonner and Jones left. Chris spent a few minutes on the radio with Perry and his team, who were rapidly exfiltrating from the scene. "Great work guys," he told them. "Get some rest. We'll touch bases later."

"Tommy, it's nearly four o'clock," Stuart called out. "We all need some sleep. Let's shut it down for the night, okay?"

"Okay. Think I could have one more Coke?"

"Sure. They're in the fridge."

Grumbling, Tommy got up and steered in the direction of the kitchen.

"Get it myself," he griped.

Reeves looked at Chris and Kate and rolled his eyes. Kate struggled to suppress a snicker and Chris just smiled.

Chapter 29

Safe House
Monday, 10:45 a.m. MST

Chris had returned to the guest house after meeting Perry at his motel for a debriefing. He'd instructed the chief to keep the SEALs on a low profile in civilian clothing. Dawson and Reeves were still sleeping and Jones had disappeared shortly after dawn.

Chris gratefully accepted a cup of coffee from Kate. A few minutes later there was a knock and Bonner entered, looking tired and haggard. He helped himself to a cup and sank into a chair.

"We did a preliminary autopsy," he told Chris. "Didn't find much. Red Falcon was a pro. The guy's fingertips were surgically erased. I'd heard about things like that, but in all my years in the Bureau I never actually saw it."

"No identifying documents, of course."

"No. The serial numbers had been removed from his pistol and silencer, along with the tags from his clothes. Everything was generic, stuff that can't be traced."

"What about travel records? There should be a ticket trail, surveillance footage at airports, auto rentals, that kind of thing."

"Like finding a needle in a haystack. Nothing so far. We don't know what he was driving, or if he even had a car. Someone might have dropped him off. Grise maybe. I'd guess he traveled in disguise and probably didn't take any direct routes. He could've come in from anywhere."

"Yeah." Chris thought for a moment. "What about his face? Surely he couldn't hide his appearance. There have to be photos somewhere, Interpol, the CIA?"

Bonner gave Chris a sour look and tapped his fingers on the arm of the chair for a moment.

"Let me put it this way: Your guys are pretty good. What were your orders to them?"

"Well...to shoot to kill at the least sign of threat..."

"You got it. Red Falcon took eight Black Talon hollow-points in the head."

"Oh!"

"Yeah. Right now he could pass for just about anybody including Madonna and O.J. Simpson. Well, probably not Madonna."

Chris smiled ruefully. "Those Heckler & Kochs do have a way, don't they?"

"Yeah, at 15 rounds per second they sure do. I've got a specialist working on reconstruction but chances are his features were surgically altered just as his fingerprints were. Say, is there anything to eat? I haven't had any sleep and I missed breakfast."

"Sure, let me fix you something," Kate spoke up. "Want some bacon and eggs?"

"Got any more of those ham and cheese sandwiches?"

"Yeah, there's one left," Kate reported, getting it for him from the fridge.

Contemplating the events of the previous night, the two men sat in silence. Bonner chewed disconsolately on his somewhat soggy sandwich. Chris sipped a third cup of coffee. After a while the doorbell rang and Jones entered. He, too, looked tired.

"Anything?" Bonner inquired.

"Nothing. I've had a couple of guys looking around. Didn't find squat." He sat down on the couch and looked enviously at the sandwich that Kate was handing to the security chief. "Got any more of those?"

Just then Dawson came out of his bedroom, yawning and stretching.

"I'll call the canteen to send over more sandwiches," Kate said, glancing at the physicist who was wearing a scarlet dressing robe featuring a black dragon. His Quantum Cowboy cap was set backwards on his head. Catching Chris's eye she winked.

"Good morning, Tommy," Chris said. Dawson yawned again, nodded in the general direction of the three men then turned toward his keyboard.

"Hey, Tommy, don't wander off," Chris implored. "Come sit down with us. We've got some more sandwiches on the way. Let's wake up Stuart so he can join us." He jumped up, strode down the hall and pounded on a bedroom door.

"Hey, Stuart, up and at 'em." There was a muffled expletive from behind the door, and a few minutes later Reeves came into the living area. His hair was mussed and he obviously had slept in his clothes.

"Hi," he said weakly, rubbing his eyes. "Anybody made coffee?"

Kate handed him a cup. Dawson got two Cokes and the other four men drank coffee. When the sandwiches arrived they all dug in, each taking a sandwich and Dawson claiming three as usual.

"It was a pretty exciting night," Chris said at last.

"You think?" Dawson said bitterly, glaring at Chris as he took a bite of ham and cheese. He chewed for a moment then mumbled, "If that bastard had been trying to kill you I bet you'd think it was more than just 'exciting'." He took another bite and followed it with a long draught of Coke.

"You're right," Chris admitted. "But we outfoxed him." He edged forward on his seat, meeting the physicist's eyes. "Tommy, you said something last night. We were too tired and occupied to follow up, but we've all been wondering about it." The others nodded and looked at Dawson. He belched and used his fingernail to work a piece of meat from between his teeth.

"Need another Coke," he said to the room in general. He was halfway through the second one. Kate sighed and started to get up. Chris waved her back down.

"Later, Tommy," he said. "First we need to know what you were talking about. What did you mean?"

Tommy looked surprised, then grinned uneasily.

"Yeah, sure," he said, setting down the remainder of his third sandwich. "I remembered I never got around to telling you the rest of my theory. Things have been so distracting. People getting murdered all the time, trying to kill me…"

"Well, why don't you tell us the rest now?" Everyone leaned forward intently and Dawson looked confused.

"I told you last night didn't I?"

"Not exactly."

"Oh, well, I thought I did. What it is, with three dimensions of time we can 'see' the past. Isn't that what I told you last night?"

"Well, yes, something like that—but we don't know what it means. Have you discovered time travel?"

Dawson broke out laughing. "Don't be an idiot," he said. "That's not possible."

Chris sat back in his chair with a defeated look and Reeves picked up the thread.

"Well, if we can't travel back in time, how can we 'see' the past?"

Dawson frowned.

"Don't any of you understand English?" he demanded. "I said we can 'see' the past, not that anyone can go there. Are you deaf or what?" He picked up the sandwich and took another large bite.

Patiently Reeves continued to probe.

"Okay, we've got that. No time travel. So, exactly how do we 'see' the past? What does that mean?"

"You don't get it, huh?"

Chris sat forward in his chair and engaged Dawson with a stern look.

"No, Tommy, we don't 'get it' because we are so far behind you that you're clear over the horizon. Can you explain this from the beginning? In simple terms?"

Dawson looked surprised. He devoured the last of the sandwich and wiped his mouth with the back of his hand. He thought for a moment.

"Well, I think I see the problem now. Look, when I said we can 'see' the past I meant we could *view* it. You've seen my software that lets us see graphic images of computer code?"

Chris nodded. "That was brilliant, but what does it have to do with time."

"Why, everything!" Dawson declared. "I designed the software so we'd have it when the quantum computer was up and running. To visualize time."

There was silence in the room. Jones leaned forward and put one hand over his eyes. Reeves looked stunned and Bonner nervously fingered his bolo tie.

"Let's get this straight," Chris said after a while. "You're saying we can actually see an event from the past, as if we were there?"

"Exactly!" Dawson replied. "Now you've got it."

"But...how?"

Tommy drained the last of his Coke and looked at Kate.

"May I have another Coke now?" he asked. "Pretty please?" She laughed and jumped up to get him one. He settled back in his chair and rested his chin on an index finger.

"I've explained about the three kinds of time, each one representative of a dimension," he began. "You saw that the past is frozen, a permanent archive of events that have happened. *Every* event that has *ever* happened." He looked around and the men nodded. Kate handed him the Coke. He popped the tab and took a drink.

"Now I also told you that the present is infinitely short, at least from our perspective," he continued. "In fact, we're physically stuck in the present. We can't escape from it. The present will carry us forward into the future for as long as we live, and carry the atoms of which we're made into eternity.

"But now that we know there are six dimensions, and that three of them are dimensions of time, we can use the quantum computer to process data from the past and future. We can create visual images. We can view a particular time and place."

Silence. Then: "Jesus." It was Jones. "Tommy, did you say you could process the *future*? You can 'see' the future? Is that what you're saying?" He looked worried.

Dawson was briefly taken aback.

"Well, sure," he said. "But it doesn't do much good. Chris, can you explain this to Mister Jones?"

"Um, not very well I'm afraid. Are you referring to the uncertainty of the future?"

"Of course."

"So if we were to process the future we'd only see a vague cloud of possibilities, not a clear image. I suppose that the further into the future you tried to look, the more diffused it would become. It would be like looking at a beach and trying to figure out ahead of time which grain of sand was going to end up between your toes."

Jones nodded. "So the future can't really be viewed."

"No, you're wrong there," Dawson insisted. "You can see the possibilities. As the future draws near to the present, the picture starts to become clear. But Fisher's right, it's uncertain right up to the instant it passes through the present."

Chris picked up the thread. "Okay, let's go back to viewing the past. Does the same problem apply, that the farther you go the more uncertain it becomes?"

"Oh, no, my dear boy," Dawson said with a proud grin. "Not at all, not at all. It just requires more computing power as you move into the more distant past."

"You mean we could look back and see events from long ago? We could see the shooting of Abraham Lincoln? We could watch the Boston Tea Party?"

"Oh yes, and much more," Dawson explained with a satisfied look. "We'll be able to watch Brutus and his friends stab Julius Caesar and count the number of times. We'll witness events in Jerusalem two thousand years ago."

Jones gave a deep sigh.

"Do you realize what that means?" he asked. "Do you really realize it?"

"Oh, sure," Dawson replied offhandedly. "It'll mean a lot of history books will have to be rewritten. A lot of historians will have egg on their face." He chuckled with satisfaction.

Nobody said anything. Jones looked at Chris and shook his head, then back to Dawson.

"It'll do a hell of a lot more than that," he said in a flat voice. "It'll unravel the whole fabric of history as we know it."

"Of course."

"And you don't see any problem with that?"

Dawson sat back in his chair, surprised.

"Whad'ya mean? What problem?"

Jones looked at the others incredulously. Chris looked stunned. Reeves was staring vaguely at the wall. Kate picked up her pen and began to write furiously in her notebook. Only Bonner seemed relatively unaffected by the latest revelation.

"Yeah," he said, turning to Jones. "What problem?"

Jones slapped his knee and stood up. He began pacing around the room.

"Paul, our whole civilization is built in part upon cover-ups, deceit, plain outright lies. Everything that's happened all down through history has been colored by myths, false blame laid on innocent people or nations, red herrings by the millions. A good deal of our entire political and economic world is based on untruths. You were in the Bureau, you know what I mean."

"Oh shit. Yeah." Bonner looked alarmed.

"Think what it would mean. It'd be like digging up a cemetery, finding all the ugly bodies under the pretty grass and flowers."

Everyone looked at Dawson. He stared at Jones for a moment then looked around wonderingly at the others.

"I never thought of that," he said in a small, wan voice. "I never even considered that." Then a smile returned to his face. "But isn't it wonderful? It'll make the world a better place. An honest place."

Nobody spoke.

"Well, won't it?"

"Maybe it will..." It was Kate. "A lot of people won't like it, but in the end it could make the world, well, at least different. No more lies and cover-ups."

Jones looked at her in amazement.

"But don't you see that it would destroy the entire structure of proclaimed 'truth' on which civilization is based? And not just received truth—belief! Could religion survive the true revelations of past events?"

"But if our civilization is based on lies..." Kate began. She stopped abruptly. "You really think it would be that bad?"

"Oh, it would!" Jones said, slashing the air with an open palm. "It would completely undermine belief and trust in every institution in the world. Hell, it would destroy them. It would wipe out the foundation of everything."

Kate stared at Jones, her mouth half open as if about to say more. Then she closed it and laid down her notebook and pen. Bonner's forehead glistened with sweat. He swiped at it with his palm. Reeves seemed to be somewhere else, continuing to gaze blankly at the wall. Untroubled, Dawson took a long drink of soda.

"Well," Chris began, "there's no doubt it could be a big problem but first we need to know if it's even possible. Maybe Tommy's wrong."

Dawson scoffed.

"Jesus, I hope he is," Jones said. "What are we gonna do about it?"

Dawson spoke up. "Why, we're going to look back to the night the guards were murdered and see who the killer was. That was my idea all along. You said you wanted to know what he looked like. We can see who he is. We'll catch that bastard."

The others looked glumly at each other. Bonner nodded and Jones sat back down in his chair. He sighed.

"Okay," he said at last. "Let's do it."

Chapter 31

Quantum Center
Monday, 1:15 p.m. MST

The Quantum Center was coming back on line. This time a skeleton crew of a few scientists and techies were in place, just sufficient to fire up the quantum computer and process the images from past time.

Dawson was prancing around on the raised control platform like an excited kid. Reeves sat at his usual place and three others sat in the outlying workstations. They had checked the Qubit Farm and the computer array. Now Dawson was supervising the integration of his visualization software into the interface.

"Come on, come on," he chastised one of the technicians. "Get that routine loaded!"

"It's loading as fast as it can," the man said defensively.

"Well, all right," Dawson backed off. "Just don't let it hang up. We haven't got all day you know?"

"You've got all the time in the world," the technician muttered under his breath.

"What?"

"I said we'll have it up in no time," the technician said.

"Well, stay on it." Tommy turned to the next workstation where another piece of software was being installed. Chris came over beside him and looked at the big overhead screen.

"Can you tell me what this program is for?" he asked.

About to launch into a tirade against the operator, Dawson paused and glanced at Chris, then followed his eyes up to the screen.

"Oh, yeah, sure," he said, taking two steps back to get a clearer view. "This is our space-time navigator."

"Space-time navigator?"

"Yeah, kind of like an autopilot to let us guide the visualization." Dawson put his hands on his hips and gazed proudly at the screen. "It took me a year to write this code. No one's seen it until today."

Chris studied the display. There were six information fields, three on each side of the screen. The three on the left were familiar, displaying GPS coordinates in three physical dimensions of space: Latitude, longitude and altitude above mean sea level. The other three were enigmatic.

"Okay, I see the space dimensions. These others on the right, labeled Alpha, Beta and Gamma, are the time coordinates, right?"

"Exactly. We need to navigate in all six dimensions, telling the system where in time and space we want to set each viewpoint."

"So this application tells the quantum computer where to 'look'?"

"Yes, that's more or less it."

"What are these extra trackballs and joysticks?" Chris asked, pointing to Dawson's workstation. "Looks like an awful lot of redundancy."

"No, we need every one of those. It's just a jury-rigged solution for now. Eventually we'll have a console like a jet fighter plane to control the viewing process."

"So each of those controls links to a different dimension?"

"That's correct. The three trackballs on the left control the three dimensions of space. Look." Dawson reached over and touched one of the balls while pointing up to the screen. The field indicating longitude was scrolling upward. "See, this moves the viewpoint to the West."

A map in the center of the screen began to scroll to the right. Dawson touched another trackball and the field displaying latitude began to change and the map scrolled upward.

"Okay, I see how that works. And I presume the software lets you preset locations and control the speed of tracking."

"Sure. It's pretty basic, just like a video game or one of those computer flight simulators."

"Great, but I don't understand one part. Since the Earth is moving around the Sun, and the Solar System is orbiting the Galaxy, and the Galaxy is flying to who knows where…how can your system know the true location in the past?"

Dawson grinned. "I wondered about that myself," he admitted. "But the theory accounts for that. You see, the vectors of time and space are intertwined. Even though the Earth has moved far away from where it was in the past, the space-time coordinates remain linked. Don't ask me to explain further; it took a couple hundred pages of calculus to make the proof. It's related to entanglement. In fact, it probably *explains* entanglement."

"And you're sure of it?"

"Oh, yes, there isn't any doubt."

Chris chuckled to himself as he noticed Dawson surreptitiously crossing his fingers. He gazed into the physicist's eyes for a moment before turning back to the control center.

"Now how about these joysticks?" he asked.

"Same thing, only for the dimensions of time. That's a little different, because we never knew about those dimensions before." Dawson touched one of the joysticks and moved it. What appeared to be a calculus formula appeared in the Alpha field and began to change as he moved the control.

"Oh, Lord, that looks complicated." Chris stepped closer to examine the readout. "You mean you actually have to work with formulae, one for each time dimension."

"Yep. I had to invent a new kind of calculus to do it, too."

Chris stared in amazement for a moment, then looked at Dawson with new respect.

"And you can actually do this?"

"Oh, yes." Dawson flashed a smug smile and waved his left hand in the air dismissively. "But I cheat. The software actually does the work. Those formulas are just readouts for the time dimensions. I just have to enter the date and time."

Chris rubbed his chin, then turned and called to Kate, who was standing near the central workstation watching Reeves power up the system.

"Kate, you need to see this," he said. "You aren't gonna believe it."

In a moment Dawson was proudly demonstrating the space-time navigator as Kate took notes. Laying down her pad and pen she got out her digital camera and made a few pictures of the controls and readouts on the screen.

"Hey, take a picture of me," Dawson demanded, stepping in front of the display. He struck a pose and she snapped several shots. "And Fisher, get one of me with Fisher."

"Oh, I don't think..." Chris demurred.

"Yes, Chris, get over there," she prodded, holding the camera in one hand and pointing with the other. Dawson put his arm around Chris's shoulder and beamed.

"Say cheese," he said.

"Cheese," Chris said, trying to look comfortable. Kate snapped the shutter.

When the photo shoot was over she picked up her note pad again and turned to examine the control center.

"You know what this reminds me of?" she mused. "Three dimensional chess, that game they played on Star Trek. Only, it's in six dimensions, isn't it, so it's got to be twice as hard, right?"

Dawson smiled indulgently.

"Well, no, not exactly."

"No?"

"Uh, I think I know what Tommy means," Chris interjected. "Each additional dimension increases the difficulty exponentially. A six-dimensional problem is three orders of magnitude more difficult than a three-dimensional one."

"Oh. Then it's a *lot* harder..."

"Yeah. About a thousand times harder."

Kate looked at Dawson with deep skepticism, but he waved his hand dismissively.

"As I told Fisher before you came over, the software actually does all the heavy lifting. All I did was figure out how to make it do it."

"Oh, so that must have just been really easy then?" Kate grinned mischievously.

Dawson chuckled and blushed at the same time.

"Well, if you must know the truth, it took me about twenty years to figure it out—and I only sleep four hours a night. You can draw your own conclusions about how easy it was."

Kate laughed and Chris joined her.

"Now we've gotta get back to work," Dawson told them. He turned away, muttering, "I swear if this lazy so-and-so hasn't finished loading that software..."

As Dawson stalked toward him the technician in question threw his hands up in the air in imitation of a rodeo roper who just finished hog-tying a calf in record time.

"It's done!" he shouted, cutting off Dawson's harangue. "All ready. The diagnostic routine is running now."

Dawson looked over the techie's shoulder to make sure he wasn't being tricked then patted him on the back.

"Good work, great job." He turned away saying to no one in particular: "Took him long enough..."

The technician looked up at the ceiling and rolled his eyes. Kate and Chris exchanged a secret smile as Dawson barged off to check on another aspect of the start-up.

"Chris, let's get some coffee," Kate said, taking him by the arm and leading him in the direction of the break area. "This is too much to absorb at one time."

"You got that right."

It was about an hour later when Dawson decreed it was time to view the past for the first time in history. He picked up a wireless mike and stepped to the front of the stage-like control center.

"Attention everyone," he cried, shouting needlessly into the mike. The speakers mounted around the lab blasted his words back at him, causing feedback. A loud screeching noise filled the room, causing everyone to put their hands over their ears. "Oops, sorry," he said, turning off the mike to stop the racket. Turning it on again he said in a cautiously low voice, "Can everyone hear me now?"

"We could if we hadn't suddenly become deaf," one of the technicians said to his neighbor, digging in his right ear with an index finger. He spoke a little too loudly for Dawson turned toward him with a pained expression.

"I said I was sorry. I'm not the one that set up this sound system."

"Sure, Doctor Dawson. Pardon my comment."

"Oh, that's all right. It just about blew out my eardrums too."

A nervous ripple of laughter spread through the room and the tension began to thaw.

"We're ready to begin, but first I want to tell you what we expect to achieve," the physicist told them. "You all know at least some of the story, and I can't explain everything all at once because what we're doing here today is unprecedented and very complicated."

He paused to look back at the control center. A new, even larger flat screen monitor had been installed above the existing arrays where it could be seen throughout the center. It was a 120-inch plasma display similar to the ones that flash advertising messages on Times Square.

"What we're going to do is to view an event from the past. What we see will be processed by the visualization program and displayed up there." He gestured at the huge video screen. "We won't have sound—that's something that needs some more work—but we hope we'll see what happened when the two guards were murdered.

"At present the quantum computer can only process one instant of past time at once, so we're not going to see motion. It'll be more like a stop-action surveillance video, a series of still shots. We'll have to run a new batch of data for each instant in time.

"Now, let's see what we can do. Pay attention everyone."

He stepped over to his station, sat down, shifted in his chair a few times to get comfortable, then put his hands on the controls. Jones and Bonner were standing at his shoulders. Dawson spoke to the security chief.

"First we'll lock in the spatial coordinates," he said. "Where should we 'look' first?"

Bonner thought for a moment and glanced around the lab.

"We know the killer came in the main door. But let's start before he comes in, see what happened to the other officer outside." He glanced at Jones. "You okay with that?" The Naval Intelligence man nodded.

"Okay," Tommy agreed. He tapped on some keys and the central map on his screen zoomed in to display a high-resolution diagram of the lab building and its surroundings. He moved a cursor to a point just outside of the main entrance door and clicked to mark the spot. A blinking green icon in the shape of a blunt arrow appeared on the screen.

"That arrow sets our viewing point. Now I can move it around and change the direction with these trackballs," he said, clicking and scrolling to move the green icon away and to the right of the door. He rotated it until the arrow was pointing toward the area where the squad cars had been parked. Using a second trackball he set the viewpoint about ten feet above the ground.

"Now we need to set the time. When do you think we should start?"

"We know that Carroll was due to be replaced by Officer Chavez at two o'clock," Bonner said. "Chavez was found beside his car, so we should start to look a little before he arrived. Let's make it one-thirty and work forward."

"Okay." Tommy turned to the time controllers and the formulae in the time display fields began to flicker, then steadied.

"There, that should be it. That identifies the initial data points. Now we need to process the image through the visualization software. That's going to take a moment." He clicked on an icon and rolled his chair back to look up at the huge display screen.

For a few seconds the screen displayed a swirl of colors and patterns then an image suddenly popped into view. A collective gasp rippled through the room as nearly everyone simultaneously drew a breath. The sound was palpable against a background of silence.

The screen showed a clear, detailed nighttime view of the outside entrance to the center. Jesse Carroll's squad car was parked just to the left of the door. The image seemed almost three-dimensional in its clarity.

"Jesus," Jones exclaimed, stepping back to get a better view.

Dawson grinned, spun his chair around and jumped to his feet.

"Houston, we have liftoff!" he shouted. He did a little victory dance and threw his Quantum Cowboy cap into the air with a whoop. All around him faces were staring in awe at the screen. Then, somewhere in the room someone began to slowly clap. Others joined in and the room rang with applause.

Dawson halted his manic dance in mid-stride. For a moment he looked like an embarrassed kid caught raiding the cookie jar. Then he smiled broadly and rewarded his audience with a short, courtly bow as the applause faded.

"Thank you," he said softly. "Thank you very much."

The time viewer had been skipping forward at two-minute increments when suddenly the front of a squad vehicle appeared on the left side of the image.

"There!" Bonner pointed. "That's gonna be Chavez arriving. Let's slow it down."

"I'll jump forward five seconds," Dawson said, manipulating the controls. The next image showed the truck parked with its nose against the building.

"This is on a five-second span," he said. "Want it slower?"

"Yeah. Let's go to two second intervals."

Dawson grunted and worked the controls. The next image showed the vehicle in the same place. Then the driver's door on the far side was partway open. Bonner stepped closer.

"You're recording all of this?" Dawson nodded. "Okay, go to the next one."

The next image showed the door of the truck wide open now and an arm could be seen pushing it. They waited impatiently as the next image was processed. When it flashed onto the screen Chavez was standing up beside the truck. Another shift and he was holding up his left arm and looking down toward his wrist.

"Checking the time," Chris murmured. "He was early."

In the next image Chavez had pulled out a pack of cigarettes and was putting one in his mouth. Then there was the flare of a lighter. Two more grab shots and they watched him exhale a puff of smoke.

"He had time to have a smoke," Bonner remarked. "We found some butts in the area, but there were too many to make anything. A lot of people step outside that door to smoke."

"Yeah," Reeves said over his shoulder. "The indoor smoking ban has that effect."

The stop-action replay continued. For several frames they watched Chavez in various poses enjoying his last cigarette.

Then, behind Chavez a figure emerged from the shadows near the corner of the building.

"There!" Bonner exclaimed. "Tommy, slow it down more, to one second intervals."

The replay continued. They saw a tall man move up behind Chavez. They saw the killer raise both hands above Chavez's head. They saw the security officer began to turn toward the attacker.

Because of the refresh delay between each image the past event unfolded slowly.

They saw the garrote falling around the officer's neck.

They saw the killer's hands beginning to sweep outward to snap the wire loop closed.

They saw Chavez attempting to raise his hands to his neck.

They saw the loop tighten.

They saw Chavez's face contorting in surprise and fear.

They saw his eyes beginning to bulge.

Somewhere in the room a woman screamed. From another direction there was the sound of retching. The horrifying sight of violent death filled the giant screen.

"Stop it!" Bonner demanded and Dawson hit a key. The image was replaced by a screensaver.

"Okay, it's time to clear the room!" Bonner shouted. "Everyone except key personnel, please leave now.

"This is a crime scene."

Chapter 32

Quantum Center
Tuesday, 8:15 a.m. MST

How do you work a crime scene viewed in past time? That was the question facing Bonner. Forensic scientists already had studied the site in detail after the discovery of the murders, but it was an entirely new concept to be able to actually witness the crime, in detail and from any angle.

"They're going to have to rewrite the FBI handbook," Bonner declared. He and Jones were sitting at one of the tables in the lab's break area, tossing around ideas about how to proceed. Also present were two specialists from the FBI forensic laboratory at Quantico, Virginia who had flown out to join the investigation.

Dale Bentley was the older of the two, late 40's with the beginning of a paunch and crow's feet around his eyes. He looked the epitome of an aging federal agent. Neat haircut, dark blue suit, white shirt, black necktie and polished brogues—the semi-official FBI uniform decreed years before by J. Edgar himself. Bentley was an old-school crime scene investigator.

His companion, Abe Keller, might be mistaken for a civilian. He was in his late 20s and affected a Geekish look. His hair was just a bit too long, his suit was brown, and his shoes were comfortable looking. A yellow tie with a pattern of small purple arrows completed the anti-Federal agent look. Trained in accounting and computer science, Keller specialized in coaxing evidence from hard drives, email accounts and web sites that crime suspects thought had been wiped clean.

The federal agents questioned whether Chris and Kate should be allowed to sit in on the discussion, but Bonner, himself a former senior FBI agent, insisted.

The subject at hand was how to take advantage of the wealth of new information potentially available to the investigation.

"We obviously can't apply the usual crime scene guidelines for trace, chemical, fibers, prints, all the usual physical evidence," Bentley mused. "Locard's principal doesn't apply."

"What's that?" Kate asked, glancing up from her note pad.

"Edmond Locard was a French criminologist," Keller explained. "He introduced the idea that when a crime takes place, something is always exchanged between the perpetrator and the victim or the scene. Trace evidence of some kind that can identify the guilty party."

"Oh, yes, I think I've read something about that."

"A lot of fiction writers mention the concept," Bonner told her. "Unfortunately, Locard's theory doesn't always work out. In this case we found no trace evidence that we could link to the crime."

"Ah, but that doesn't mean Locard was wrong," Bentley pointed out. "It just means the clues were either not found, or weren't recognized."

"That's true," Bonner admitted. "Now we may see things we missed. For example, we know that Chavez was smoking a cigarette when he was killed. We collected all the butts from the area, and we can identify the one he was smoking. Who knows, there may be some trace evidence right there."

"Could be," mused Keller skeptically. "There could be a lot more to learn as we work through the entire scene. Heck, we've only seen a few seconds of the actual crime."

"Right," Bonner agreed. "When we originally collected evidence it was like working with blindfolds on. Now we know exactly where to look and what to look for."

"This is gonna be a pretty big job," Keller pointed out. The others nodded agreement.

"There's almost no limit to what we can discover from the crime scene through past viewing," Chris remarked "We can see every detail, from any angle we want, and record it all as evidence."

"Yeah, I see that," Keller said. "We need to draw up a plan on what to follow up or we could be at this forever." He pulled a yellow pad in front of him and began to sketch a flow chart.

Bonner opened a folder and spread out some prints of the scenes that had been witnessed a few hours before.

"Unfortunately, we still don't have a clue who the killer is." He pointed to a print that clearly showed the man's face, an image captured just as the garrote began to tighten. "We ran this image through our database of employee files here, and the FBI lab at Quantico did the same. We even sent it to Interpol but no joy there either."

Bentley picked up the print. He reached into his briefcase and pulled out a jeweler's eye loupe, screwed it into his left eye socket and began to examine the print.

"I'm seeing this at fifteen times magnification and I'm amazed at the detail," he said.

"The output was done on a high-definition laser machine at 2880 dpi," Keller informed him. "The data is just incredible." He picked up another of the prints and began to study it.

"You know, I'm wondering how this guy could be slipping past us," Bonner said. "I still think he's someone who has access to LANL. The security here is about the best in the world. After all, this is where the atomic secrets are."

"I wonder if he's disguised somehow," Keller suggested. "Maybe one of those latex face masks. In the dim light the officers probably wouldn't have noticed."

"That's a thought," Bentley replied. He spent more time gazing at the printout, then removed the eye loupe and put it back in his case. "I'd like to look at this at higher power. Can we get a microscope in here?"

"Sure." Bonner clicked on his radio and placed a call. About ten minutes later a security officer appeared carrying a wooden case. Bonner signed a receipt and set the box on the floor next to Bentley.

Opening the box Bentley pulled out a Nikon binocular microscope and set it on the table. He adjusted the stage and asked Keller to plug in the power cord. He clicked on the halogen light built into the instrument. Then he picked up the print, took a pair of scissors from his kit and cut out the section where the killer's face appeared. He slid it onto the stage and rotated the objective turret to select 50X magnification. He bent over the microscope and began to move the stage using the micrometer controls.

"Very interesting," he said after a while. "It's just amazing how much detail there is here." He looked up at Keller. "I think you're right…it sure looks like a latex mask. No telling what this guy really looks like. We can't ID him from his appearance."

Bonner groaned.

"Crap!" Jones slid his chair back in disappointment.

For a long moment the six just sat there thinking.

"Well, we can hope to find some other evidence when we review the whole sequence of events," Bentley hazarded. "Call Doctor Dawson over and we'll brief him about how we want to proceed."

Bonner stood up and walked through the lab to where Dawson was bent over his workstation. In a moment he returned with the physicist in tow.

"Doctor Dawson, your images are incredible," Bentley told him. "We've already been able to learn something just from these first images."

"Ah! Then you know who the killer is?"

"Well, no, but we have learned that he was disguised. He was wearing a latex mask. Thanks to the high-definition of these images I could see where the latex was slightly wrinkled around his neckline, just above the collar. There's no way we can ID him from his face."

"Oh, that's too bad. What can we do then?"

"We want to use the time viewer to work the whole crime scene," Bentley told him. "We may see something else that will give us the clues we need. It's our only hope."

Dawson pulled up a chair and sat down. For a moment he appeared to be thinking, then he turned to Bonner and raised an index finger in the air.

"Do you still think it was someone who works here at the lab?" he asked.

"Yes. Actually, that would be almost the only way to explain how he could gain access to the facility. He would have come in on his usual pass, then put on the mask to change his appearance."

Dawson thought for another moment.

"You know, I seem to remember that when I was getting my security clearance here they made some kind of picture of my eye. Do you do that for everyone?"

Bonner sat back in his chair. Bentley and Keller leaned forward intently to hear the answer.

"Well, yes," the security director responded. "Our database has iris recognition scans for everyone cleared to work here..." He glanced at the two FBI agents. "Iris patterns are as distinctive as fingerprints, aren't they?" Keller nodded.

"Well, then, there's your answer," Dawson said, standing up and getting ready to walk away.

"Hey, wait," beseeched Bonner. "What do you mean? How does that help us?"

Surprised, Dawson turned back to the table. Six pairs of eyes were locked on him like laser beams.

"Why it's obvious," he said, spreading his hands out palms up. "We'll just look at his eyes."

Silence, then: "We can do that?" Bentley had a stunned expression on his face.

Dawson gave him a funny look.

"Of course. Don't you realize that past time is extremely detailed, right down to the atomic level? Absolutely everything is there. It's just a matter of looking closely enough. We can see right into the man's eye and blow it up on that screen as big as a jumbo-sized pizza. Hell, we can go in and blow up an individual blood vessel as big as this room if we wanted to, count the corpuscles."

Silence again. Bentley and Keller exchanged looks then the younger agent spoke.

"Investigation just ain't gonna be the same, Boss."

"You got that right." Bentley shook his head. He looked skeptically at Dawson. "Okay, do we need to fire up the quantum computer, or can we get the image from the data you've already captured?"

"Oh, that's an interesting idea. Actually I hadn't thought about it but I think you're right—we already have the data so it's just a matter of massaging it."

"Well, then, what are we waiting for?" Bonner declared, pushing his chair back and standing up. The others followed him to the control center at the far end of the room. They gathered expectantly around Dawson's workstation. The physicist had lingered to grab a cup of coffee before wandering after them.

"What's the hurry?" he shouted. "We've got all the time in the world!" He chuckled, earning a glare from Bonner.

"Listen, Tommy," Bonner told him through gritted teeth, "people have died. Until we catch the killer no one is safe, not least you. Have you forgotten what happened just yesterday?"

Dawson's expression froze.

"Oh!" he said with a note of surprise. "Yes, I see." Without another word he sat down, powered up the workstation and brought the computer array on-line. Bonner, Jones and the two FBI agents crowded close behind him.

Chris and Kate watched from a few feet away. Kate took out her digital camera and made a few pictures, then turned to Chris with a whispered question.

"How could this be possible? That man's eyes are just tiny specks on the screen…"

Chris shrugged.

"I guess the quantum computer captured a lot more than that display screen was showing," he suggested in a low voice. "In fact, it makes sense. The interface simply processed the data to generate a wide view. Now Tommy will reprocess the data to see more detail."

"You're right, Fisher," Dawson called over his shoulder. "We captured everything in that instant of time. It's all a question of processing power. The server array will take longer to bring out what we want." He glanced at Bonner. "Is this the screen shot you're interested in?" He pointed to the overhead display where the image of the killer loomed above them. Bonner nodded.

"Okay, this'll take some time." Dawson typed a string of code and crop lines appeared on the overhead screen. The crop box zoomed in toward the killer's face. "Which eye, right or left?" he asked.

"Doesn't matter, we scan both eyes," Bonner replied.

"Okay. Looks like the left eye is more square-on to the point of view, so I'll go with it." The crop box moved tighter and centered on the eye of the killer. Dawson pulled down a menu and clicked PROCESS, then sat back in his chair. "Now we just have to wait." He picked up his cup of coffee and took a sip.

"How long?" Both Bonner and Jones spoke at once.

"Don't know."

"Well, shit," Bonner said. "Give us a clue here, Tommy. Minutes, hours, days?"

"Really don't know, but I'd say at least minutes, maybe hours but I don't think so. We're dealing with some heavy processing here, and we've only got those conventional servers." He waved vaguely at the hundreds of linked computers. "Sometime in the future we'll be able to process the files with quantum computers and that will be almost instantaneous, but for now we gotta be patient. Look, I'll set up a progress bar. It won't be very accurate, but it will give you some idea of what's happening."

He typed away for a moment and a window appeared on his monitor. A blue bar was slowly inching across a white field from left to right.

"There," he said. "Watch the bar. When it gets to the end it means it's done."

"Looks encouraging," Keller commented. "Already about twenty percent. Shouldn't take more than ten or fifteen minutes at that rate."

Bonner and Jones relaxed. The Naval intelligence officer stepped to a nearby swivel chair and sat down. Bentley and Keller stepped back and explored the scene with their eyes.

"It's like something from one of those movies about mad scientists," Bentley murmured.

"Ha!" Keller exclaimed. "This *is* a movie about mad scientists."

"I heard that," Dawson said haughtily. "We resent the implication, and I'm speaking for all my associates. And besides, this is no movie. If it were, one of the Three Stooges would be playing your role. And not Larry. I like Larry."

Keller grinned. "Sorry, nothing personal. Just that I can't help thinking James Bond should be arriving any minute."

"He won't be coming," Dawson informed him. "He's not on the approved list."

Kate and Chris looked at each other. Kate was struggling to keep from laughing. Chris rolled his eyes upward.

Several minutes passed as the blue bar moved slowly to the right. Sometimes it paused and seemed to have stopped, then jumped ahead. At last it neared the end. Dawson slid his chair forward in front of the keyboard. The bar paused just short of the end, then made one last jump and disappeared.

"That's it," Dawson said. "Took about twelve minutes. Not bad for such a high-rez image. Now let's see what we've got."

He worked his controls and the large overhead screen flickered then displayed a close-up of the killer's left eyeball. It was startling in its detail and scale. On the ten-foot screen the eye was three feet in diameter, looming over them like some apocalyptic vision. Even the smallest detail was sharply defined.

"Well, there you are."

Everyone stared for a moment, mesmerized by the effect. Chris's mind flashed to a scene from the movie *Lord of the Rings*, with the eye of Sauron looming over the landscape. Bonner was the first to speak.

"Copy that image and email it to the security lab," he commanded, pulling out his radio. "We'll be able to have a match in minutes." He stepped away, already speaking to someone at the security office.

As they waited Keller studied the image carefully, then turned to Bentley.

"Yep, this investigation bidness of ours is definitely gonna change," he remarked grimly.

The senior agent nodded without taking his eyes from the screen.

Kate was making more pictures. She turned to Chris with a look of elation on her face.

"Just think what this will mean! Nobody will be able to get away with anything. It'll make crime obsolete."

Chris shook his head.

"No, it won't—it just means criminals will have to be smarter from now on."

Kate's face fell.

"Oh, yeah, I see what you mean," she mused. "Like, from now on everyone will be wearing reflective sunglasses."

"Yeah, something like that," he agreed. "And maybe glasses with one of those artificial noses and mustache that makes you look like Groucho Marx."

They both laughed and even Keller joined in. Bentley and Jones didn't seem amused, but Dawson joined in the levity.

"I always wanted one of those," he said. "It'd go nicely with my Quantum Cowboy cap, don't you think?"

Now even Jones and Bentley smiled. Bonner, waiting for word to come back from the security office, looked over at them and raised one eyebrow.

"What's so damned funny?" he asked. Then as everyone looked embarrassed, "Never mind..."

A few minutes passed, then Bonner smiled and signed off of his radio link.

"They got a match. They're emailing the file over now. Tommy, refresh the inbox and display the contents please."

Dawson nodded and clicked a few keys. The icon for an email attachment appeared. It was labeled CONFIDENTIAL followed by a code number. It asked for a password and Dawson looked at Bonner inquiringly.

"Here, let me," the security director said. He leaned over and typed a string of letters and numbers. The digital folder opened to reveal a personnel file. First were images, face-on and side view portraits, a set of fingerprints and iris scans.

"This is him," he said triumphantly. "I even remember seeing him around. Let's see who he is." He scrolled past the images. Text appeared identifying Doctor Harvey Pearson, a senior physicist.

Everyone was silent for a moment as they read the synopsis of Pearson's record. Finally Bonner looked around at the others.

"I think we've just found ourselves a killer," he said. "And a mole as well, no doubt. Jesus, this guy's been here for twelve years! Heaven only knows what he's done."

"I think you're right," Jones added. "This isn't what I expected at all. I think he's been an agent-in-place for all those years. Now he's been activated."

"Well, he's going to be an agent in a place he won't like," Bonner said grimly. "He's going down, hard."

Chapter 33

Los Alamos
Tuesday, 5:45 p.m. MST

Dr. Harvey Pearson had been placed under discreet observation by one of Bonner's investigative teams. The suspect had been observed returning from the lab to his house, stopping to pick up a few groceries and to top up his gas tank at a Shell station. Now he was inside.

An FBI special weapons and tactics team had been brought in from Denver and was prepared for action. The net around Pearson was drawing tight.

Everyone had returned to the safe house except for Bonner and Chris. They were briefing the SWAT team leader. They sat in an unmarked panel van parked about two blocks away from Pearson's house. The vehicle, armored and equipped as a mobile tactical control center, had been driven up from Albuquerque.

"We know this guy's killed at least two men and possibly more, but we want to take him alive if we can," Bonner said.

The special weapons and tactics leader nodded. "We've issued flash-bang and gas grenades," he said. "If they have to shoot my men have been ordered to aim for the legs."

"Okay, great. Let's run over the plan one last time."

"Sure. Your officers have closed off the street from each end. Our three vehicles will pull up to the house, this one in front and one at each side. Two teams of three will deploy from the vans on the left and right and secure the sides and back of the house. This van will block the driveway exit."

Bonner grunted in agreement and turned to Chris with one eyebrow cocked. "Any comments?"

"Sounds good so far. He'll be trapped like a rat."

"Hell, he *is* a rat," Bonner said with feeling. "Okay, then?"

"I and my team will deploy from the van and take cover behind the vehicle," the FBI man continued. "My negotiation specialist," he indicated another agent sitting nearby, "will use a bullhorn to announce our presence and offer Pearson a chance to surrender. We'll give him two minutes to come out."

"Okay. If he does it'll be over. Let's say he doesn't."

"Then the deployed strike teams will take him out. Team leaders will go in through the front and back doors simultaneously, full body armor and flash-bangs. The other four will cover the corners to pin down the location."

Bonner looked at Chris, who seemed to be thinking.

"I remember something that happened when I was in Afghanistan," Chris related. "A group of Army Rangers had a Taliban stronghold surrounded. When they broke in the enemy set off a huge explosion. Killed all of the Mujadaheen, but took two Americans with them and injured several more. I think you need to consider the possibility this man could be willing to sacrifice himself."

"That's a good point" the SWAT leader mused. "We think he may have a connection to Iran, so he could be a holy warrior, eh?"

"Oh yes, if he's Iranian I would absolutely expect him to be a jihadist."

"All right, let's think about this. Instead of deploying my men close to the house, we'll pull them back. And let's send in only one team leader at first, through the front door. That way if Pearson has any intention of becoming a suicide bomber we'll limit the danger."

Bonner and Chris nodded. Lewis looked at his watch. "Okay, are we set?"

"Set," Bonner told him.

Five minutes later the vans converged on the Pearson house. There was no sign of activity inside as the paramilitary teams took their positions.

"Okay, let's go." The leader opened the van's sliding door on the far side and jumped out. Carrying a bullhorn the negotiator followed him along with two tactical officers. Bonner and Chris remained in the vehicle, observing through one-way bulletproof glass. The bullhorn came to life.

"Doctor Harvey Pearson," the negotiator called out. "This is the FBI. Your house is surrounded. Come out with your hands on top of your head and you will be safe. You have two minutes to surrender."

There was no response. The negotiator glanced at Lewis then checked his watch. He raised the bullhorn and paused, counting silently to himself.

"You now have ninety seconds to surrender," he announced.

There was no response. A third warning came at sixty seconds and a fourth at thirty seconds. The house remained silent. The time ticked down to zero and the team leader spoke into his comm mike.

"Go, go, go!"

Dressed in heavy armor and a massive helmet and goggles that made him look more like a robot than a man, a tactical officer stepped to the front door, kicked it down with his heavily booted right foot and tossed a flash-bang grenade underhanded into the living room. He stepped briefly aside as the grenade exploded then launched into the house, weapon raised.

He was met by a rising cloud of greasy smoke. Inside, the house was on fire. Now smoke could be seen issuing from two open windows on the sides of the house.

"Careful," the leader warned. "Go to oxygen."

"Roger," the team leader responded. "Oxygen on." The SWAT gear included a small OX canister good for about ten minutes.

There was a pause as the black-garbed figure moved deeper into the house.

"Can't see much. Smoke's coming from downstairs. Don't see anyone."

"The fire started in the basement?"

"Roger. I'm checking the upstairs rooms."

"Strike teams, hold back," Lewis ordered.

"No one up here. There's flames coming up the stairs now."

"Get out of there, now!"

"Roger."

In a moment the SWAT officer emerged from the front door. Thick smoke billowed after him. Flames could now be seen inside the house. The strike teams moved back and encircled the building as the fire continued to spread.

"Fire trucks are on the way." Bonner had been busy with his radio. "Looks like you were right, Chris, he's after those 72 virgins but no booby traps."

Chris nodded and stepped out of the van. Bonner followed him and they stood on the sidewalk watching the spectacle. Up and down the street people were coming out of their houses to see what was going on. A man emerged from the back door of the adjoining house in the rear. He was a rough-looking individual wearing biker leathers and boots. For a moment he watched the burning house, then started to walk toward it waving his arms.

"Hey," he shouted, apparently concerned that the fire might spread to his own house. "What the hell're you doing? Call the fire department!"

One of the SWAT team agents moved to cut him off.

"Sir! Please step back," he shouted. "This is an active crime scene. Clear the area."

The biker hesitated, then shrugged and walked back into his house. A moment later Chris noticed a pickup truck pull into the back street and drive sedately away.

"Huh, strange," he mused.

"What?" Bonner turned to him.

"Oh, nothing I guess. Just wondering why that neighbor decided to leave."

"Well, we both heard him being ordered to clear the area. Guess he took it literally."

"Yeah, I guess so." They turned to watch as the fire shot from a window and spread to the roof of the house. In the distance the hooting of an approaching fire truck could be heard.

Chapter 34

Los Alamos
Tuesday, 9:25 p.m. MST

It was several hours before the fire was out and the ruined house had cooled enough to be searched. It was fully dark now and the area was illuminated by several sodium work lights running off of a generator aboard one of the fire trucks. Inside the house a fire inspector dressed in a bright yellow Nomex suit with goggles and an oxygen mask carefully negotiated the charred stairway leading into the basement. He was carrying a fire axe in his left hand and a four-cell flashlight in the right.

"There was an office down here," he reported on the comm net. "The fire started here all right. There's a pile of charred books and papers piled in the middle of the room. I smell accelerant, probably gasoline."

"Any sign of the body?" Bonner was standing in the front yard holding a fireman's portable radio.

"Negative."

"Oh crap," Bonner said to the sky, then pressed the TALK button. "Nothing at all?"

"Negative. I don't see where he could have been hiding."

Could he have escaped? How?

"Keep looking."

"Affirmative."

Bonner lowered the radio and turned to the fire chief who was standing nearby.

"You ever see anything like this?"

The chief shrugged. "Could be a hidden compartment," he suggested.

"Yeah." Bonner pressed the TALK button again. "Check the walls for any hidden compartment," he instructed the investigator.

"Affirmative. Three walls are poured concrete. The back wall was covered with bookcases, pretty badly charred. That's the only possibility."

"Check it out."

"Affirmative." The sound of axe blows came from inside the house. "Wait, here's something..."

"What you got?"

There were more sounds of axe work, then a shout.

"There's an opening behind the bookcases," the fireman reported.

Bonner turned to the SWAT team commander. "Get a couple of your men down there. We need to move fast. He might still be alive and could be dangerous." Bonner hit the TALK switch again. "Hold off. We're sending two SWAT guys down. Wait for them."

"Affirmative."

Quickly slipping on their helmets and goggles two men jogged into the ruined house and disappeared in the gloom. As they picked their way down the stairs they switched on high-intensity lamps mounted on their helmets, lighting up the charred room. Raising their submachine guns they took positions to each side of the bookcase. The investigator had already broken through one section revealing a dark opening beyond.

"Okay, stand back and go to oxygen," one of the tactical officers told the fireman. "We're gonna use teargas." The fireman nodded and stepped to the far corner of the room. The second SWAT officer pulled a gas grenade from his utility belt, pulled the pin and tossed it inside the opening. There was a loud "pop." After a moment gas began to trickle out of the opening. There was no other sound or response from inside.

"Okay, take it down," the tactical leader instructed the fireman, gesturing at the bookcase. With a few strokes the fireman enlarged the opening. The SWAT man waved him back and moved closer to peer inside.

"It's a tunnel," he reported. "It leads directly away from the back of the house."

Outside, Bonner groaned. "Jesus, don't tell me he escaped."

Chris looked thoughtful.

"I think he did," he said. "Remember the biker guy from the other house? Could have been him, in disguise."

A few minutes later the team inside the tunnel reported in. They'd discovered the secret room with its prayer rug. The picture was clear—Pearson had a second legend and used the cover to escape right from under their noses.

"And we let him just drive away!" Angry and disgruntled, Bonner handed the fire department radio to the chief and walked away, hands thrust into his pockets. A pair of heavy firemen's gloves had been left on the ground. He swore and kicked one of them across the yard. Then he kicked the other. It sailed into the air and fell near a puddle of foam. He turned and marched back to Chris.

"Now what are we gonna do?" he demanded.

Chris smiled wanly.

"I may have an idea," he said.

Chapter 35

Quantum Center
Tuesday, 11:47 p.m. MST

It was nearly midnight and the Quantum Center team had reassembled. Dawson and Reeves were at their control positions and the quantum computer was on-line. Jones and the two FBI forensic men were standing nearby. Dumbfounded, three of the SWAT officers looked on. Chris and Bonner were standing behind Dawson's chair as Kate recorded the scene with her video camera from a few feet away.

"Are you telling me this thing can look into the past?" the SWAT team leader asked Bentley.

"You don't know the half of it," the evidence specialist told him.

"And you think we can actually, well, follow this guy, see where he went even though it was hours ago?"

"Yep."

"Well, if that don't beat all. How's it work?"

"Top secret, I'm afraid, but basically it lets us take snapshots from specific times and places in the past."

"So it's not like a movie?"

"No, it's a series of still shots. It takes some time to process each one, so this will take a while."

"I get it," the man said. "It's kind of like tracking a perp by examining surveillance camera footage."

"Yes, only we can decide where and when each 'camera' is located and what it sees."

"Jeesh. That oughta be helpful."

"Yeah," Bentley responded, glancing meaningfully at his partner. Kelly grinned.

Chris and Bonner were helping Dawson set up the first time view. The map had been centered on the house behind Pearson's and Dawson had clicked to set the viewpoint. Now he was using the trackballs to position it.

"Move it right out into the street," Chris told him. "Good. Now, he came out of the driveway and drove off to the South, so move it North a little more and point it down the street. There, that's good. Hold it there."

Dawson looked up over his shoulder.

"Now what about the time?"

"We need to guess about that, but we can come close," Bonner said. "We know the exact time I placed the call to the fire department, and it was a couple of minutes later that the suspect came out of the other house. He was probably in the yard for two minutes before going back inside." He turned to Chris. "How long would you guess it was from that time until we saw the truck drive away?"

"Not long, maybe another two minutes."

"Okay, let's give it six minutes all told." Dawson nodded and adjusted the time coordinates. He hit the ENTER button on his keyboard and they heard the interface array begin to spool up as instructions were sent to the remote quantum computer.

A moment later an image appeared on the giant screen. It showed the street in clear detail. There was no sign of the truck.

"Would you look at that!" one of the SWAT officers exclaimed. "It's like a frame from a high-def movie."

"It's a lot more than that," Dawson told him. He was adjusting the time coordinates for thirty seconds later than the first shot. The computer bank hummed into action and presently the screen refreshed. Again the street was shown in sharp detail. This time they could see the back of the red pickup about a block away.

"Got him!" Bonner pumped the air.

"Indeed," Dawson said. "Shall we see where he goes next?"

"Definitely."

Painstakingly Dawson began the task of adjusting the space and time coordinates to track the pickup, one step at a time. Sometimes he overshot, losing the truck when it made a turn or stopped, but each time he merely stepped back to pick up the trail again. He followed the truck through a maze of streets, around the center of Los Alamos, back to residential areas and in and out of the city park.

"He was definitely trying to lose any tail," Bonner pointed out. "I'm starting to wonder, though, if he takes off cross country we might lose him."

"Not going to happen," Jones informed him. "We've got an APB out with the State Police and all Sheriff's departments and local patrols in the area. He knows he's gotta go to ground. He's holed up somewhere and we're gonna root him out like a badger in its den."

"Yeah, I think you're right," Bonner said with a relieved expression. "It's just a matter of time—and we've got all the time in the world, right Tommy?"

"Oh yeah," agreed the self-proclaimed Time Genius.

It was getting late now, past midnight. As predicted the process of trailing the killer was taking a long time. They had followed him to Highway 4 leading South out of town and Dawson was jumping from one bend in the road to the next to keep the truck in sight. Unfortunately the road twisted and turned like a snake's track.

After a while Chris suggested they could speed things up by moving ahead to a straight section of road several miles ahead and waiting for the truck. He pointed out that if it didn't show they could backtrack and find where it had stopped or turned off.

At about 1:20 a cell phone went off in Bentley's pocket. The senior FBI forensic man pulled it out and stepped away to accept the call.

A moment later another ringtone was heard and the SWAT commander answered it, stepping in a different direction.

Phones held to their ears, the two men stared at each other across the room as they listened. Bentley ended his call and beckoned to

Kelly. He began to whisper urgently. The younger man tensed and glanced around nervously.

The SWAT guy concluded his own call, signaled to the others, and soon they, too, were engaged in a taut discussion.

Then Jones's cell rang and Chris knew something serious was happening. As Jones stepped away to take the call, Chris moved closer to Bonner and signaled to Kate to join them. Reeves looked up questioningly from his station and Chris raised one eyebrow and shrugged. Bonner shifted uncomfortably on his feet.

Bentley approached the SWAT team with Kelly following. The FBI men huddled together in a whispered conference.

"What's that about do you think?" Kate asked. Chris put his finger to his mouth.

"I don't know," he whispered. He reached over and touched Reeves on the arm. The project engineer glanced around at the FBI men who now seemed to be engaged in a low-key argument. Kelly was obviously upset and Bentley was trying to calm him.

Dawson continued to track the pickup down the highway, oblivious to what was happening around him.

In a moment the conference broke up and Bentley walked over. He looked nervous and was trying to cover it up. Kelly stayed back with the SWAT officers. His face had gone pale.

Chris looked up as Bentley approached.

"What's up?" he asked innocently.

"Oh, nothing important." Bentley's expression said otherwise. "We just got some orders from above. We're being given another assignment..."

"Really?"

"Yes. FBI headquarters is pulling us out of Los Alamos."

"All of you or just you and Kelly?"

"All of us."

Chris looked at him and Bentley wouldn't meet his eyes. Bonner stepped forward aggressively.

"When?"

"Immediately."

"What! Right in the middle of this operation? Why?"

"Don't know. Something came up. All FBI personnel have been ordered to redeploy. Tonight." He looked around helplessly. "Right now."

"And you don't know why?"

"No. We weren't told."

"You didn't ask?" Chris interjected.

"When a Deputy Director tells you to jump, if you value your career you don't even ask 'How high?' you just jump." Bentley ran a finger around the collar of his shirt. Beads of sweat had broken out on his forehead.

"Okay. So you're all leaving?"

"Yes, every agent in the area is being pulled back."

Chris glanced at Jones who was still talking on his phone a few yards away. He, too, appeared to be upset. He was snarling into the phone.

Chris caught Bonner's eye and signaled with a little sideways glance. *Let me handle this.* Bonner gave a tiny shrug of assent.

"Well, then, I guess you better get going," Chris told Bentley. "Thanks for your help."

"Yeah, thanks for everything," Bonner said, trying to contain a hint of sarcasm.

Looking relieved, Bentley turned away and gestured to Kelly to follow him. Without a word the SWAT team followed the forensic men.

An uncertain moment of shock passed between Chris, Bonner and Reeves. Kate looked confused. Even Dawson recognized that something unexpected was happening. He swiveled around in his chair to watch the agents walk away.

"Where the hell are they going?" he asked. Nobody answered.

Jones finished his conversation, snapped the cell phone shut and dropped it into his pocket. He wore an expression of anger but overlain with a touch of something else.

For a moment the tableau froze. Chris and Kate were standing close behind Dawson with Reeves seated at the right and Bonner to the left. Jones was standing about thirty feet away, watching with contrived nonchalance as the FBI agents made their way to the exit.

As soon as the door closed behind them he gestured to Chris and walked quickly toward the lunch area.

Chris followed and caught up as Jones reached for a rack of coffee cups.

"Coffee, Fisher?" he asked. Chris nodded and Jones poured two cups. He handed one to Chris and took a sip from his own.

"What is it?" Chris asked.

Jones looked around to make sure nobody else could hear.

"I take it the FBI has been ordered out," he said quietly. "That call was to give me the same instructions."

"We're being isolated," Chris observed. "I don't like it."

"You think I do?"

"Of course not. What do you think it means?"

Coffee cup held to his lips and his eyes far away, Jones thought for a moment. Once again his eyes scanned the vast space of the Quantum Center.

"I don't know," he said at last. "But I've got some ideas and I don't like them one bit."

"We're being set up aren't we?"

Jones nodded solemnly.

"Looks like it. But you're a serving Navy officer and the order to withdraw applies to you. I got specific instructions that you're to pull out with the rest of us."

Chris shook his head emphatically.

"No way," he stated. "SEALs don't run. These people need us, Jones. You can't leave now."

Jones set the coffee cup down on the counter. He gazed at Chris approvingly.

"I've got to," he grunted. "I might be able to do more good that way. You're on your own, but I want to tell you what I think." He guided Chris by the elbow to a nearby table and chairs. They sat down, leaning close with their elbows on the table. Jones spoke in a low, serious tone.

"My guess—and it's only that—is that somebody in high places has got wind of Tommy's time viewing trick. We've already discussed how dangerous this is. Every nasty secret, every dirty little fuckup that's been covered up, every career that's built on a foundation of lies—all those things are under threat. Powerful men and women, maybe all the way up to the top, could see their reputations ruined. Some could be charged with treason, sent to prison for life, or worse."

Jones paused and stared into Chris's eyes.

"That's how serious this is." He slapped the table angrily with his open palm.

Startled, Chris sat back in his chair as if Jones had struck him. His mind whirled as he considered the possibilities.

"You're saying it's not just the Iranians we should be worried about," he said.

Jones nodded.

"You're saying our own government…"

Jones cut him off.

"I'm not going to speculate, but let's say that certain elements *within* the government might not want this to come out. A lot of people have watched this quantum computer project with great interest, but no one expected this. Tommy's discovery changes everything."

"Yeah."

"It's not just our government, but the whole world. This could be the biggest event in history."

"Yes, I can see that." Chris flexed his hands. He sensed the danger of the situation. A ripple of tension went up his spine.

"Look," Jones told him, "I'm gonna level with you. As you already guessed, my name isn't Jones. It's Chase, James Chase. I'm a rear admiral. Deputy Director of Naval Intelligence. Twenty-eight years of loyal service. No dirty secrets, no lies, no fuckups."

Chris blinked and nodded.

"I've got to follow my orders, but I'll cover for you," Chase continued.

Chris nodded again.

"No one in Washington knows about your reactivated SEAL team. Use them well."

"I will sir."

"I'm about to 'lose' my encrypted satellite phone." Chase reached in his pocket and handed the phone to Chris. "It's very secure, as you know. Use it with caution. You'll find some useful numbers in the directory."

Chris took the phone and slipped it into his pocket.

"I've got to go but I have one more tip for you, and it consists of three letters."

"Let me guess," Chris said. "It's Charlie Indigo Alpha isn't it?"

Chase nodded solemnly and rose to his feet. He held out his hand and Chris stood up to take it.

"Calm seas and fair winds, sailor."

"Aye, sir," Chris responded. "Thank you, Admiral."

He braced to attention and saluted. Chase returned the salute, turned on his heel and left.

Chapter 36

Quantum Center
Wednesday, 2:18 a.m. MST

Staring at his coffee cup, Chris sat at the break table for a long moment. *What have I gotten myself into?* There were so many things happening that he couldn't keep them straight in his head. *Iranians. Central Intelligence. Tommy's continuing revelations.* And now it was all falling on his shoulders.

Bonner walked up, poured himself a cup of coffee and sat down heavily. He looked tired, and more than that he seemed stricken, a man betrayed.

"Jesus," he breathed. "The feds pulled out. What's going on?"

Chris said nothing.

"I assume Jones is gone too?" Bonner inquired. Chris nodded somberly. "And what about you?"

The security director sipped his coffee and waited for a reply. Chris's mind was churning. *The Admiral said I could trust Bonner.*

"You're right," he verified, "my orders are to leave." He paused and swirled his nearly empty coffee cup around in a little circle to stir up the dregs. "I told Jones that SEALs don't run."

"Ah!" Bonner gave up a thin smile. "So it's you and me against, well, whatever's going down. Wonder how that's gonna work out?"

Chris shrugged and looked around the center. Dawson and Reeves were still tracking the killer through the forest. Kate was standing near them, watching their progress but glancing uneasily toward Chris and Bonner from time to time. Except for Dawson, who seemed oblivious to what was happening, the place held a palpable atmosphere of tension and uncertainty.

"Okay, let's take a look at this," Chris said. "We need to make a plan and we need to do it right now. Jones told me some things. What really concerns me is that someone within the CIA may be behind the orders to pull out the support."

"Shit." Bonner sighed and set down his cup too forcefully, causing the last of the coffee to splash onto the table. "The freaking Company? I should have expected it." He mopped up the mess with paper napkins, picked up both cups and went to refill them. "What do you think they'll do?" he said, speaking over his shoulder as he poured.

Chris shrugged with a negative twitch of his head.

"I don't know. Tommy's past viewing can reveal too many dangerous secrets. There are probably people within the CIA who've sent more bodies to the grave than Genghis Khan. They know about Tommy's discovery now and at least some of them don't like it one bit."

Bonner nodded glumly as he sat back down and handed Chris a cup. "Yeah, I get the picture." He suddenly slammed a fist on the table. "Why in Hell didn't Dawson warn us about what he was up to?"

"Paul, he doesn't think that way. You know how he is. He saw only the good, the positive things about his work. He's smarter than all of us put together, but in some ways he's like a child."

"Oh, yeah, I know." Bonner stirred two packets of sugar into his coffee. "You think we're in serious danger?"

"Yeah."

"Somebody already tried to kill Tommy to stop this. If something did happen to him and his papers were lost, could the rest of you figure out what he was onto?"

Chris pondered the question for a moment. "Well, maybe, now that we know it's possible. We're all in danger. They'd have to completely eradicate all traces of the project."

Bonner sighed. "That's kinda what I figured too." Full of nervous energy he kicked back his chair, stood and began to pace around the break area. "We're like sitting ducks here."

"There's nowhere for us to go," Chris pointed out. "As long as Tommy and the rest of us are alive, we'll all be in danger until this situation is resolved."

"Jesus!" Bonner sat back down and slumped forward, elbows on the table. "We're well and truly screwed, aren't we?"

"Well, maybe not."

"What do you have in mind?" Bonner looked up with interest.

"There're some good guys out there, too. Jones for one. He told me he's on our side. By the way, his real name is Chase. He's deputy director of Naval Intelligence."

"Yeah, I know. He's been working undercover."

"I thought you might be aware of that, since he gave you as his contact," Chris remarked. "Look, there are others we can count on, assuming we can hold out long enough and get the word out. I know there are plenty of decent people within the CIA, too. I think we're looking at just a few really bad apples. The people that might be coming after us are acting outside of the law, and out of desperation. This isn't going to be something official."

Bonner perked up a little. "Yeah, I can see that. They're not going to send in the Marines, are they?"

"No. Probably not even their own staff, but contract people, off-the-books agents, maybe even double agents. People like Red Falcon. Try to make it look like it was the Iranians or the Israelis or whoever is handy. Maybe aliens."

"Okay, that makes sense. So we might not be up against the sharpest pins in the cushion."

"Right. Chase told me the Company is better at making messes than it is at cleaning them up."

"That's gospel in the FBI, too. So what do we do?"

"We start with the five of us here. Tommy and Stuart are essential to the program. Kate already knows too much, so we need to keep her close for her protection. Plus, as a journalist she could be a real asset. If we can get some publicity out through the media it could make all the difference."

"What about you and me?

"Well, I guess we'd just be what they call collateral damage."

Bonner winced. "I wish you hadn't said that."

"But don't forget my SEAL team," Chris continued. "That's our hole card. The Admiral told me nobody in Washington knows about them."

Bonner nodded. "So, it's we few against the whole world. An emotionally handicapped genius; an engineer; a science writer; a burned-out FBI agent; a retired Navy officer; and six over-the-hill commandos. That sure makes me feel a whole lot better." He grimaced. "Sounds like the plot for a bad movie."

"But think about it," Chris told him. "Besides what you mentioned we've got the power of Tommy's invention. We have knowledge that nobody else has. We have the secure environment of the lab and your security staff. We can use all of that to our advantage."

Bonner looked around glumly at the empty center, the three distant figures at the control center, the humming wall of servers. It was a dismal scene.

"Doesn't look like we have much choice, does it?"

Chris shook his head. "Let's get to work. First, I'm pulling in my SEAL team. Can you alert your staff to provide immediate entry to the Lab for the six of them?" Bonner nodded. "While I'm calling them, start making a list of your most capable senior staff. We need to put together a cadre and button up this facility as quickly as we can."

Chris reached over and placed a hand on Bonner's forearm.

"Look," he said gently, "this is your bailiwick. I'm a take-charge kind of guy. It's just my nature and training. If you resent my getting in front of this, just say so."

"Oh, good Lord no. I'm just thankful you're here. I'm an old FBI gumshoe, a plodder, Hell, a damned paper-pushing bureaucrat. An antiquated lawyer for Christ's sake. What this calls for is tactical and command skills. You're the best hope we have of seeing this through."

Chris squeezed Bonner's arm. "Thanks. We'll make a good team."

"Reeves has to be part of the command, too," Bonner said. "The top boss is away in Europe, some conference at CERN, so Stuart's acting director."

"We'll keep him in the loop. In fact, he's got to be the public face of the lab."

Bonner seemed to cheer up. "Right, let's do it. I'll alert my officers." He stood up and stepped away, turning on his radio. Chris fished out Admiral Chase's secure sat phone and dialed. His call was answered on the second ring.

"This is Perry."

"Leonard, it's Chris. The shit's hitting the fan. Get the team together and hustle to the main gate of the lab. Bring the vehicles with everything. You'll be cleared to enter, no questions asked. Someone will escort you to the Quantum Center."

"Aye, aye sir." Perry clicked off.

Bonner came back to the table and sat down. "I've instructed the main gate to let your guys in and issue them class A passes."

"They're on the way. I told them to come straight here to the center. I suggest you get your hand-picked people here, too. And, Paul, what do you think about calling in your off-duty security staff? We might need every living body we can get and we need to lock down the facility."

"I agree. In fact, we should shut down the entire lab, send everyone but security folks home. It's the middle of the night so there aren't many people on the premises. I'll set it in motion to warn everyone else to stay away. We've got a procedure for that."

"Good. You should have some plausible cover story to give them. Perhaps a radiation leak, something like that. A little misinformation now could pay off later. While you're organizing all that, I'm going to brief the others."

"Right. I'll call in our Hazmat personnel, too. Gotta make it look real." Bonner reached for his radio.

Chris walked down to the control center where Dawson and Reeves were still busy tracking the killer. Kate stepped forward to meet him.

"What is it?"

"Trouble," he told her. "Big trouble."

Reeves turned from his workstation. "How big?"

"Big as it gets. We could be under siege here. Tommy, may I have your attention please?"

Dawson turned in his chair with a broad smile.

"I've got the bastard pinned down," he announced proudly. "He's gone through a locked gate in the Jemez forest and holed up in a cabin. He's sitting in there now. We've got him!"

"Great, good work. But Tommy, we need to be concerned about something else now. It's not just the Iranian spies. Elements from our own government might be after us."

Dawson looked confused.

"No," he said. "Why?" he added. Confusion turned to concern. "What the hell?"

Chris held up a hand for attention. "Tommy, your past viewing could be a threat to some people, some of them in high places."

"Why?"

"Because, Tommy, it might get them in trouble."

"Oh, yeah. I guess it could. You mean, if someone had done something bad and they don't want anyone to find out."

"Exactly."

"But, that's a good thing," Dawson exclaimed. "No more secrets. It means no more lies, no more dishonesty, no more crime."

"It wouldn't be good for the people who are hiding the secrets," Chris explained patiently. "Some of them could be high ranking officials in Washington—in the Central Intelligence Agency, the State Department, the Congress or even the White House."

Dawson's eyes went wide.

"You think so?"

"Yes, Tommy. It's a certainty."

"Oh, yeah, then I guess I see what you mean." He spun his chair idly from one side to the other. "So what do you think they'll do? Sue us or something? There are laws you know. We have rights..."

Chris sighed. "It's a lot more than that," he explained. "We think they may try to destroy your discovery, cover it up for good."

"But, that means..." Dawson froze as full realization struck home. "You think..." he stammered. You think I'm in danger again."

"Yes, Tommy. And not just you—all of us."

Reeves and Kate moved closer. Kate had a strained look and Reeves appeared to be struggling to control a tic in his right eye.

"That's why the FBI was ordered out?" Reeves asked. "To leave us open for an attack?"

"I think it's a real possibility. I could be wrong, but somebody may try to shut us down. If so, we'll be ready for them. Paul and I are already putting a plan into action. We'll need your help and support. We've got a jump on things and we need to move. From this moment, this will be our command center. We may have a fight ahead of us, but we have a chance.

"My SEAL team is on its way here now, " he continued. "Paul's pulling in some of his key people and putting the whole lab on high alert. We're going to be in a lock-down mode here, and we need to stay one jump ahead of the bad guys."

The administrator nodded bleakly. "Okay. Let me know what I can do."

Chris turned to Kate. "You can be part of this, too," he told her. "If trouble does start, we need to get the story out to the media. That's up your alley, right?"

She smiled wanly and nodded.

"And Tommy," Chris continued, "you're the most important element in our plans."

Dawson looked surprised. "Not me! I'm no G.I. Joe."

Chris smiled at the thought. "No, you're the Time Genius. You told us so yourself. Do you think the CIA goons have anybody like that on their side?"

Tommy stood up straight, pulled off his cap and held it up to show off the logo.

"No, they sure don't," he exclaimed. "And I'm a Quantum Cowboy, too. Says so right here. Hell, those guys won't stand a chance."

Now everyone smiled, even the morose Bonner who had walked up just in time to hear this last exchange.

"You know, Tommy, I think you're right," Bonner told him. He stepped up to Dawson and they exchanged high fives. "Bring 'em on!"

Chapter 37

Quantum Center
Wednesday, 6:12 a.m. MST

The quantum center had come back to life. In the break area Bonner was briefing a small group of his senior officers, all lieutenants and above. Gathered around the large planning table in the middle of the lab Chris and his SEALs were poring over a detailed map of the LANL and the surrounding area. At a corner desk Kate was busy drafting a report on her laptop. At the control center Dawson and Reeves were working with the time viewer.

Bonner's meeting broke up and the officers filed out of the lab. The security director walked over to where Chris and Perry were marking out fields of fire on the map. Chris looked up.

"How'd it go?"

"Better than I could have hoped," Bonner reported. "I'm proud of those guys."

He pulled up a chair and filled the SEALs in on the security plan.

"We're setting up groups, like military platoons, with a lieutenant and two sergeants assigned to a couple dozen patrolmen. We're going to post a platoon at each gate and two more in reserve. We'll have more than the usual number of men and women on duty, so we're doubling up on the patrol vehicles, two or three in each.

"As you know our patrolmen ordinarily carry only sidearms, but one of my captains remembered the old armory from the days when this was the center of atomic secrets. There used to be a solid military presence here, Army and Marines. We found 30 riot guns. They're 12-gauge Ithaca pump shotguns from the Vietnam era."

"That's good firepower for close-up work," Perry remarked. "How's the ammo?"

"It's old, and that's a problem. I'm sending a couple of men over to Santa Fe and as soon as the stores start to open they're going to be buying up all the double-aught and solid slug 12-gauge shells they can find."

"Good," Chris said approvingly. "What else was in the armory?"

"Well, there are some old-style flak jackets and we're equipping all the gatehouse staff with those. Also tin-pot helmets. And, oh yeah, we've got a couple dozen M-16s. We'll issue the rifles to the second line of defense and put the shotguns up front."

"What about air defense weapons? I don't suppose we have any Stingers?"

"Well, no. There are a couple of 25 millimeter guns of some kind but I don't know what we could do with them."

Sharp's eyes lit up.

"Chain guns?"

"I guess. They're M242s, whatever that is."

"It's a truly kick-ass automatic cannon that fires high-explosive shells," the sniper replied. "I gotta see those. That's some real firepower—could be just what we need to defend the center."

"Great. I'll have someone take you over to check them out."

Chris nodded approval. "We're going to use the SEALs as the last line of defense here at the Quantum Center," he advised Bonner. "We're going to set up sandbag pillboxes on the roof, one at each end. Those chain guns might be perfect for that. We've also got M-4 carbines and one Barrett 50-caliber rifle."

"You're the experts on that," Bonner said. "On a more mundane subject, we've got a supply of folding cots that we're going to set up in various places so that people can get some rest between shifts. I'll have some brought over here for you guys. Also, we've brought in a skeleton staff for the cafeteria. We'll be providing hot meals there, and delivered to you here."

"Great, no MRE's," exclaimed Garrett. "Already ate my share of them in Afghanistan."

"Mouse, you ate more than your share," quipped Tuttle. "I think you ate enough to feed the whole platoon."

"Well, I am a growing boy you know," Garrett explained with a grin. Everyone chuckled.

"All right, we have a plan in motion," Chris said. "Anybody got anything else to suggest?"

De la Vega raised a hand. "Teach and I've been talking. We've got this Iranian spy character hiding out in the woods, and we're thinking we should do something about him."

"Whatta you have in mind?"

"Oh, just a little 'welcome to Paradise, here's your 72 virgins' kind of thing, like we used to do for the al-Qaeda lads. Teach and I can handle it, no sweat."

Chris turned to Bonner. "What do you think?"

"I'd say go for it. No sense letting that rat off the hook. Plus, for all we know he could cause us some more grief. But we need him alive to see what he can tell us."

"Think you could take him alive?" Chris asked De la Vega.

The commando's face fell. "Well, anything's possible I guess. We've got flash-bangs and gas…"

"That reminds me of another thing that we have in our inventory, and that's a supply of Taser shock weapons," Bonner chimed in. "Any use for those?"

"Oh, yes," De la Vega replied eagerly. "That could help us catch your rat. Can you get us a couple?"

"I'll have some brought over right away."

"Okay, let's get to work." Chris slapped the table and dismissed his team. He and Bonner walked over to the control center where Dawson and Reeves were working. For a moment they stood watching. A view of the main entrance of the LANL was displayed on the big overhead screen. Chris studied it curiously.

"What the hell is that?" he inquired.

"Oh, hello Fisher," Dawson greeted him. "You're just the man we want to see." He gestured proudly at the screen. "Behold the latest achievement of time viewing."

Chris studied the image more carefully. There was something very strange about it. The gatehouse, road, nearby buildings and other fixed objects were clearly shown. But overlain against that sharp backdrop was a kind of ghostly cloud of images that seemed to be made up of people, cars, even an eagle soaring overhead.

"Tommy, what is it?"

"You can't figure it out?"

Chris scratched his head and looked again. It was as if a light bulb went on over his head.

"You're viewing the future!" he exclaimed.

"Bingo. You're seeing an instant from future time captured thirty minutes before it reached the present. Those fuzzy images represent alternate possibilities."

Dawson stood up and pointed at one area of the image.

"See there, what seems to be a man walking toward the gate?" Chris nodded. "Notice that he's blurred, like there are multiple pictures of him, superimposed on each other. Each of those images represents a possibility in future time. Only when the event reaches the present will it become clear. The man could continue to walk toward the gate, or he could turn and go somewhere else. Here, look." He pointed to a set of images halfway across the screen. "There he goes, in another possible scenario, where he just remembered he had to be someplace else."

Bonner cleared his throat. "Tommy, what would it look like if you viewed the scene closer to the present time?"

"Oh, the picture becomes clearer. Here, I'll show you. We did this at fifteen minute increments up to an hour ahead." He typed on his keyboard and the picture changed. "Here's the view an hour before. As you can see, the uncertainty is much greater. You can barely make out anything. The images are all over the place."

He typed some more and the picture changed again.

"Now this shows the instant fifteen minutes before it happens. It's a much clearer picture."

The man was still blurred, but the multiple images were tightly spaced. Everything was becoming recognizable. Chris could even make out the license plate number on the back of a ghostly vehicle that was frozen on the screen as it approached the gate.

"What exactly are we seeing here, Tommy?" he asked.

"Well, like I said, it's an instant captured in future time," Dawson explained. "It's just like when we view a past instant, but this is one that hasn't happened yet. It's still in a state of uncertainty. When we view it closer to the point where the event actually happens, the picture becomes clearer. After it reaches the present, that is passes into past time, we see it in perfect detail."

"Tommy," Chris said quietly. "I take it this mean we could see the enemy coming before they get here?"

"Well, of course!" Dawson exclaimed. "Why do you think I'm going to all this damned trouble?"

"Jesus," Bonner swore under his breath. "This could save our asses."

"Sure," Dawson chortled. "Ass-saving is my new specialty. I'm thinking of having a T-shirt made. It pays to advertise, you know. Somebody said that once."

Chapter 38

Jemez Mountains
Wednesday, 10:22 p.m. MST

De la Vega and Chen parked their SUV among some bushes off the forest road about two miles from the locked gate. Navigating with tactical GPS units they circled through the forest to approach the cabin from the rear. Each was wearing camoflage gear including helmets and armor. They carried M-4 carbines and Glock pistols plus the Taser disabling guns Bonner had provided.

They worked their way cautiously through the Ponderosa forest, their footsteps muffled by deep layers of pine needles. Several times they were startled by the frantic chattering of gray squirrels offended at their intrusion. After about an hour they reached a position 200 yards behind the cabin, crouching and crawling the last distance before taking cover behind a fallen log. Drawing Nikon 16x32 image stabilized binoculars from their cases they began a careful reconnaissance.

The cabin was quiet with no sign of habitation. No lights, no smoke from the rustic chimney. Each rear window was sealed with heavy wood shutters. The cabin appeared to be closed down for the winter—but the red pickup gave it away. It was parked in the rear, out of sight for anyone approaching up the lane.

The two commandos watched with the patience cultivated during long days and nights of combat reconnaissance. Methodically they scanned every square foot of the scene with the powerful binos.

At last Chen reached out to gently touch De la Vega's shoulder. He pointed to his eyes, then off to the right. "The tree next to the boulder," he whispered. De la Vega scanned the tree.

"Uh oh," he breathed as he spotted the tiny camera concealed in the branches.

Carefully De la Vega edged closer to his partner, putting an index finger to his lips.

"He probably has sound detection mikes, too," he whispered.

The situation was more difficult than they had anticipated. Besides visual and sound monitors inside the cabin there were likely to be automatic warning devices in place. Intruder recognition software was in its infancy, but someone such as Harvey Pearson might have the latest technology.

They continued their surveillance. After about twenty minutes they'd discovered two more cameras, quartering the approaches to the cabin.

"We're going to have to make a full assault," De la Vega concluded. "And hope to Christ he doesn't have the place mined."

"Yeah, that's a roger," Chen agreed. "I've got my doubts about taking him alive, unless he wants to give up."

De la Vega pondered for a moment.

"We're gonna have to do this at night," he whispered. "Sun doesn't set for eight hours. I say let's come back later and bring Shooter, maybe Mouse too."

"What if he decides to leave in the meantime?" Chen asked.

"I don't know. Maybe we should keep the place covered."

"Yeah, we can use the tactical radios to call in reinforcements. But what if we're needed back at the center?"

"I hope not. Let's check in with Perry."

"I like that idea. We don't want our transmission to set off any alarms. You go up to the other side of that hill to make the call. I'll stay here to keep him bottled up."

De la Vega nodded and began to crawl away from the cabin.

Twisting his body to make a bed in the pine needles, Chen settled down to watch and wait.

Twenty minutes later De la Vega had moved three quarters of a mile from the cabin and was on the north side of a steep hill. He activated his radio and paged Perry. Quickly he outlined the situation.

"Let me get back to you," Perry told him and signed off. De la Vega sat with his back against the trunk of a towering pine tree and watched the white clouds that were almost ubiquitous in the deep blue New Mexican sky. An eagle was soaring over the rugged valley that opened ahead of him.

The radio buzzed in his ear and he touched the earpiece.

"Commander Fisher says to pull back. We can't spare backups and we might need you here. We'll take care of the rat later."

"Aye aye," De la Vega responded and clicked off.

Shit. He began to work his way back to where Chen was on point. An hour later they were in the SUV headed back to Los Alamos.

Chapter 39

Los Alamos
Thursday, 1:24 p.m. MST

Rumors were rife. Early that day the national lab had been closed down. Civilian employees were turned away at the gates and armed guards were seen around the perimeter. The lab's public affairs spokesman had issued an explanation that might be said to have raised more questions than it answered. The announcement went out as a statement from acting director Reeves. Wording was vague. A possible radiation leak was mentioned and the public statement assured there was no immediate danger. Nonetheless, the coffee shop chattering classes were busily spinning webs of speculation.

Bonner, Chris and Reeves were seated at a table in the break area of the Quantum Center, lunching on roast beef sandwiches with coleslaw and potato salad delivered fresh from the cafeteria. All three men had been able to grab two or three hours of sleep.

"The shutdown is raising a lot of questions," Bonner remarked. "I don't think we can keep a lid on it for very long. The national media are already noticing and it won't be long before the talking heads will be here in force, turning rumors into news."

"I'm afraid you're right," Chris said. "But look at it this way: the increased attention may work to our benefit. The more TV cameras and microphones on the scene, the more difficult to organize an attack on the project."

"There is that. But an influx of outsiders will also make our security harder. What kind of timetable do you expect from the supposed attackers?"

Chris pondered that for a moment.

"Well, they're going to know that the longer they wait, the worse their odds. The public attention is the last thing they want. It might even be to our advantage to go public with the whole scenario. That would certainly change the situation."

Bonner grimaced and Reeves cleared his throat as if about to speak then fell silent, toying with a dill pickle spear.

"Sure, I understand the problem," Chris continued. "This is a top secret project and we've all sworn to uphold that secrecy. But, look, if we're in danger from elements in our own government, doesn't that give us the right to consider every option to protect ourselves—not to mention the project itself?"

"Well, maybe…" Bonner muttered. "The problem is, until we're actually attacked we don't even know if the danger is real. When you think about it, we might be over-reacting. Maybe it was just a coincidence that the FBI units were pulled out when they were."

Chris shrugged skeptically.

"Anything's possible I guess, but I have a feeling we're right. Admiral Chase sure took it seriously. When you think about it, there have to be some people who are really worried about this and might do almost anything to stop Tommy's discovery."

Bonner nodded morosely. "I think you're right," he murmured.

"Even if we're wrong, we need to prepare for the worst," Chris continued. "Remember the old Roman adage, 'If you want peace, prepare for war.' That needs to be our motto now."

"Ah, yes," said another voice, "*Si vis pacem, para bellum.*" The three men turned to see Dawson approaching. "That's a very old idea, gentlemen. Those were the words of Scipio Africanus, the Roman general who defeated the Carthaginian Empire. From all accounts he was an uncompromising sort of fellow. I always liked the version that goes 'If you want peace, prepare for peace'." He reached for a beef sandwich and took a bite.

"That sounds like something a Hippie anti-war protester would say," Chris remarked. "Things don't work that way because strength discourages attack, while peaceful intentions are all too often seen as weakness."

"Well, there is the middle ground that actually embraces both," Dawson replied, setting down the sandwich and filling a cup with coffee. "Teddy Roosevelt understood that. 'Speak softly but carry a big stick,' he used to say. There's also the image of an iron fist in a velvet glove." He tore open three packets of sugar and emptied them into the cup. "I know you're going to say that even those examples involve force. I guess I'm rambling, aren't I?'"

"No, not at all," Chris assured him. "These are ideas we need to consider." He leaned back in his chair, crossed his legs in front of him and clasped his hands behind his head in thought.

"You're really turning the motto only halfway around. It may not seem to make sense to say 'If you want war, prepare for peace'— although there's a lot of truth in that. Remember the appeasement the British tried with Hitler, and how that turned out. We have to accept the underlying principle that peace must be backed up with force. It comes back to the same thing Scipio said."

"Yeah, it does, but the Cold War demonstrated the principal of balanced power," Bonner mused. "The greatest arms buildup in history failed to result in all-out war."

"Well, that was probably due to the policy of mutually assured destruction," Chris replied. "With each side facing the possibility of total annihilation, it was easy to maintain a state of, well, perhaps you should call it not-war. But is that the kind of peace anyone really wants?"

"True. Too much like the peace inside a high security prison," Bonner said. "Real peace couldn't be something enforced by fear and threats of destruction."

"But what other kind of peace is there?" Reeves interjected. "What you're imagining doesn't even exist except in some Utopian dream. Strife is universal. Look at that novel about young boys marooned on an island, *The Lord of the Flies*. Hardly a day goes by and they're at each other like rabid wolverines."

While he was talking Kate walked up carrying a stack of papers. She set them down on a table, helped herself to a sandwich and sat down to listen.

"We're talking about the yawning chasm between humanity's higher aspirations and our baser instincts," Reeves informed her. "We humans are capable of acting on a high moral plane, but then you can't deny the raw inhumanity of the Holocaust, the Gulags, the Cambodian genocide."

"Ah, yes," Dawson mused, "the old argument that started when Darwin demonstrated that men and women are merely higher animals rather than celestial beings. Having been thrust back into nature—which Tennyson described as 'red in tooth and claw'—has had profound effects on the way we humans view ourselves. It shouldn't have surprised anyone who's read *Genesis*."

Everyone sat quietly for a moment as they absorbed Dawson's words. At last Bonner stirred.

"Well, that's all very interesting but we have some serious things to consider." He started to stand up but hesitated when Dawson pulled out a chair and sat down.

"It's not just philosophy you know," the physicist told him. "Everything that we've done here and everything we will do are intimately associated with these ideas about human morality, ethics and responsibility."

"Well okay, I guess," Bonner said uncomfortably. "But how can that help us now?"

Dawson lounged in his chair. He took another bite from his sandwich and sipped his coffee, spilling some on his T-shirt. Chris noted with mild amusement that the shirt featured The Grateful Dead. It was hard for him to picture Dawson at a rock concert.

"I've been doing some thinking," the physicist said wistfully. "When I figured out that the Universe has three dimensions of time, I got really excited. It was the most enormous idea I had ever encountered—and it came right out of my own head. I realized that I was going to be famous. Well, okay, I already had the Nobel and all, but I mean this was going to make me *really* famous, right up there with Einstein."

He paused to take another bite of the sandwich. Chris, Bonner and Reeves were watching with intense interest. Kate had her steno pad out and was taking notes. Dawson took another drink from his coffee cup.

"Then things started to happen," he continued with a morose expression. "Someone tried to kill me. Others actually died. The project appears to be in danger. And why? I didn't expect anything like that, and it's my fault. It's because I think like a robot, not like a human being. I didn't see what time viewing would mean to the way things are."

"The way things are?" Kate inquired.

"Yes, the way things are," Dawson repeated, turning toward her with a little smile. "Not the way things *should* be. Not the way things *might* be. The way things are. The fabric of our entire civilization, as we perceive it, woven over centuries like a fantastic tapestry—a tapestry made up as much of myths, half-truths and outright lies as fact."

He finished off the sandwich and got up to get another one.

"I never gave a thought to what time viewing would do to that tapestry," he said, talking over his shoulder as he poured another cup of coffee, "until some of you pointed it out to me. Now I'm not so sure…"

No one spoke. They watched as he stirred three packets of sugar into the coffee, sipped then added a fourth packet.

"Look," he said, returning to his seat, "I'm afraid I've made a mess of things. I'm sorry that you're all in trouble because of my grand idea. I never imagined it would come to this."

Chris placed a comforting hand on Dawson's shoulder.

"It's okay, Tommy," he murmured. "We're going to see this through to the end."

Dawson looked gratefully at him over the coffee cup.

"Well, that's all right then," he said. "Meanwhile, we've got to decide how to prepare for peace or we're going to have a real war. I swear I don't know what to do about the political, social, and religious implications of this. I'm just a physicist, what would I know about all that stuff?"

"Quite a bit, actually, if you just set your mind to it," Chris assured him. "The idea of preparing for peace makes sense to me, as long as it could be a true peace. Obviously there are going to be a lot of people who aren't threatened by time viewing. Good people, people like us. They're potential allies. Those others who have dirty laundry to hide deserve to be uncovered. That tapestry of yours is due for a thorough cleaning."

"Actually it needs to be completely re-woven I'm afraid," Bonner chimed in. "But that's not our problem right now. We need to focus on what's happening here and now." He turned to Reeves. "How's the preview program going?"

Reeves looked over at Dawson who waved to him to answer.

"Tommy's set up a rotating program to check the future views of the main gate and several other sites, plus a 360-degree panorama of the horizon to alert us to approaching aircraft. We've got a couple of techs watching the screen like hawks."

"Great," Bonner exclaimed. "I can't believe that we might be able to actually 'see' an attack coming before it even gets here."

"Yeah, it gives us a major advantage," Reeves said. "But future viewing is slow, a lot slower than past viewing because there's so much more data to be processed due to the uncertainties. About three minutes per batch."

Bonner grunted in acknowledgement "For my part, our security teams are on full alert. I've got patrolmen in plain clothes out in the town, just acting normal but keeping their eyes open. They're our first line of defense. The perimeter guards are armed and on their toes, working eight hour shifts. We also have teams in vehicles circulating around the grounds and observers with binoculars and night vision scopes in strategic locations."

He glanced at Chris before continuing: "The final defense of the center is in the hands of the SEALs. I must say I've been really impressed."

Chris nodded in appreciation and picked up the thread.

"As you've seen, with the help of Paul's staff we've established rooftop pillboxes at each end of the center. We cut holes in the roof and put ladder-stairs in place for access or retreat. Those 25mm cannons are installed up there and manned 24 hours. I have four men on duty at all times, on a four hour cycle for rest and refreshment."

"What about the coverage from the roof?" Bonner asked.

"It's almost ideal. We can see up and down the approach roads from every direction, and the aerial view is excellent. We've got coverage in all directions. With advance warning, if anyone shows up we'll be ready to rock and roll."

"We're well supplied with food and we have our own turbine generator," Bonner noted. "We've brought in the civilian medical staff, too, so the clinic is up to speed in case we have injuries."

"Hope we don't need that," Reeves commented. He turned to Chris. "How long can we stay on this high alert status?"

"Probably at least another day, but I don't think we'll have to wait that long. My guess is that something is going to happen soon, probably tonight. We've got to be ultra ready."

"What about that spy that your guys scouted?" Bonner asked. "I sure would like to get that bastard."

"I would too," Chris agreed. "We'll see about that, but right now we can't spare any of the SEAL team. Storming that cabin calls for experienced commandos."

Bonner thought for a moment.

"You know," he mused, "I've got some former Special Forces guys on my staff. Those Green Berets know how to do that stuff."

"Well, that's a thought. Why don't you pull their files and we'll take a look. I might be able to spare one of my SEALs to lead the mission."

"I'll do it." Bonner looked around at the others. "Anything else we should discuss?"

Kate looked up from her notes with a little ironic smile. "Pardon me, but if you've finished making preparations for war, what about Tommy's ideas about peace?" She smiled at the physicist.

"Oh, that was all just talk," Dawson said. "Just me letting my mind wander."

"No, it wasn't," she retorted. "You said yourself that these are important considerations. I for one took it quite seriously. Surely there are ways to assure peace other than through military force."

"Well, thank you." Dawson smiled uncertainly. "But, I don't really know what to do..." He shrugged and looked around at the others with a helpless expression.

"Well, I think I've got a little piece of the idea. I've been creating a dossier on the project and the events that have taken place. When the time comes, that can be pretty useful."

"Oh, yes," Chris joined in. "You're right. Having our story written down and ready to present can be a big plus. That gives us at least a start on preparing for peace. What else can we do?"

"The written word is one thing, but I want to document this on video, too," she replied. "I've shot some scenes with my little hi-def camera and it's not too bad thanks to the image stabilization and auto focus. For example, I've got a lot of background footage from my interviews with Stuart, and I shot Tommy's talk over at the New Mexico Institute. Now I'd like to expand on that and document the entire project and what's happened here."

The others responded enthusiastically.

"Great," Bonner exclaimed. "That's a really sound idea. Where do you want to start?"

"Why not with you? I'd like to record the whole story of the security situation here, the murders, the discovery of the spy, the whole thing. We can move around the scene while you describe the events."

"Okay, you're on. I need to take care of some business, but check back in a couple of hours."

"Fine. Meanwhile, do you think you can find me a tripod? It would let me get in the picture to interview you."

"Sure, we have one at my headquarters. I'll have someone bring it over. I seem to remember seeing a set of photo lights, too. Would you like those?"

"Yes, super. Meanwhile I'll continue to work on the dossier." Kate shut her notebook and put it back in her messenger bag. She looked around at the others. "Well, what're you staring at? Let's get to work."

"Aye aye, ma'am," Chris replied and threw her a lazy salute. Kate gave him a strange look, then nodded and turned to her computer.

Chapter 40

Jemez Mountains
Thursday, 6:42 p.m. MST

Dr. Harvey Pearson, in actuality the Iranian spy Sanjar Jalaly, was on edge. For more than a day he'd been waiting in his getaway cabin hoping for word from his handler, the man he knew only as "Uncle." He had shed his biker legend along with the padded leathers and elevator boots. Before leaving his house he'd sent a coded message to alert the Iranian spymaster.

The cabin was a secure safe house, equipped with a surveillance system, enough food and water to last for weeks, and even weapons. Still, Jalaly was unsettled.

How did they get onto me? I was so very, very careful.

Again and again he went over the events of the past few days, seeking an explanation for what had happened.

In the Quantum Center. It must be something they found. But what? I wore the face mask, gloves, special clothing to prevent shedding clues. I was so careful!

Checking his watch he set his thoughts aside. As it neared the time for *Maghrib*, the sunset prayer, he made his preparations. Pouring fresh water into a bowl he murmured "Bismillah ar-Rahman ar-Raheem." *In the name of Allah, Most Gracious, Most Merciful.* Dipping his hands he began the cleansing ceremony by washing them three times.

When he was done he spread a prayer rug on the floor and aligned it in the direction of the Kaaba, the sacred place in the Holy City of Mecca toward which all prayers must be directed. Satisfied, Jalaly kneeled, pressed his forehead to the floor and began the prayer of submission.

"Allahu Akbar…"

Minutes passed as he performed the ritual, speaking the Arabic words of supplication required of every Muslim at Sunset. Still, he was not at peace, struggling to focus on the words of the prayer. Thoughts of his situation intruded in his mind. When he was finished he remained prostrate for several minutes more, pressing his forehead fervently to the floor.

At last he rose, rolled up the prayer rug and put away the cleansing bowl. Through an open window on the front of the cabin he saw the last fading gleam of day. Stepping into the bedroom he opened a concealed panel to reveal a compact surveillance center. Six digital recorders were mounted in a rack, LEDs glowing to indicate they were active. Six-inch flat-screen monitors displayed views of the forest around the cabin. There had been no automatic warnings, but he would make sure. Jalaly pulled out a keyboard on a sliding tray. One by one he ran each recorder back by two hours then skip-framed the images forward. Satisfied at last, he returned to the main room and sank into one of the worn easy chairs.

Will he come? What should I do?

Jalaly had been told that in case of trouble he should make his way to this cabin and wait for further instructions. No details had been provided.

His mind whirled.

I am devout. I am pure. I have served Allah against the Great Satan. If I die I shall leave this life as a Holy Warrior, a Jihadist, deserving of the delights of Paradise. That is reward enough. And yet…

And yet Jalaly could not find peace. As he sat there in the gathering darkness he recognized that most ancient of enemies: Fear.

All his years he had lived the role of an elite, first as the son of a leading cleric, then as a scientist, posing as an Infidel American. Not for him the heat and dust and blood of violent Jihad—that was for lesser men, the ignorant rabble blindly following in the thrall of equally ignorant mullahs or warlords. Men who would throw away their lives like so much dust or desert sand. His own existence was on a higher plane, a dutiful toiler in the Great Work of creating the new world, engaged in the holy task of preparing for the final triumph of Islam over all other faiths.

He froze in horror at the thoughts that had just come forth unbidden from his mind. The terrible memory that he had become a murderer sprang into his consciousness. Was he no better than the lowest terror bomber or killer of children?

No! Shaitan has placed these thoughts in my brain! Every man who follows the path of Jihad is pure in the eyes of Allah. Each is cleansed and made holy by his submission to the words of the Prophet.

He stirred uneasily and glanced suspiciously at the deep shadows that were gathering in the cabin. His many years of contact with Americans had planted seeds of doubt. He could see it now. *But the Infidels deserve the wrath of Islam. How can I doubt this? Am I no longer worthy in the eyes of Allah?*

Jalala began to weep. As the last twilight faded from the window he sat rigidly in the tattered chair, tears rolling silently down his cheeks and dripping from his chin. At that moment he felt his comforting faith begin to waver, passing far beyond his spiritual reach. He was overwhelmed by uncertainty.

As the tears streamed from his eyes like a cleansing desert rain his long-enduring confidence in Allah began to leak from his heart.

"Allah," he croaked. "Allahu akbar..." His voice, too weak to echo, dissolved into the lonely room. Jalaly wrestled with the demons of his belief as the cabin fell into darkness.

Some time later he stirred, confused. *Where am I?* He tried to stand up but fell back in the chair, his legs trembling. Slowly he emerged from the fugue state. Panic had receded but fear and doubt still loomed in his thoughts.

He tried again and was able to stand, leaning with one hand on the chair for support. He shook his head gently from side to side, trying to get the blood moving to his brain. Strength gradually returned.

Unless his watch was betraying him he had been in a state of not-sleep for an impossible time. It was nearly midnight, past the hour of the late evening prayer. Glancing at the rolled prayer rug, Jalaly shrugged and turned away.

He shuttered the open window and turned on a dim light. He stepped into the cabin's tiny kitchen and prepared a cold meal of hummus, dates and olives. He sat at the small kitchen table and ate. His eyes were focused somewhere impossibly distant, far beyond the walls of the cabin.

In the forest a moment before Jim Chen had whispered into his tactical headset.

"He's inside. The window was just shut."

"Roger, Teach," came a reply. Several microphones clicked. Ranged around the cabin with Chen were six others, members of the LANL security force who had enthusiastically volunteered for the mission. Each was a former Special Forces soldier, including five Army Green Berets and one Delta Force veteran. Jim Chen was preparing the attack on the cabin.

"Time check, twelve-oh-six," he whispered into his mike. "Be careful. He's moving around in there now."

The men were positioned around the cabin beyond the range of the six surveillance cameras, each of which had been identified and marked on their GPS tactical displays. The field of view of each camera was shown in red, areas to be carefully avoided.

The six men were deployed in teams of two, with two teams covering the front quadrants of the cabin and the third ready to move in from the rear. They were equipped with flak jackets and M-16 rifles from the LANL armory plus their standard Beretta 9mm sidearms. One member of each team also carried a fully charged Taser.

Chen was concealed directly in front of the building, about 200 yards out and with a view up the approach road. As suggested by Fisher, he planned to allow Pearson a chance to surrender peacefully before moving in, then to attempt to take him alive. Chen was uneasy with those orders. Having seen buddies regret having hesitated in the face of danger, he was an advocate of swift and deadly response to mortal threat. But on the other hand, Chen appreciated the intelligence value of capturing the spy.

The attack was set for 3 a.m. Chen settled down behind an ancient Ponderosa log, monitoring the guarded frequency and watching the cabin through the night vision scope mounted on his M-4 carbine. There was no further activity. Waiting for o'dark thirty, the commando team laid low, concealed in the silent forest.

Chapter 41

Quantum Center
Friday, 1:24 a.m. MST

Chris Fisher woke and sat up on the folding cot. He'd been able to grab four hours of much needed sleep. Nearby Perry was catnapping, still dressed for combat and with an M-4 laid out beside him on the bunk.

Chris yawned, picked up his own rifle and headed for the coffee urn. With a steaming cup in hand he walked to the control center. Reeves was awake and watching the future viewing screen along with two techs. Nearby, sprawled in a swivel chair, Dawson was snoring.

"All clear?" Reeves whispered. Chris nodded and continued to the far end of the center. One of the rolling work platforms used to maintain the server array was positioned beneath a ragged hole that had been cut through the roof. He climbed the expanded steel steps and poked his head through the hole. It opened inside the sandbagged pillbox, which was about ten feet in diameter and three feet high.

Sharp was seated like a Buddha at one side of the bunker, happily fingering the controls of the 25mm chain gun. His Barrett 50-caliber rifle lay alongside him. Night vision goggles covered his eyes, but in the light from below Chris could see the smile on his lips.

"Shooter's in heaven, sir," Carlos de la Vega said from the other side of the bunker. "He's like a teenager in love."

"Della, you have no idea," Sharp responded. "I love my Barrett, but this cannon is like a sweet angel come down to Earth." He leaned forward dramatically and kissed the cold steel.

"Well, at least you're on top of things," Chris said, grinning. "Any problems?"

"No, sir," De la Vega replied. Sharp nodded agreement and lazily swung the big gun from side to side on its mount.

"I just wish we'd been able to test fire these babies," he said wistfully. "These hot loads are more than twenty-five years old and I never trust anything but fresh ammo."

"Well, we have to hope it's okay." Chris took another look around. "We'll be on full alert shortly. I'm going to check the other bunker and then wake Perry."

"You won't have to," grunted the chief, climbing up beside Chris on the platform. "You guys okay?"

"Aye sir."

"All right, stay smart."

Perry and Chris climbed back down and walked to the other end of the lab where Floyd Garrett and Joseph Tuttle were on duty in the second sandbag bunker. Garrett was hunched over the chain gun, night vision goggles in place. Tuttle was using a pair of binoculars to scan the skies.

"Everything okay men?"

"Oh, yeah," Tuttle replied, taking the binos from his eyes. "We're ready for them."

"Bring 'em on," Garrett added. "This sure beats the heck out of selling cars."

"Carry on then." Chris and Perry went back down and strolled over to a cubicle where Bonner and two of his senior officers were conversing. Kate was sitting nearby recording their planning session with her video camera.

"All quiet?" Chris asked. Bonner nodded.

"We're going to go to full alert now," Chris continued. "Perry and I will back up the guys in the bunkers. How are your outside teams doing?"

"Fine," Bonner reported. "They keep asking questions I can't really answer, but they're performing their duty."

"That's all you can ask." Chris turned to Kate. "How about you? Did you get any rest?"

She turned off the camera and smiled at Chris.

"Yes, I did. One of Paul's men set up a folding bed for me in one of the offices—not one of those horrible cots."

"Yeah, they are kind of primitive, but you should try sleeping on frozen ground with rocks as big as goose eggs."

Kate winced. "Well, I did get a good nap. Earlier I spent about a half hour interviewing Paul. I've gotten some footage of him in action. Later I want to talk with Stuart and Tommy together. That should be really interesting."

"Great," Chris said. "Have you had any more thoughts about Tommy's idea of preparing for peace?"

"Nohing solid, but I've been thinking about it."

"Okay." Chris glanced at Bonner and his officers, nodded and turned to Perry. "Leonard, join Mouse and Snapper. I'll back up the others."

"Aye aye." Perry picked up his M-4 and strode away.

"Keep alert guys," Chris said. Slinging his rifle over his left shoulder he walked to the break area and refilled his coffee cup. He picked up a six-pack of bottled water. Next he checked in with the control center. The big screen was showing rotating views of the main gate and other scenes, changing about every ten seconds. Chris gaped in surprise.

"How are you doing that?" he asked Reeves. "I thought it took two or three minutes to process each future time view."

Reeves grinned. "Yeah, it did. But time marches on, not to make a pun. Tommy figured out how to process several views at once. It still takes three minutes to complete a run, but we're batching the data. While we check these views, the quantum computer's processing the next batch. We've got the system multi-tasking."

Dawson stirred in his chair and looked up bleary eyed.

"Time marches on, indeed," he said. "When we really get up to speed in a year or so, we'll have fully detailed views, with sound, motion, 3-D—everything just as if you were standing right there."

"Oh," Chris said without enthusiasm, thinking about how much more dangerous that would make the time viewer. "Fine. That's great."

He walked over to the ladder stair and climbed back to the top. He stuck his head through the hole and tossed the bottles of water to Sharp. He handed up his M-4.

"Chief Perry's taking the other end and I'm going to hang with you guys." He scooted up onto the roof and sat down with his back against the wall of sandbags. "It's getting on to o'dark thirty."

"Aye," Sharp acknowledged, handing back the rifle and popping a bottle of water out of the pack. "It's quiet as a tomb out there."

"Let's hope it doesn't turn out to *be* a tomb," De la Vega remarked. He reached out to catch the water bottle Sharp had pitched to him, unscrewed the lid and took a drink.

Time did march on, until at about 2:10 an excited voice came on the comm net. It was Reeves.

"We've got something!"

"I'm coming," Bonner's voice.

"Me, too," Chris said. He scrambled down the ladder stair and sprinted over to the control center. Puffing, Bonner met him at the edge of the raised dais. They stared up at the screen. Reeves jumped from his chair and joined them.

The image showed the main entrance to the lab. In the freeze-frame view the security station was in the process of blowing apart. The roof had lifted about ten feet into the air, the walls were ballooning out and bright flames had engulfed the entire structure. As with all future views the scene was blurred due to uncertainty.

"Holy shit," Bonner exclaimed. He turned to Chris. "What's happening there?"

Chris studied the image for a moment. He could see vague shadows that he knew were the bodies of guards being blown to pieces. *Jesus!*

"It looks like it's been hit with at least one RPG," he told Bonner. "We call them rocket-propelled grenades, but that doesn't do them justice. The Russians developed them as anti-tank weapons. They can also take out lightly fortified buildings, as you see there."

"Those bastards!" Bonner exclaimed. "They're killing my men with Russian weapons."

"Well, not yet," Chris reminded him. He turned to Reeves. "How much lead time do we have until the event?"

Reeves pointed to a digital time readout on the screen.

"About twelve minutes. This just came up from a scan to fifteen minutes into the future and it takes nearly three minutes to process."

"Paul, I'd suggest you advise your men to evacuate the guard post," Chris suggested. "Have them pull back and keep the gate under observation, but not to engage. We need to know what we're up against."

Bonner nodded, stepped away a few feet and began speaking into his radio.

Snapping on his own tactical unit Chris rallied his SEALs, putting them on top alert. On the roof Garrett and Sharp tracked the 25mm guns in the direction of the main gate.

"When will we see the next views," Chris asked.

"Couple of minutes."

"And they'll also be looking ahead fifteen minutes, so that about three minutes will have elapsed since the last one?"

"That's correct."

"All right."

Bonner stepped back beside Chris.

"The guards are evacuating," he reported. "I told them to take their patrol vehicles and join the backup platoons. They're deploying to the left and right of the route in from the gate."

"That's smart. If the attackers come on foot or in conventional vehicles, your guys should be able to take them down. Those 12-gauge slugs and double-aughts will make mincemeat out of them."

"Fine by me."

Impatiently the two men watched the screen, waiting for it to refresh.

"Here it comes," Reeves said softly. Dawson had come fully awake now and was standing beside the others as a new view appeared on the screen.

The first shot showed the main gate, which was now a flaming ruin. Bodies could be seen scattered around the remains of the guardhouse.

"What the hell," Bonner exclaimed. "Look, they've killed my men!"

Dawson chuckled and Bonner turned on him furiously.

"What's funny?" he demanded.

Dawson held up his hands in supplication.

"Hey, calm down. Those men aren't dead. Remember this is the uncertain future we're seeing. We're processing possibilities, not reality. By the time it gets to real time, your men will be safe. They're already safe."

"Oh, yeah," Bonner breathed. "Sorry I got a little edgy there…"

"That's okay," Dawson told him. "Time viewing takes a different way of thinking. The future view shows what would have happened if we hadn't known what was going to happen—and that can get pretty confusing even to me."

"Yeah, I get it now," Bonner agreed, turning back to the screen and addressing Reeves. "So, there's nothing more to see yet?"

"Nothing to see at the main gate. Let's check the other views."

The images appeared one after another as the screen refreshed. None of the other vistas showed anything.

"About a couple minutes until the next view," one of the techs called out. "It's processing now. About nine minutes until the explosion takes place."

Chris thumbed his mike.

"SEAL team, be advised: In about nine minutes the main gate is going to be hit by RPGs. Guards are preparing an ambush along the route in. The trap's set and we're the bait. Out."

Time seemed to slow down as they waited for the next set of future views. Bonner began to pace impatiently and Chris picked up his M-4 and checked it for about the tenth time that night. He noticed that Kate had set up her video camera on its tripod and was recording the scene. He smiled grimly at her and she responded with an unenthusiastic wave.

Perry came on the comm net.

"Sir, if they have RPG-7s with HE warheads they could take this place apart like a house of cards."

"Those weapons aren't much use over about fifty or a hundred yards," Chris reminded him. "They need to get real close, and we're ready for them."

"Aye."

The digital clock seemed to be going even slower now. Bonner was muttering under his breath. Dawson stood up and was beginning one of his manic dances.

"Calm down Tommy," Chris told him. "We're going to be okay."

The physicist stopped in the middle of a pirouette, walked back to his swivel chair and sat down. He started rocking back and forth. The chair began to emit a regular screeching noise as it absorbed his weight. Bonner glared at him, then turned away to watch the slowly moving clock count down to the next view.

"This is excruciating," he said to nobody in particular.

The screen refreshed and once more there was no sign of anything at the main gate.

"Do you think we've missed them?" Bonner asked. "Could they've slipped in between the time shots?"

Chris studied the scene.

"No, I don't think so. Look at all the rubble scattered in the road. They'd have to drive over that. Anyway, your men would have seen them and reported in." He turned to the tech. "Give us another time check."

"Yes, sir. The view we're seeing now is still from fifteen minutes ahead. It took about three minutes to process so what we're seeing is about twelve minutes ahead of the present. It's now about six minutes to the explosion."

"Damn! And we can't see what's happening now!" Bonner exclaimed.

Dawson looked up at him in astonishment.

"Paul, if you want to see what's going on now, just step outside and look around. This isn't *now* that we're dealing with here."

"Oh, yeah," Bonner said. "I keep forgetting that."

"The RPG attack will take place in about six minutes. But we already know that no one came through the gate for at least another six minutes after the explosion."

"Hmm, that strikes me as strange," Chris said. "Scroll those other views." The screen refreshed as it cycled through the viewpoints, ending with the wide panorama of the horizon. As it appeared Chris sucked in his breath.

"That air view—can we enhance and enlarge?"

"Sure," the tech responded. He bent over his keyboard. In a moment the view seemed to zoom out toward the horizon.

"Scan to the West," Chris instructed him. The image changed again, panning across the rugged mountains that ringed the valley of Los Alamos. Chris watched carefully. Then…

"There! Stop!" The image jittered and froze. Chris stepped closer and pointed. "Look, just above that outcropping of rock."

The others peered. There was a tiny speck in the sky. The tech zoomed in closer. The image was blurred from uncertainty.

"Could be a bird, a hawk or an eagle," Bonner suggested.

"Yeah, it's a hawk all right—a Black Hawk helicopter," Chris declared. "I know one of those when I see it, even ten miles away. They're coming by air."

"Oh, Jesus," Dawson cried. He began to rock harder.

Glancing at his GPS display unit, Chris switched on his comm.

"SEAL team, be advised. Air attack imminent, coming from West. Black Hawk will approach from about two niner zero degrees, low over rock outcrop. ETA to visual approximately eleven minutes. Gunners align cannon to defend. Others keep watch on the ground. No further activity from the gate. Out."

"What can we expect from that copter?" Bonner asked.

"Not sure," Chris replied. "If it's in full military rigging it could have up to sixteen Hellfire missiles and heavy chain guns."

Bonner's jaw dropped.

"Sixteen Hellfire missiles! Jesus, Mary and Joseph! They can take out the whole town."

Dawson moaned. His motion accelerated and the entire chair was rocking with him now.

"Maybe," Chris agreed. "But we've got the lead. The idea is to take them out first. Thank goodness for those chain guns. With them, we have a chance if they make the mistake of coming in too close before attacking."

"How close is too close?"

"The cannons will reach out as much as two or three miles but we'd like them closer than that. My guess is they'll come right in over the valley before attacking. If they do that, they're in for a surprise."

He clicked on the comm mike again.

"Shooter, Mouse, you're on the firing line. Hold until my order, or until you see them launch."

The digital clock scrolled on and finally refreshed. First up was the view of the main entrance. Nothing was apparent. On the sky view the Black Hawk was looming over the horizon, clearly visible now. As seen in the future view it was coming straight toward the Quantum Center.

"Christ, we might actually see the dirty son of a bitch fire even before it happens," Bonner realized. Chris and Reeves both nodded. "Maybe we have a better chance than I thought. But why isn't anything coming in the gate?"

"I think that was just a diversion to keep us from noticing the copter. They don't know we can see ahead in time."

Dawson stopped rocking and looked up.

"You know, I hadn't thought about it, but future time viewing has a hell of a lot of potential for use in combat, doesn't it."

Chris rolled his eyes toward the ceiling. "Yeah, Tommy," he murmured patiently. "That's one of the reasons why your invention is so dangerous. It'll change just about everything. Meanwhile, this is the first field trial—and we're the experimental lab rats."

As he listened to this exchange Bonner had been thinking.

"Chris, I heard you say those RPGs are only good up to fifty or a hundred yards, right?"

"Yes."

"Well, that means whoever fired on my gatehouse was pretty close." He turned to the technician. "How many minutes do we have until the rocket attack takes place."

Glancing at his timer the tech held up four fingers.

"Then we still have time to prevent the attack, right?"

"Well, uh, yeah," Reeves admitted with a strange expression on his face. Chris grinned. Bonner chortled and slapped his thigh.

"Hot damn!" Grinning, he reached for his radio.

Chapter 42

*Jemez Mountains
Friday, 2:12 a.m. MST*

Jim Chen was sweeping the area around the cabin with his night vision scope when a whispered voice came through his earpiece.

"Hey Teach: We got company. Delta One over."

Startled, the SEAL laid the rifle aside and pulled down his night vision goggles to provide a wide-angle view. Delta One was the code for one of the commandos stationed behind and to the East of the cabin. Chen touched his mike.

"Whatta you got?"

"Somebody coming through the woods, from over the hill behind us. He's moving stealthily. Presently at about 65 degrees from my position and 75 yards out."

Chen looked at his GPS display, drawing an imaginary line from Delta One's position. It led across the nearby ridgeline to the next valley. There was a forest service road about two miles away, cross-country. Chen figured the intruder must have walked over from there.

He grunted under his breath. *We don't need this.* Chances are the intruder was a confederate of Pearson's. He ran through the options. They could simply hold their positions and watch to see what happens. That could complicate their chances of taking the spy alive if the intruder joined up with Pearson. Or they could intercept him or her. That could result in a struggle and give away their presence.

Either way, it was both a blessing and a curse. If they could take a second prisoner the intelligence haul could be even bigger than from Pearson alone.

Chen made his decision.

"Hold back and let the intruder pass through," he instructed. "Divide your teams, one man observe the cabin, the other the intruder."

Tap.

Tap.

Tap.

They waited. Delta One kept up a whispered commentary on the stranger's progress. It was clear that he or she was heading directly for the cabin, pausing every few feet to scan the area.

"I can see him now. It's a man. Dressed in a coverall but doesn't seem to be combat ready. No rifle. Delta One over."

"Roger."

About five minutes passed.

"He's reached the rear of the cabin. Moving around the East end. Delta One over."

"Roger." Chen raised the goggles and brought the riflescope to his eye, pointing it toward the cabin. In a moment he saw a tall, slender figure slip furtively around the corner onto the little porch, pause to look around, then step up to the door. With a quick movement he unlocked the door, opened it, stepped inside and gently closed it behind him.

"He's inside," Chen announced. "Bravo Team, begin to move around toward the West end of the cabin. Charlie Team, hold in place. Delta Team, work around to the East end."

Tap.

Tap.

Tap.

Inside the cabin Jalaly had been sleeping in the easy chair. He was emotionally drained by his confrontation with fear and shame.

His faith shaken, he was on the verge of a mental breakdown. There was a little sound and he jerked half awake. Then the door swung open and a dark silhouette appeared briefly against the starlit forest before the door closed.

Jalaly's AK-47 was leaning in a corner across the room. Panicked he tried to jump up but a hand pressed him back.

"Peace, Brother," a familiar voice intoned.

Jalaly fell back into the chair. *Everything will be all right now.*

"You've come...?" he croaked, his voice weak. "I thought..."

"Yes, Allah be praised I am here." His controller turned on a table lamp beside Jalaly's chair and looked intently into his face. "You are well?"

Jalaly shrugged. "I've not been injured," he said, sidestepping the question of his mental state.

"You've been through much," his handler told him. "Now it's time for your reward."

"Ahh," sighed Jalaly. "My return to Mother Iran...?"

The tall man made a noncommittal gesture and sat down in the second chair. "In time, Brother, in time." He crossed his legs and looked around the room. "First you must tell me what you have done."

Jalaly's face fell. He rubbed his hands together as if washing them. Bowing his head he placed his dry palms over his face. He seemed to shrink into himself.

"I have done nothing wrong, nothing," he murmured in a hopeless tone.

"You must have made some mistake," the handler encouraged. "Think."

Lowering his hands Jalaly looked up with anguish on his face.

"There was nothing, nothing that could have given me away," he pleaded. "I followed every precaution..."

His interrogator gazed steadily at him and Jalaly dropped his eyes. His hands began to writhe together once more. He emitted a little sob.

"Sanjar, you must tell me where you failed."

"I did not fail!" The denial came as a shout. "I did not! I swear by the Prophet."

"You insult Mohammed. Have you become an apostate?"

"No!" Jalaly froze. *Have I?* He shrank further back in his chair.

"Calm yourself Brother. Now think: How did you fail?"

Jalaly did not speak for a long time. The handler waited patiently, his gaze steady. At last Jalaly raised his eyes and whispered.

"It must have been at the center. When I killed the guards. They must have found some clue…"

"You were careless then?"

"No!" Again Jalaly dry washed his hands. Fat tears welled in his eyes, sparkling in the dim light. "I was careful. I did everything correctly." His voice rang in despair.

"Hmm." The interrogator pursed his lips. "Well, it cannot be helped," he said at last.

Jalaly glanced up in relief. *It's going to be all right.* He tried to smile but found it was impossible.

The commando teams had closed in around the cabin, being careful to stay hidden from the surveillance cameras. Now Chen had moved to within fifty yards of the front porch. Bravo and Charlie teams were closing in from the sides.

Hearing a sound, Chen froze. It had come from inside the cabin. A voice, raised in anger? No, more like fear. *Gotta move.*

"Go! Go! Go!"

Chen ran forward, jinking left and right to keep to the cover of tree trunks. About 20 yards from the cabin he stopped and took cover. He checked left and right and saw the Bravo and Charlie teams close in on the ends of the cabin. He knew the Delta team was covering the rear.

Cupping his hands over his mouth he shouted.

"Yo! You in the cabin. You are surrounded. Come out with your hands on top of your heads."

Silence. The wind stirred the upper branches of the Ponderosas, mimicking vague human voices. Somewhere a pinecone fell, bouncing through branches and landing with a soft thump on the mat of pine needles that made up the forest floor. Chen waited for a count of twenty.

"You have ten seconds to come out with your hands on top of your heads."

Inside the cabin the interrogator had jerked his eyes from Jalaly. He swore under his breath, jumped up and drew a pistol from its hidden holster.

"What have you done?" he challenged. Jalaly moaned and sank even further in the chair. "You've given us both away!"

"No! I did nothing."

Grunting, the handler strode across the room and grabbed the AK-47.

"Stand up!" he ordered. Slowly, Jalaly complied. The rifle was thrust into his hands.

"Here is our plan. You are going out first, firing as you go. Shoot anyone you see. Go left to draw their attention that way. I will be behind you and go right. Allah will protect us—or receive us in Paradise."

"Please..." Jalaly tried to thrust the assault rifle away but the handler pressed it into his hands and propelled him toward the door.

"Go with Allah. This is your chance for redemption, Brother."

Whimpering, Jalaly stepped forward as his handler swung the door open. He stepped onto the porch and turned left, the AK sagging at his side.

"Shoot, you fool!" hissed the controller, still inside. Jalaly took two more steps and let the rifle fall to the porch with a clatter. Just then he was hit in the chest with the darts of a Taser. Shocked by fifty thousand volts, his body jittered like a marionette then fell to the ground.

Behind him the Iranian handler ran out the door firing his pistol wildly. At the corner of the porch a commando appeared, aiming a rifle and shouting at him to halt. Another commando was trying to get in position to use a Taser.

The Iranian raised his pistol. Before he could fire there was a stuttering sound and he felt his legs give way. He fell sideways, rolled and again tried to bring the pistol to bear. There was another stuttering sound. This time the commando had aimed to kill. The pistol clattered on the wooden floor and was immediately kicked away. The Iranian's body shuddered and lay still.

"We're clear!" Chen announced. Jalaly had already been cuffed.

The commando team gathered around the porch, alert in case any other parties might be present. Moving in from the side, one of them kicked the door wide and tossed in a flash-bang grenade. It exploded and he and two others stormed in after it. In a moment they were back outside reporting that the cabin was empty.

It was a little past 2:40 a.m. and Pearson had been taken alive. Chen breathed a sigh of relief as he gave high fives to each of the fighting men.

"Great job, troops. Really great."

"Thanks," said Delta One, who had fought in Operation Iraqi Freedom and Afghanistan. "Brought back some memories, both good and bad." He looked down at the huddled bodies on the porch, one dead, one alive. "This would rank as one of the good ones," he observed.

"Shit," exclaimed another, "any time you walk away with your balls still attached is a good one."

"Yeah, man, I hear you on that," Delta One replied.

Chapter 43

Quantum Center
Friday, 2:47 a.m. MST

Bonner shouted into his radio, instructing one of his platoon captains to send three teams armed with shotguns out through the gate to look for persons unknown armed with grenade launchers.

"They'll be within a hundred yards of the security gate, and they're due in less than four minutes," he said.

"Huh? Howd'ya know?"

"We have information," Bonner told him curtly. "Immediately upon identifying the subjects use lethal force, repeat, lethal force."

"You mean kill them?"

"Yes, you idiot! Load with double-aught. Get going!"

"Yes, sir." The confused security officer clicked off. In a moment three squad vehicles were rolling toward the gate from inside the grounds, each carrying four men armed with Ithaca 12-gauge shotguns. The platoon captain was in the middle car, his radio set to a guarded frequency.

The squad vehicles shot through the gate, fanned out and stopped about twenty yards from the gatehouse. One car was to the left, the other two to the right spaced about thirty yards apart in a loose semi-circle facing out. Following the captain's orders the drivers turned off the lights and stopped. The men jumped out and took cover behind their vehicles, shotguns at the ready.

"Nobody here," the captain reported.

"Not yet," Bonner told him. "Just keep on your toes. About two minutes."

The captain turned off his comm. "Jeesh," he grumbled to the sergeant beside him. "I think the Bossman's lost his marbles. This is a snipe hunt. There's nobody out here."

"Yes sir," the patrolman replied in a neutral tone.

The captain switched back on and addressed his team.

"Okay, men. There's supposed to be some guys coming here in a minute or two. They'll be armed with rocket grenade launchers, you know, those long things you see on TV, that terrorists use against tanks. When you see them, shoot to kill."

"Would you repeat that?" someone asked.

"You heard me! Soon as you see a grenade launcher, shoot. Take out everything, their vehicle, anyone with them."

Another voice came on the net. "This is Bonner. Proceed as ordered."

Confused, the guards hunkered behind their squad cars, shotguns ready. A minute went by. Another ten seconds passed. Then, in the distance came the sound of an engine racing.

"Probably some teenagers out for a joy ride," the captain advised the sergeant. *Bonner must be crazy.* He flicked on his comm. "Hold until I fire first," he told his men, modifying his earlier order.

Around a nearby corner came a metallic silver civilian Hummer, driving at least 20 miles over the speed limit. With a screech of rubber it slid to a stop straight in front of the guard station and about 75 yards away. Both side doors opened and two men dressed in black jump suits and balaclavas leaped out. One of them opened a rear door and pulled out what looked like…

Holy shit! They are *terrorists!*

"Jesus, Bonner, they're here. Just like you said."

"Kill them!" Bonner ordered. "They're going to attack the lab."

"Roger."

Clicking off the push-button safety at the front of the trigger guard of his Ithaca, the captain aimed toward the Hummer and pulled the trigger, pumping another shell into the chamber on the recoil. Around him the others began to fire. They continued to pump and shoot. The night was filled with thunder and the smell of gunsmoke.

A 12-gauge 00 shell fires a spray of nine pellets each measuring a third of an inch in diameter, roughly equivalent to a 32-caliber bullet. Each tactical shotgun held four shells it its tubular magazine plus one in the chamber for a total of five. The twelve guns delivered a storm of pellets that struck the vehicle and its hapless occupants from three directions.

The two terrorists fell like sacks of wet sand. The RPG clattered to the pavement. The Hummer's tires exploded and it dropped onto its rims. Its windows blew out. The bodywork became a metallic version of Swiss cheese. The firing continued until every gun was empty. In a matter of less than four seconds the twelve guards had fired 540 individual buckshot pellets at the would-be attackers and their vehicle, a wave of hot lead striking at more than 1300 feet per second.

The echoes rolled back from the mountainsides and subsided and then there was only the sound of liquids leaking from the shredded Hummer. Coolant, brake and power steering fluid, windshield washer mix, battery acid, oil and gasoline mingled with blood, urine and lymph in a spreading pool around the shattered vehicle and its former occupants.

The guards stared at the scene, then as one they turned to the captain. Washed in the light of a nearby streetlamp, his face was ashen. For a moment he just stood there, his empty shotgun hanging from one hand. At last he stirred and looked around.

"All right, what're you looking at?" he shouted. "Secure the area. Grab some fire extinguishers.

Back in the Quantum Center Bonner had been monitoring the comm net. He gave a triumphant shout.

"We did it! We got the rats!"

Even Dawson joined in the brief celebration, throwing his Quantum Cowboy cap into the air.

"First victory for time viewing!" he shouted. "Woo hoo! We're kicking ass."

"So far," Chris said dryly, glancing at the big overhead screen where the image of a Black Hawk helicopter was looming in the sky.

Chapter 44

Quantum Center
Friday, 2:51 a.m.

"**All right, settle down everyone,**" **Chris ordered.** "When's the next view?"

"Coming up," the tech responded.

"Go straight to the copter," Chris told him. The image flickered onto the screen. Now, captured fifteen minutes in the future, the copter was passing over the edge of Los Alamos. It was descending, now just a few hundred feet above the ground.

"Okay, that's still about twelve minutes away, right?"

"Right. The first sight of the copter should be coming up on real time in a couple of minutes."

Just then De la Vega's voice came over the comm net.

"I've got the helo in sight," he reported. "Just as predicted. It looks to be about twenty miles out."

"I'm coming up," Chris replied. He sprinted to the ladder stair and charged up the steps two at a time. In the bunker De la Vega had a 40-power night vision spotting scope set up on a compact tripod.

"She's coming," he told Chris, pointing to the scope. Chris bent down and took a long look.

"Too far away to see too much," he concluded. "Keep watching and give us reports."

"Aye aye."

His combat boots clattering on the steps Chris hurried back to the control center. The other grab shots had shown nothing of interest.

"Stuart, can we bring the time coordinate closer to real time? I'd like to concentrate on that Hawk for the next few minutes."

"Actually that wouldn't be a good idea." Reeves turned to Dawson. "We're better off using a longer time, but not too long because uncertainty increases."

Dawson nodded agreement. "When we take a future shot we see what might happen exactly when the shot was taken," the physicist explained to Chris. "Between each shot real time moves forward, so we see the progression. There's no point in looking too close because it would just give us less time to react."

"Okay, got it," Chris agreed. "How much time do we have now until the actual ETA?"

"The view you saw a moment ago shows where the copter will be in about ten minutes in real time," the tech reported. "The next shot's processing now."

"Okay. What I'm interested in is seeing the final attack, if there is an attack. We can't just start shooting unless we're sure we're in danger. For all we know that copter could be friendly."

Nobody said anything. The atmosphere in the Quantum Center was electric. Besides the SEALs, Bonner had brought in a squad of his security patrolmen. They were scattered around the large room in groups of two or three. Although uncertain about what was happening, even the guards were staring at the giant screen in fascination. Some of them were showing signs of uneasiness as they absorbed the idea that the Quantum Center might soon be the target of Hellfire missiles.

Although tension was high everyone was calm, even Dawson who had now reclaimed his ball cap and wandered off to get coffee. Chris, Bonner and Reeves were waiting impatiently for the next grab shot from the future.

Time seemed to drift slowly. Dawson returned carrying a cup in one hand and swinging a six-pack of Cokes in the other. He offered the soft drinks around and the technician accepted one, taking a drink then carefully setting it safely aside.

The process ground slowly on. At last Reeves announced, "Here it comes."

The picture flashed on the screen. The Black Hawk was frozen in mid-air less than a mile away. It had dropped further, flying almost at treetop height.

"Okay, guys, there's your bird. That's where it'll be twelve minutes from now."

"Zoom in," Chris ordered. The tech operated a trackball and the picture steadied in on the helo. Chris stepped closer to inspect the image.

"It's got a couple of missiles on hard points," he said. "Can't tell what else."

"Well, it's not sixteen…" Bonner said with relief.

"No, but one of those Hellfires could take out this whole building and make toast of us. Each one contains about twenty pounds of high explosive."

"Wish you hadn't said that."

"At least we can see that the Hawk's going to come in close enough to give us a chance," Chris announced. "I'm concerned about the low altitude though. Doesn't look like the far chain gun will be able to track it."

He activated his radio and alerted Sharp that he would probably be the only gunner able to hit the target.

"I'm on it."

The next grab shot was processing. It would give them the view three minutes further on. Chris figured that would provide the evidence he needed to assure that a deadly response was appropriate.

Time crept. Chris walked in a little circle, scuffing the floor with his heavy-soled boots and stopping at intervals to glare at the digital time clock. At last the screen refreshed and the tech scrolled through to the view of the helicopter.

Chris's breath caught in his throat. A cry of dismay spread through the room. The image showed the helicopter hovering just above the treetops. Twin lines of fire led from the weapons racks, pointing directly at them. Frozen in the air not more than a hundred yards from the lab were two missiles, a scant instant from striking the building.

"Oh my God!" Kate said. "They've fired the missiles!"

"No," Reeves assured her, 'not yet. Remember, that won't happen for twelve minutes, actually a little less now. Just like Bonner stopped the attack on the gate, we have time to stop this one."

Chris studied the image carefully. Turning to the tech, he asked for an accurate fix on the copter and the exact time when the grab shot was made.

After a moment the tech flashed GPS coordinates on his screen and started a digital count-down clock.

"There's your time and place," he said. The clock showed about eleven minutes remaining until the rockets would be fired.

"Okay, great." Chris entered the coordinates in his tactical GPS display, set the elapsed time indicator on the little flat-screen device, and headed for the far bunker where Garrett and Tuttle were on watch.

"We know when the copter will shoot," he told them, handing the GPS to Tuttle. "That's a huge advantage. But we also know its position and it looks like this gun won't be able to target it."

"Let me see that." Garrett reached for the GPS and studied it, looking up from time to time to identify landmarks. "You're right," he agreed. "The copter will be too low for us to see."

"Can you move the gun?" Chris asked, although he was sure of the answer.

Tuttle shook his head regretfully. "The gun alone weighs 250 pounds and the mount we cobbled together must be at least that heavy. Plus it needs to be secured or the recoil would probably kill the gunner. It's bolted to the structural I-beams under the roof."

"Yeah, that's what I figured. Okay, you guys stay alert for anything on the ground. It's going to be up to Shooter and the other gun."

Chris disappeared down the ladder stair and ran back to the control center. The countdown clock had advanced to just over nine minutes until the real-time missile launch.

"We've got another grab coming up," Reeves told him. They turned to watch the screen.

"I've set the coordinates for a view showing this building from a couple hundred feet away," the technician told them. "Here it comes."

The screen refreshed. Nobody said anything. The scene was like something from a nightmare. The center was in ruins, flames shooting out of the broken roof. The image had caught debris flying into the air from explosions. The entire wall containing the server array with its hundreds of linked computers had collapsed. Bodies could be seen scattered among the wreckage.

Dawson moaned softly and began to rock again. Several others showed signs of severe stress. One of the guards started to run for the door but was stopped by a sergeant.

Grabbing a PA mike from its socket on the control panel, Chris thumbed it on and addressed the room.

"This isn't going to happen!" His voice echoed through the room. "Whoever they are, this is the future they want for us, not reality. They will not succeed. We're literally ahead of them and we're going to kill them before they can kill us." He spoke with his command voice and the room settled down. Dawson stopped rocking and looked relieved.

Chris switched off the mike and handed it to Reeves.

"Keep a lid on things here. I'm going up top." He headed for the ladder stair and in a moment was briefing Sharp and De la Vega. Using Bluetooth, both men transferred the GPS and time readings into their own units. They began to plot the location on the real-world scene.

"The copter will be 900 yards away when it fires," Chris pointed out, watching the approaching Black Hawk through night vision binoculars. It was coming steadily on, showing no running lights. He glanced at the timer. "It'll launch in about seven minutes."

Sharp took an azimuth reading from his GPS and tracked the chain gun to point at the spot where the helicopter would appear.

"You okay, Shooter?"

"Aye, tip top, sir." Sharp swiveled the big chain gun in its mount, peering through the optical sight. "We're ready to rock and roll."

Chris said nothing. Across the bunker De la Vega was sighting his M-4 over the sandbags, scanning the ground below for intruders. Chris picked up the binoculars and watched the Black Hawk skim nearer. It was about six miles out now and beginning to descend just as the earlier grab shots had shown.

"We're going to have just one chance," he said, edging over behind Sharp. The sniper mumbled something. Chris sat on the sandbag wall. Just as he raised the binoculars again he noticed De la Vega surreptitiously cross himself.

"Five minutes, men."

The helicopter came closer. It was sliding over the ridge of the mountains that ringed Los Alamos to the West. Now they could faintly hear the distinctive, heavy thump-thump-thump of its rotors. The pilot was slowing now as he approached the point shown in the future preview.

Using a range finder, De la Vega began to read off distances and time remaining. Sharp checked the action of the chain gun again and wiggled his fingers to make sure they were limber. He rotated his shoulders to loosen the muscles.

"Shooter," Chris instructed, "the time is key. I'll give you a countdown to thirty seconds before the launch. That's your cue to take him out."

"Aye aye."

They watched intently as the black craft drew nearer. "I see just the two missiles on the rails," Chris reported, peering through the spotting scope. "It's a bare bones copter, standard door-mounted light machine guns. Looks like two men manning them, one on each side." He glanced at the timer. "Three minutes forty seconds to launch."

"He's almost in range now," Sharp said. He was obviously itching to fire.

"As long as she's moving closer, let her come and wait for my mark."

"Aye aye."

The seconds ticked down.

"Two minutes to launch," De la Vega announced.

The copter slowed almost to a crawl now, advancing like a gigantic black shark negotiating a reef. It was less than a mile away and settling lower toward the ground. Its nose was pointed directly at them. In the night vision binos Chris could see the pilots now, outlined in the windscreen.

"Distance 1400 yards," De la Vega said. "Still coming."

Chris moved over beside Sharp and laid a hand on his shoulder.

"Steady now, let her come. She fires from under 900 yards."

"Hell, sir, I could shoot the tits off her right now," Sharp exclaimed.

"Just wait a moment longer. Steady. She's still coming." Chris picked up the binos and raised them to his eyes.

"One thousand yards," De la Vega reported. "Slowing...slowing. I think she's about to hover sir!"

"Okay."

"Nine hundred yards. She's stopping sir. She's going into hover. Fifty-two seconds to launch."

Chris took a last careful look at the looming machine through the image-stabilized binoculars. Even at this distance he could feel the power of its twin GE turbines, each cranking out 1800 horses. The ship rocked a little then settled into a steady position. As the attack helicopter prepared to fire he laid aside the binos and looked down at the timer on his GPS. Forty-eight... forty-seven...forty-six...

"Get ready," he told Sharp. "On my mark"

"Aye."

Chris began to give a countdown. "Thirty-five... thirty-four... thirty-three... thirty-two... thirty-one... Now!" Chris tapped Sharp on the back and the sniper pulled the trigger.

The chain gun emitted a loud *Pop!* The drive motor whined torturously.

"Oh crap!" Sharp leaned forward and tried to cycle the action. "You ungrateful whore!" He glanced at Chris. "It's a squib. A round burst in the chamber. She's fucked."

Chris's heart fell. *So this is how it ends.* Then he watched with surprise as the man who had been the Navy's top-rated sniper three years in a row went into action. Leaping up from behind the chain gun Sharp snatched the 32-pound Barrett 50-caliber rifle as if it were a Daisy BB gun. He slammed it down on the edge of the sandbag bunker and racked the action to load a cartridge.

The Barrett semi-automatic rifle is nearly five feet long. It's fed from a 10-round magazine and fires M33 50-caliber ammunition driving a 661-grain bullet at 2750 feet per second. At 900 yards, the distance to the hovering copter, the bullets would fall by about 22 feet in their arcing flight, with the drop compensated for by adjustment of the scope.

"What are you doing?" Chris asked.

"Gonna kill the sum-bitch!"

"Very good, proceed." *There's no way he can take down a Black Hawk with a damn rifle. Well, what else we got?*

Sharp snapped the covers off of the front and rear elements of the night vision scope he'd fitted onto the gun earlier. He turned the range knob, letting the integrated computer in the scope calculate the adjustment needed to compensate for the bullet's fall.

Unable to think of anything to do, Chris watched as the master marksman prepared to attempt an impossible shot. He felt his fingers clasping the silver medallion that hung around his neck. *Coyote.* He stroked it gently then let it fall back on its chain. He glanced at the timer. It was passing ten seconds to launch.

Sharp was steadying on his target. Somewhat bemused, Chris watched him prepare by breathing, in, out, in, out…hold.

Blam!

The rifle bucked. Sharp slammed it back down. Almost immediately he fired a second round.

Blam!

Chris watched as Sharp repositioned the rifle and stared intently through the scope then started in surprise as he saw the man relax and draw his eye back. *What the hell?*

De la Vega whooped and Chris raised the binos. He saw an incredible sight. The Black Hawk was whirling out of control, its huge blades beginning to chainsaw into the top branches of a tree. As Chris watched it whipped over on one side. Rotating blades struck the earth and shattered, fragments flying high into the air. Then the whole body of the craft slid heavily to the ground. As the last of the blades broke away the huge turbines spun up the stubs, causing the wreck to heave and jump, twisting and churning up the earth like a dying beast before settling down with a final crash. Smoke began to pour from the remains, then the wreckage burst into flame.

Chris was awestruck. He turned to gape at Sharp who was standing up beside his Barrett with a grin on his face, watching the helicopter self-destruct.

"Shooter, how did you do that?" he inquired. "Those helicopters are armored…"

"Hell, sir, those pilots were just sitting there looking right at me through the Plexiglas. A Barrett can penetrate an inch of steel, so there's not a windscreen made that will stop it. Damn fools."

"You shot the pilots?"

"Best way to kill a snake is to cut off its head, sir."

Chris looked back at the helicopter. It was engulfed in a ball of flame. There was no sign of survivors. One of the Hellfires blew, scattering pieces of the wrecked bird.

He looked at De la Vega who met his eyes and shrugged. He turned back to Sharp who was still watching the wreckage of the Black Hawk with a delighted smile on his face.

"Shooter, I gotta tell you, that's one for the record books."

"Hell, sir, they were only nine hundred yards away. For a Barrett, that's like shooting fish in a barrel." He stroked the rifle, then scowled down at the inert chain gun. "I guess it's true what they say," he mused, "you gotta dance with the one that brought you."

He patted the Barrett again as Chris and De la Vega broke out in relieved laughs.

Chapter 45

Quantum Center
Friday, 4:23 a.m. MST

The excitement over the successful defense of the Quantum Center was settling down at last. The heroes of the hour were Martin Sharp and Tommy Dawson. Sharp's fellows had hoisted him onto their shoulders and carried him around the perimeter chanting the SEAL motto as a march cadence. Everyone clapped and stamped their feet. Sharp looked embarrassed and affected an "aw shucks" expression.

When the SEALs offered to do the same for Dawson he quickly turned them down.

"Hey, hey, you're not getting me up there," he squeaked, hustling away to the safety of his workstation.

"Aw, come on Tommy," one of the technicians chided him. "You're a hero too."

"Well, I still say no," exclaimed Dawson, sitting down in his desk chair and crossing his arms over his chest. "But someone could get me a Coke," he added. Then looking around cannily and seeing his opportunity, he added, "Two Cokes. And a couple of those Danish." He looked around expectantly. "Please?"

Amid laughter someone hustled off to get Dawson his reward.

Chris and Bonner were huddled at one of the break room tables with cups of coffee and stale donuts. Both looked exhausted. For a long time they just sat looking around without saying anything.

Chief Perry walked up, poured a cup of coffee and asked permission to join them. "Christ, yes," Bonner told him, pushing out a chair. "Grab a donut before they're all gone."

Perry eyed the stale pastries suspiciously. "Thank you, no, sir."

"More for us."

"Welcome to them, sir."

"Relax, Leonard," Chris told him. "This isn't the Navy, not really. Sure, technically we're active duty, but after tonight I don't know which side we're on."

"You're right," Perry agreed. "Just a while ago Mouse, Snapper and Della were asking me about that very question."

"You mean about their official status?"

"Yes sir."

"The temporary duty documents I gave each of you came directly from the Department of the Navy and signed by me. You're fully covered with pay in grade plus bonuses."

Perry looked away for a minute, tapping his fingers lightly on the table.

"So who owned that Black Hawk? Who were those guys we killed?" His forehead wrinkled with concern. "Are we gonna get blowback?"

Chris looked at his former second-in-command. Perry gazed steadily back. His expression was earnest.

"Leonard, I don't know who they were, and frankly I don't give a damn. We have proof that they were here to kill us and destroy the project. We acted in self-defense."

Perry looked unconvinced.

"Yes, sir, but what kind of proof?"

Surprised, Chris sat up straight in his chair.

"Why the future time views," he replied. "They clearly show the attack."

Perry was silent, contemplating his superior officer with a neutral, no-nonsense expression.

"But the attack never happened," he said at last. "How do we explain that?"

Chris thought for a moment, took a bite of donut and washed it down with coffee.

"It would have happened. The record shows that."

"From what I remember of military law, only facts can be submitted in courts martial," Perry told him.

Chris realized Perry was serious about his concern.

"Well," he hedged, "the time views *are* facts…" He paused. "Sort of. God damn it Perry, we saw what would have happened right on the screen. You saw the re-play later. Any court of justice would take one look at that and declare our actions to be defensive and fully justifiable. And anyway…"

He paused to take a deep breath before continuing.

"…You were acting under my direct orders."

Perry regarded him solemnly.

"You're suggesting the Nazi guard defense, sir? Orders that must be obeyed?"

Exasperated, Chris shrugged.

"Look, Leonard, I don't know the answers. I only know we did what we had to do and if there's a penalty for it then there's something rotten in the system."

The chief relaxed slightly. He took a sip of coffee, but peering above the rim of the cup his eyes remained brooding and intense.

"Okay," he allowed at last, "I think you're correct. I just need some help with what to tell the men."

"Ah!" Chris smiled. "Tell them they did one Hell of a good job here, that the country owes them a huge debt of gratitude, and that if any son of a bitch tries to lay anything on their heads I'll take it to the Joint Chiefs and the Supreme Court if I have to. That good enough?"

Perry nodded curtly, stood up and prepared to leave.

"Thank you sir. I'll let them know."

"Tell them this, too," Chris added, his voice rising. "There'll be decorations in this. I'm recommending the whole lot of you for the Navy Cross. Oak leaf clusters for those like you who already have it, and presented by the God damned highest-ranking admiral in the whole fucking Navy. No, by the President himself or I'll raise a stink so strong they'll smell it on Mars."

Perry stared at him for a moment.

"Aye sir," he said, trying to suppress a grin. "The men will be happy to hear it." He spun on his heel and marched away.

"Wow," Bonner exclaimed when Perry was beyond hearing range. "Is he always that uptight?"

"No," Chris replied, staring morosely at the remains of his stale donut. "Leonard Perry was, no strike that, *is* the best master chief the SEALs ever had. He's a man's man and he takes his responsibilities more seriously than you can even imagine. Put in his twenty years and picked up a degree in political science from UC-San Diego along the way. Summa cum laude. Sharp as a tack."

"Oh, I can see that," Bonner said. "I wish I had a few like him..."

"From what I've seen, you've got a pretty good outfit."

"Well, yeah, I'm proud of the force. But this is a government operation, part of the Department of Energy, so there's only so much I can do to prune the tree, if you know what I mean."

"Well, that's pretty much the case with any public organization," Chris responded. "Even at Stanford we've got tenured staff that have long outlived their usefulness and there isn't a thing anyone can do about it."

"Yeah," Bonner mused, staring into his cup. "Frankly, I'd dump about a quarter of my people if I could. The others put out the effort to more than make up for the deadwood, but that's really not fair to them..."

Just then there was a racket from the front door. Chris and Bonner glanced up then jumped to their feet. Jim Chen had entered the building. Behind him came six men in commando gear escorting a prisoner in cuffs and ankle chains.

"It's Pearson," Chris told Bonner. "They got him."

"Hot damn! And, see those guys around him? Those are some of the real stars in my outfit. Looks like they and your guy pulled it off. I can't wait to get that Pearson bastard alone..."

"Whoa," Chris said with a chuckle. "We don't allow torture in this country."

"Who's talking about torture? I just want to beat the crap out of him."

Bonner grinned to show he was kidding. It was a weak, sort of half-grin that bore little assurance of its sincerity. Chris returned the smile and winked. They went to meet Chen and his prisoner.

Chapter 46

Quantum Center
Friday, 9:42 a.m.

The center was quiet. Nearly everyone had caught some sleep and most of the staff had been dismissed. Sending away the rest of his security officers, Bonner asked for two volunteers to guard Jalaly who was cuffed to a steel desk in one of the small offices along one side of the center. The SEALs had changed back into their civilian clothes and gone to their motels for some well-deserved rest.

Caterers from the LANL cafeteria arrived with a hot breakfast. They set up two segmented serving trays on stands and lit alcohol stoves to warm the contents. Yawning and stretching, the few remaining people began to stir in response to the smells of bacon, sausage and fresh-brewed coffee.

Bonner was first in line, loading a plate with pancakes, scrambled eggs, sausage links and hash browns. Dawson was right behind him, taking the same but twice the amount. They sat down together and began to eat, watching the others line up.

"Tommy, as much coffee and Coke as you drink I'd think you'd be a mass of nerves," Bonner observed.

"Nah, they're stimulants," Dawson replied. "How do you think I keep my mind working on overtime, eighteen hours a day?"

"Er, I guess that's right." Bonner didn't look convinced. He tried to keep his own coffee consumption down to four cups a day and wouldn't touch a cola unless dying of thirst.

Dawson drained his coffee cup and rose to refill it. He stopped to say hello to Kate who had just emerged from the office that had been designated as her private bedroom. She'd made an effort to freshen, wiping her face with cold cream and running a brush through her hair.

"My you look alert this morning," Dawson remarked.

Kate grimaced. "Tommy, you're so full of bull." Slinging her messenger bag over her left shoulder she picked up a plate and helped herself to a cheese omelet, four rashers of bacon and a small serving of hash browns.

"Come join us," Dawson invited, returning to the table where Bonner was still digging into his pancakes.

Before long Chris had joined them as well. Later, empty plates pushed aside they were deep into Monday morning quarterbacking the events of the previous night.

"I still can't get over the audaciousness of the attacks," Bonner pondered. "We still don't know who it was, or where they came from."

"We're working on that," Dawson informed him. "I've got my assistant running past views on that copter. We should soon know its point of origin. We'll try to track the Hummer, too, but that's going to be a lot harder."

"That may help some," Chris said, "but it's my guess that 'Hawk wasn't military, but from a black ops source. The CIA has always used fronts to provide air transport and attack. Remember Air America? And, they're the ones flying the Predator drones in Afghanistan, taking out Taliban and al-Qaeda. They've done lots of good work, but that doesn't mean there aren't some bad guys among them."

"I think you're right," Bonner noted. "During my time in the Bureau we always suspected there were things going on that just didn't smell right. Hell, the Company could have been penetrated by foreign agents—it's been known to happen. You surely won't find any clean trace back to Langley. And, for all we know, the Company may not have had anything at all to do with it."

"Oh, too bad," Dawson said pensively. "Shall we stop the search then?"

Bonner gave him a strange look.

"No, as long as your tracking doesn't interfere with our future time surveillance, keep working on it."

Dawson nodded solemnly. "No problem there. We're just adding the copter views to the future view batches. We can take it back about ten miles at a time, so it shouldn't be long until we know where it flew out of."

"What we need to worry about now is whether they'll try again," Chris informed the others. "My thought is that last night was a serious setback for them, whoever they are. It'll take time for them to organize a new operation, but I think we should be back on full alert by this afternoon."

"I've placed my staff on condition orange until noon to give them time to rest," Bonner reported. "But we've still got the lab buttoned up. I don't think we can keep it in lock-down for much longer though."

"After last night, I should think not," remarked Kate. "How are you explaining the Hummer and the helicopter?"

Bonner shook his head and shrugged.

"We issued a news release that's ninety-nine percent pure BS. After this the media will never believe a thing we put out."

"They don't anyway. What did it say?"

"Heh, you'll probably laugh..." Bonner looked embarrassed. "We didn't have much time to think about it, and the fact is we were pretty tired. Okay, for the Hummer we explained that a couple of escaped cons stole the vehicle and planned to rob the Los Alamos National Bank when it opened. They'd made a bomb loaded with ball bearings and planned to threaten to set it off unless the bank opened up their safe. Then the bomb accidentally went off in their vehicle, killing both of them and turning the Hummer to Swiss cheese."

Nobody said anything for a while. Kate cleared her throat and Chris chuckled.

"What happens when somebody notices that all those holes in the vehicle came from the outside?" inquired Chris.

"No problemo, Senor," Bonner grinned. "We confiscated the remains of the Hummer and brought it inside the lab. The city cops are truly pissed, but I asked the chief to keep it quiet and so far he's going along with it."

"Okay, not too bad," Kate admitted. "How did you explain the copter that was shot down? Fortunately it crashed in an open area instead of taking out a bunch of houses. Nothing like civilian deaths to stir up trouble."

"Well, yeah, we were truly lucky about that. Actually, we told the media that the helicopter was bringing some special equipment here, to help deal with the radiation leak we claim to be cleaning up. We explained the sound of the sniper shots as explosions in the copter's left engine—I learned a long time ago that details like that make BS smell better."

Kate gave him a sphinx-like smirk. "BS with eau de cologne, eh?"

"Yeah. Anyway, the pilot tried to regain control but caught a blade in the trees—that actually happened and there were witnesses so we're good on that. Then, like the heroes they were, the pilots struggled to keep from hitting the nearby houses. Fortunately, the crash and fire destroyed the windscreen so nobody saw the bullet holes. The pilots were pretty much cremated in the fire, too."

"They'll find out eventually," Chris observed. "There's certain to be a full-scale investigation."

"Sure, but not for days, maybe even weeks."

"Hmm, okay. But meantime the lab remains on lock-down and the rumor mills are cranking out every conspiracy theory you could imagine. I heard there's even a story going around that the Black Hawk was actually a UFO and that we're hiding ET here."

Dawson and Kate laughed.

"There's enough funny stuff going on without that complication," Chris noted with a grim smile. "Unless…" He turned to Dawson. "Your new theory doesn't have anything to say about that, does it?"

Dawson looked uncomfortable. He picked up a cold sausage link and took a bite, followed by a drink of coffee.

"Well," he said nervously, "I haven't finished my calculations in that area…'

"Oh Christ!" Bonner exclaimed. "I suppose thanks to your time dimensions we can expect a visit from little green men."

"Well, no, that isn't what I meant. But six dimensional space-time probably does make FTL travel possible."

"FTL travel?" Bonner looked confused.

"Faster than light. According to Einstein nothing can go faster than the speed of light—that's about a hundred eighty-five thousand miles per second—but Albert was assuming only a four-dimensional space-time."

"Oh." Bonner still looked confused. "That seems awful fast. Why would anyone want to go faster?"

Dawson looked surprised. "Well, yes, it's pretty fast by our standards, but it still takes light eight minutes to get to the Earth from the Sun and that's only about 93 million miles. For a photon to get to the nearest star takes over four years."

Bonner shrugged. "Never mind that, we need to focus on our situation right here and now. Let's not get sidetracked."

"I agree," Chris chimed in. "As I see it we have three main priorities. First is to keep up our protective posture. We'll go back to red alert this afternoon. Second is to continue to develop Tommy's idea of preparing for peace, in other words how to get out of this with our butts in one piece." He paused to let the pun sink in and was rewarded by a sour glance from Dawson. "Kate's working on that aspect. And third—and actually this is something we can do right now—is to see if we can get any useful information from our prisoner."

"Ah, yes," Bonner said with satisfaction. "It's time to have a friendly chat with that weasel."

"I had one of the guards take a plate in to him, so he's had his breakfast," Chris reported. "I think now would be a good time to have that chat."

In a few minutes the Iranian was brought out of the office where he'd been restrained. He was frog marched to an open space in the middle of the room and pushed into a straight-backed chair. One of the guards fixed the ankle chains to the leg of the chair. At Chris's instruction he removed the handcuffs. Jalaly rubbed his wrists and nodded at Chris gratefully.

Chris, Bonner and Reeves sat down in swivel chairs facing the spy. Jalaly was flanked by the two guards, each holding an AR-16. Kate was off to one side about twenty feet away, her video camera recording the scene. Dawson wandered off to check on the time search. Wanting no other witnesses, Bonner had instructed Dawson to send his assistants away. No one else was in the building.

Bonner sat in the middle with Pearson's security file open on his lap. He spent a long moment just gazing at the Iranian. Jalaly sat straight in the chair, watching his captors. He looked nervous, particularly when he saw the expression on Bonner's face. Jalaly hadn't slept and it showed as stress lines and drooping eyelids. His hair was unkempt and there was a small tic in the left corner of his mouth.

He spoke.

"I wish to make a confession," he said in a flat, emotionless voice. "I will tell you everything."

Bonner grunted. He looked unconvinced.

"You killed two of my men," he stated. "Do you deny that?"

Jalaly recoiled as if he had been slapped. His eyes dropped to focus on the floor between his feet.

"No," he said, voice low.

"No what?"

"No I do not deny it." Jalaly looked up and they could see tears welling in his eyes.

"Okay, you're willing to help us?" Chris asked.

"Yes. I will tell you anything."

"We don't want to hear *anything*," growled Bonner. "We want to hear the facts, the answers to our questions. Complete answers."

"I understand."

"And you'll do it?"

"Yes."

Bonner looked at Chris and Reeves. Chris picked up the cue.

"Who're you working for?"

Jalaly straightened up. "Mother Persia," he said. His voice was low but had a touch of pride in it. "I have been a spy for Iran."

"What kinds of information have you stolen?"

"Anything, everything." With a hint of pride Jalaly spread his hands, palms up. "Nuclear secrets of course, but anything that might benefit my country."

"And you've done this for twelve years?"

"Yes."

"Who are you really?" Bonner asked. "I presume that Pearson isn't your real name?"

A little humorless smile passed over Jalaly's features. "There is no such person. There was once but he died as a baby and I was given his identity."

"I see. And your real name?"

"I am Sanjar Jalaly."

"An Iranian?"

"Yes. I chose to become Pearson for the similarity to the word 'Persian.' I am a patriotic son of Iran."

"I see." Bonner flipped through the security folder. "Surely not everything in your file is false? What is true?"

"I never attended MIT. My education through undergraduate levels was in Tehran. But I did earn a doctorate at Cambridge."

"As Pearson?"

"Yes."

"And what about the second identity, this Jim Rowland?"

"That was created to give me the ability to move and act in different circles, outside of the lab. Rowland is also a false identity, of course. The real James Rowland died at the age of two."

"I see. And I suppose it was as this Rowland character that you passed information to your masters?"

"Yes."

"Tell us exactly who you were in contact with?"

"Only one man, the one who died last night in the forest."

"Who was he?"

"I do not know."

Bonner raised his eyebrows. "You don't know?"

"Not specifically. He never told me his name. I believe he was an agent with VEVAK, the Iranian Ministry of Intelligence and Security."

"Was he based here, in Los Alamos?"

"I don't think so. I saw him infrequently. He would contact me through coded ads in the Santa Fe newspaper."

Bonner closed the security file and glared at Jalaly for a moment.

"So you were merely a spy, gathering information?"

"Yes."

Bonner said nothing but his face reddened with anger. His hands squeezed into fists causing his knuckles to go white. He held his temper in check but his voice was tinged with rage.

"And you killed people? My people? On my watch?"

Jalaly shrank back in the chair.

"Yes," he squealed.

"Why?"

"It was ordered. I had never done such a thing. I was told I could soon return to Mother Persia, to become a professor in Tehran. I was given the passcodes."

"Ordered? By the man who was killed last night?"

"Yes."

"Why?"

Jalaly seemed to be off somewhere else. His eyes were focused at infinity and his breathing was very slow. Bonner waited.

"I am not sure," Jalaly said at last. "I didn't want to do it. I'm sorry that I did."

Now Bonner's rage spilled over. He threw the file onto the floor and slapped the arm of his chair."

"You're sorry!" he nearly shouted. "Is that all you can say?"

Jalaly looked frightened but he gazed defiantly back at Bonner.

"I answered the call of jihad," he said quietly, as if that would explain everything.

"Ah." Bonner relaxed. "It's that again, the holy warrior crap."

Jalaly winced.

"Please, do not insult Islam," he muttered.

"Do not insult Islam!" shouted Bonner, his anger returning. "Your crazy so-called religion indoctrinates maniacs like you to murder innocent people and we're not supposed to insult it?"

He turned to Chris. "I don't know how much longer I can talk to this, this..." He stopped and pointed helplessly at Jalaly. "Why don't you have a try? You must know more about these jihadists than I do."

Chris nodded, thinking about the Taliban and al-Qaeda warriors he'd hunted in the mountains of Afghanistan. He knew them to be dedicated to their various causes, no matter how misguided and bizarre they seemed to Westerners. He watched Jalaly for a moment, trying to relate him to some of the Afghan and Arab Islamists he'd encountered. There was no easy match. Those holy warriors were unsophisticated, many of them functionally illiterate except for having memorized the Koran, considered by many of the faith to be the only "education" required of a faithful Muslim. Jalaly was trained in the Western manner, with a Ph.D. from one of the world's leading universities. Chris shook his head in wonder.

"You consider yourself a holy warrior?" he asked.

Jalaly looked at him with surprise.

"Of course. Every good Muslim is called to jihad."

"Not all," Chris contradicted him. He watched Jalaly closely then spoke in a quiet tone.

"But you've lost your faith, haven't you?"

Jalaly jerked and tried to stand up. Jangling, the leg irons tripped him up and he sat back down. He looked stricken. He turned his eyes upward and murmured something.

"What did you say?"

Jalaly did not answer at first. Tears began to roll down his cheeks. He was shaking.

"Yes!" he cried. "Yes! I have lost my faith. Allah help me!" He buried his face in his hands.

Bonner looked at Chris curiously.

"How did you figure that?" he asked.

Chris related how Jim Chen's report included a description of Jalaly's actions on the front porch of the cabin. Although the other man was heard exhorting him to shoot, he dropped the AK-47 and made no attempt to resist.

"It was his opportunity to earn a place in Paradise, to die fighting against the Infidels," he explained. "A true jihadist would never pass up a chance to earn the approval of Allah. By failing to fight and die he damned himself."

Bonner turned back to Jalaly who was still crying into his hands.

"Is that true?" he demanded.

Jalaly nodded without raising his head. "Yes, it is." The terse words were muffled by his tear-dampened palms.

"Well, I'll be," Bonner remarked. "An apostate, huh?" He turned to Chris. "We might get some good intel from this guy…"

Jalaly looked up. He wiped his eyes with his eyes.

"Yes," he said. "I know some things…"

At that moment the scene changed, suddenly and dramatically. Without warning one of the two guards standing behind Jalaly raised his M-16 rifle and fired a bullet through the back of Jalaly's head. Everyone froze as the Iranian slumped forward onto the floor.

"Stay in your seats!" the guard commanded, pointing the rifle in the direction of Chris, Bonner and Reeves. "Don't move or you're dead."

The second guard looked stunned then started to raise his own rifle.

"Never mind," the first guard told him, "I jammed the mechanism when you went to the restroom. And if you reach for your pistol I will shoot you. Drop the rifle, now!"

The weapon clattered on the floor.

"Now take your pistol from the holster with two fingers and hold it out to me very, very slowly."

The killer stepped sideways near the other man to take the pistol. Keeping Chris, Bonner and Reeves covered with his rifle held in one hand he raised the pistol, flipped the safety off and shot the other man through the head. The guard's body fell in a spreading pool of blood.

"Who are you?" Bonner demanded.

"Shut up! Open your mouth again and you're dead."

Bonner shut up.

"Now, I want the three of you to stay in your chairs and I want you to place your hands on top of your heads. Yes, that's it. Now interlock your fingers. Good. Next, I want you to tuck your toes around behind the legs of the chairs. Excellent."

He turned his head slightly toward the distant control center where Dawson stood, staring in consternation at the sudden turn of events.

"You!" the guard shouted. "The fat weirdo! Get over here."

"Tommy, stay there," Reeves warned. The guard stepped forward and slashed the acting director on the side of the head with the pistol while jabbing the M-16 into Bonner's paunch. Reeves groaned and slumped back in the chair. A trickle of blood ran down the side of his face.

"What do you want?" Chris asked without thinking. *Oh shit!* The guard stepped in front of him.

"Do I have to say it?" He looked Chris sharply in the eye. Chris shook his head and the guard stepped away.

"I've said what I want, which is for that weirdo to get his fat ass over here. You!" he shouted again. "Come!"

Dawson was on his way, anger suffusing his face.

"Who you calling a weirdo?" he demanded. "Fat, maybe, I'll give you fat, but nobody ever called me a weirdo." Dawson was angry. He stopped about ten feet away and put his hands on his hips. Then he looked down at the two dead bodies and his face turned pale.

"You are a weirdo, and you're about to be a dead weirdo," the guard told him. He began to swing the rifle toward Dawson. Chris tensed himself to act. *Not enough time!* He started to rise, realizing that before he could get out of the chair Tommy would be dead and the rifle would swing back to take him down too.

As always seems to happen in moments of extreme danger, time slowed to a crawl. Everything seemed to be happening in slow motion, like a time-lapse movie. Tommy began to react, raising his hands as if to ward off the bullet. Chris twisted his feet free from the chair and began to lean forward. The rifle continued on its steady arc toward Dawson.

Blam! Blam!

Two shots rang in quick succession and the scene reverted to real time. Chris sprang to his feet. The guard jerked like a marionette. He dropped the rifle. It hit the floor with a metallic crash. He slumped, then sprawled face down across the weapon. Chris placed a foot in the small of the man's back to immobilize him. Below the man's shoulder blades he saw two exit wounds, both bleeding profusely, both dead center in the heart-lung kill zone and no more than two inches apart.

Chris looked up at Dawson, who was shaken but uninjured. *What the...* He glanced at Bonner, half expecting that he had been the source of the gunshots, but the security director still sat frozen in the chair, only now beginning to lower his hands. Slumped sideways in his chair, Reeves appeared to be unconscious.

"Over here," said a quiet voice.

He spun toward Kate. She was standing beside her video camera with one hand resting on the tripod and a pistol dangling in the other. She wore an almost triumphant smile that Chris found odd considering that she'd just witnessed three violent deaths. And not only that...

"Kate!" he cried. "You shot him?"

"Damned straight," she said. "Nobody ever deserved it more." She began to walk toward Chris. He noticed that despite the smile her hand was shaking slightly.

"Your shooting lesson paid off!" he declared, glancing at the pistol in her hand. "That Beretta..." His voice trailed off.

Kate was holding a matte black Kimber 45-caliber automatic, one of the deadliest and most distinctive handguns made. He looked up in surprise, meeting her emerald eyes.

"No," Kate said, looking down at the pistol. "It's not the Beretta. It's my personal weapon."

"What!"

Kate sighed and shoved the big forty-five back inside her ever-present messenger bag.

"I'll explain everything," she declared, "but first we have to clean up this mess. Make sure that guy's dead, although I have no doubt he is. Tommy, sit down. You look like you've seen a ghost. Chris, see about Stuart would you? And Paul, better call an ambulance and a security team."

Bonner realized his mouth was hanging open. He closed it.

"Yes, ma'am." He reached for his radio.

Chapter 47

Quantum Center
Friday, 11:23 a.m. MST

There had been a hectic flurry of activity. Bonner called for medics and a team of forensic officers. Chris got ice and a towel for Reeves, who was groggy but regaining awareness. Dawson went to check on the progress of time viewing.

Before the officers arrived some quick decisions were made.

"Well, I guess my cover's blown," Kate announced. She reached for her messenger bag, rooted around in it for a moment and pulled out a leather wallet. Without a word she handed it to Bonner. He opened it, glanced inside and handed it to Chris with a nod.

It contained credentials identifying Lieutenant Katherine Elliott, Office of Naval Intelligence. A badge was pinned on the facing side of the wallet. It featured an American eagle wearing headphones. On the eagle's chest was an eye, the same as the one atop the pyramid on the back of the one-dollar bill. Around this illustration was the motto of America's oldest intelligence service: "In God we trust. All others we monitor."

Chris studied the ID for a moment then gave it back to Kate.

"You work for Admiral Chase," he deduced.

Kate nodded. "I need your help to maintain my cover."

Chris and Bonner looked at each other for a moment.

"Okay, then we need to work fast," Bonner declared. "We need to edit the storyline."

Kate handed her Kimber over to Chris who carefully wiped it, even removing the unspent shells and cleaning them. He handled everything including the empty shell casings on the floor to replace Kate's prints with his. He gave her his Beretta after wiping it and laid the Kimber on a nearby desk.

Meanwhile, Kate backed up her video camera and erased the sequence when the shootings took place. She moved over to the control center near Dawson. They briefed him on the change in story and he immediately agreed, looking approvingly at Kate.

When the investigative team arrived the events were described almost exactly as they happened, with the exception that Kate was not implicated. It was explained that she had been busy working with Dawson, far away from the interrogation.

"Just as the rogue guard was about to shoot Doctor Dawson," Bonner told them, "Lieutenant-Commander Fisher here managed to pull out his gun and shoot him." He glanced at Chris who corroborated the story and handed over the Kimber as evidence. Out of the corner of his eye he noticed Kate wince as the weapon was zipped into an evidence bag.

After several hours the frenzy had settled down. Statements and photographs had been taken, the bodies bagged and removed. Reeves had gone to the nearby clinic for treatment and the quantum center was empty except for Chris, Bonner, Dawson and Kate.

The security chief wiped his forehead with a handkerchief and sank down in a chair at a workstation along the server wall. Chris walked over to the next station and pulled a chair around to face the area where the shootings had taken place. It was now fenced off with yellow tape. Kate walked up looking slightly embarrassed. She set her messenger bag on a nearby desk and hopped up to sit beside it, her feet dangling inches from the floor.

Chris looked at her with interest. The security chief was keeping quiet. After a moment Kate let her eyes wander away. For a long moment no one said anything. Chris gazed steadily at Kate before opening the conversation.

"Good shooting, Ms. Elliott…if that is actually your name."

Kate's eyes turned toward him and she rewarded him with her little two-toned laugh.

"You can call me Kate," she said. "And yes, it's Elliott. FYI I really am an editor at *Science Today* magazine." She paused. "Among other things…"

Bonner gazed at her contemplatively. He said nothing.

"Ah," Chris said in a flat voice. He paused for a beat. "Those other things being spying for the Navy."

Kate shrugged and straightened her posture.

"So, you were playing me," Chris interjected. "Pretending you didn't know how to shoot, acting like the innocent girl reporter."

Kate shrugged. "I had to protect my cover. I've invested five years in establishing my reputation at *Science Today*."

"And all the time you could shoot the eye out of a gnat at twenty paces."

"Well, no…" She paused then added with a smile: "My Kimber would splatter the whole gnat."

Chris had to smile with her. "Yeah, I guess it would." He glanced at her messenger bag. "Tell me, how do you get away with carrying around a loaded piece?"

Kate grinned. "The Navy has ways. When I travel abroad, my pistol either stays behind or takes a different route, by special courier or some other means. Here at the lab, the Admiral quietly arranged for my bag not to be searched. Like you, I'm a serving officer with a top security clearance, although we don't want that to get around. Paul was the only one who was in on the secret." She glanced at Bonner, who nodded in assent. He kept his eyes on Kate, refusing to meet Chris's eyes.

"Look," Kate implored, "this business has gotten really out of hand. It all started as a routine monitoring assignment, the kind I specialize in. Nobody expected what it would turn into."

Chris stared at her for a moment. Bonner shifted uncomfortably in his chair.

"Why didn't the Admiral tell me you were his?" he asked.

"Chris, once again it was to protect my cover. We've worked for a long time to put me in a position to monitor science projects that could have security implications for the United States. I report on a lot of different stories all around the world and most of them are routine. Then again, sometimes something comes up that Naval Intel really needs to know about. It's important work for our national security."

"I can see that," Chris mused. "In my case I didn't need to have a cover, what you saw was what you got." He glanced at Kate. "I even understand why you didn't tell me. Against orders, right?"

"Absolutely." She grimaced. "I really wanted to…"

"You know, this explains a lot. The way you caught my cue in the cafe and spilled your iced tea as a distraction. It was masterful. I kind of wondered about that. And even though you tried to act like an innocent lamb, I could see a lot of steel under the wool."

Kate laughed out loud. "Baaa," she replied.

"So, look, I don't see any reason why you can't keep your cover." Chris turned to Bonner. "Do you Paul?"

The security director stirred.

"No, we should be able to keep a lid on this. You, Tommy and I were the only witnesses."

Kate gave them a grateful smile then began to tell her story.

She had graduated from the University of Indiana with a degree in journalism and a second major in general science. After two years working as a newspaper reporter for the *Cleveland Plain Dealer,* and following the breakup of an unfortunate marriage, she applied for a scholarship at Boston University's Center for Science and Medical Journalism. There she entered a three-semester program leading to an M.A. degree. She graduated with honors.

As she neared the end of her time in Boston, Kate began to cast out queries for a new job. The center's director encouraged her to go for a doctorate at BU's College of Communication, even offering her an adjunct position with a modest salary that would have made it possible. But Kate was not inclined to an academic life.

Publishers sent representatives to interview graduating students and Kate spoke with several of them. She was particularly interested in possible careers with *The Wall Street Journal* and *Science Today* magazine.

In her last week before graduation she received an unusual message from the director's administrative assistant: An un-named organization had requested a confidential interview with her. No details would be provided unless she agreed to sign a blind non-disclosure agreement. The executive assistant reached into her desk drawer and pulled out the NDA. Laughing, Kate waved it away and said she wasn't interested.

The next afternoon she was walking south on St. Mary's Street on the way from the campus to her apartment, about ten blocks away. As she neared Euston Street just south of the Massachusetts Turnpike, a Lincoln Town Car pulled over to the curb. It was a livery car with a driver and blacked-out windows in the back. The rear window rolled down as the car pulled to a stop and the passenger called out to her by name.

"Oh," Chris interrupted. "I bet I know who that was."

"Yeah, I'm sure you do," Kate replied. "But at the time I was just a little apprehensive. Momma told me not to talk to strangers in cars."

Stopping but keeping her distance, Kate asked the man what he wanted. He told her he'd asked for a private interview with her and was disappointed that she'd refused.

"I don't want to work for someone who won't even tell me who they're with," she'd responded.

"Okay, fair enough," the stranger told her. "I'll tell you this much: You'd be working for the United States government in a classified job. That's why we asked for a confidential interview."

"Why don't you hire one of the other graduates?" she asked. "I bet plenty of them would jump at it."

The man told her she was the only one he was interested in.

"Look," he pleaded, "would it hurt just to talk?" Kate admitted that it wouldn't and he suggested that she return to the director's assistant first thing in the morning, sign the NDA and keep an appointment with him at ten o'clock in a private office that had been reserved at a distant part of the campus. Kate agreed.

"So that's how it began," she told Chris and Bonner. "As you've guessed the interviewer was Admiral Chase—he was a Captain then—and the job was with Naval Intelligence. It was an intriguing deal they offered. I'd attend Navy Officer's Candidate School at Newport, Rhode Island and be commissioned as an ensign. I'd serve two years of visible active duty, training in intelligence work. Then I was guaranteed a spot at *Science Today*, working undercover. Captain Chase said the magazine wanted me, but was willing to wait while I, quote, 'undertook military service,' unquote."

She grinned wryly at Chris before continuing. "The old goat had already spoken with the publisher who as it turned out was a classmate of his at Yale. That ticked me off a little, but it was an offer I couldn't refuse. What made it truly sweet was that I'd receive both Navy pay in grade plus a generous bonus for the first two years. After that, full Navy pay plus the usual salary and benefits from the magazine. Who could pass up a deal like that? And best of all I'd be doing work I love, science journalism."

"Wow," Chris said. "And has it turned out to be as sweet as advertised?"

"Absolutely. My employers think I'm a reserve officer but I've got seven years of active service recorded in my personnel jacket. I'm on the promotions list to make lieutenant-commander next Spring." She smirked at Chris. "Just like you."

"Congratulations," he said. "I've never even worn the uniform of that rank."

"Well, you've earned it."

Chris nodded in appreciation. "Thanks. But what I want to know is why Admiral Chase co-opted me into this deal at the last minute. Obviously you were plugged into the operation well in advance. Did he know something, or just have an idea trouble might be coming down the road?"

Kate thought for a moment, considering how to answer. She idly touched her Thunderbird medallion.

"Well, the Navy funded most of this project. It's important to us." She hesitated, then continued: "There were some rumors…"

"Yeah," Bonner interjected. "Remember the file you read when you first arrived. and the briefing Reeves and I gave you? We'd picked up hints that foreign interests were poking around, and now we know it was true."

"Okay, understood, but you didn't answer my question, Kate. Why did Chase bring me into this?"

Kate shrugged and looked away.

"He wanted someone to give me backup," she said with obvious embarrassment. "I argued against it, of course, but I'm glad it was you."

"Ah, so I was just sort of an enhanced bodyguard. Hired muscle."

"Oh, no, I didn't mean it that way!" Kate looked stricken. "He just felt he needed two of us on the ground, and someone with technical knowledge I don't have. I'm sorry you weren't given the whole picture. I told the Admiral he should brief you fully, but he overruled me on the grounds that you were only on temporary assignment."

"Yeah, and it almost became permanent a couple of times," Chris murmured. "Actually three times if you count the kidnappers in the cafe, the helicopter attack, and what happened here this morning."

"Yeah, and thanks to you and your men we survived the first two," Kate reminded him. "This time it was my turn."

"Well, thanks."

"And thank you."

"I guess Chase did pretty good by teaming us up," Chris admitted. He grinned and stood up. "Now the Lord only knows what's going to happen next."

Kate hopped down from the desk and stood facing him.

"Well, sir, we're incommunicado and you're the ranking officer here, so from now on I'm under your command." She threw him a brief salute.

"At ease, Lieutenant," Chris told her solemnly. "Remember, no saluting necessary when in civilian dress. And that goes double when you're a spook."

"Aye, sir."

"And call me Chris."

"Aye aye, Chris," Kate responded with a devilish grin.

"And eighty-six the bullshit, okay?"

"Okay," Kate agreed. She wiped away the smile but a twinkle remained in her eyes.

"Now we better get back to work."

Chapter 48

Quantum Center
Friday, 12:47 p.m. MST

Bonner had returned the LANL to red alert. Chris's SEALs were back in position at the center and security squads were on patrol around the grounds. Dawson had two assistants at work viewing the future for signs of impending attack.

Chris, Bonner and Perry were discussing defensive tactics at a table in the break area when Dawson hurried over. He was holding a printout in one hand and waving the other in the air excitedly.

"We've got 'em," he exclaimed as the three looked up to watch him approach. "The helicopter, we know where it came from."

Chris waved him to a chair. Dawson started to sit down, then changed his mind and went to the cooler for a Coke. He returned, sank into a chair and popped the tab on the soda. He took a long drink.

The three others were staring at him patiently. Finally noticing, he set down the can and picked up the printout.

"This is really interesting," he told them. He was as excited as a kid on the first day of Summer vacation. "You'd never guess where that 'copter came from."

"Area 51 in Nevada?" Chris hazarded.

Dawson deflated.

"Howd'ja know?"

Chris waved away the question. "Just a wild guess."

"Oh." Dawson looked a bit flustered. "Well, it was a pretty good one. Actually it came from an underground hanger in a remote area of the base, along the northwest side of Groom Lake. We poked around inside for a while and it's obviously a civilian operation. CIA I'm sure."

Everyone absorbed that for a moment.

"It's not surprising," Bonner said after a moment. "As I recall, Area 51 is the most secret military facility in the world."

"Yeah," Chris added. "It's where they test the most advanced aircraft. No surprise the Company would have its fingers in there somewhere. It's the perfect cover for them. Some of the top-secret planes that fly out of there are engaged in intelligence gathering, like the rumored Aurora plane that may have replaced the Blackbirds."

"I've heard they have flying saucers, too," Dawson said with a happy grin. "I can't wait to do some further checking around out there."

Bonner and Chris looked startled.

"Uh, Tommy, maybe you better stay away from that," Chris suggested. "We're going to have enough trouble as it is without being guilty of penetrating the deepest secrets in the world."

Dawson looked disappointed. He took another drink of soda.

"Well, okay," he said grudgingly. "But there are two more helicopters in that secret bunker so we're going to keep a watch on them."

"That's fine, Tommy, but leave the flying saucers for another day. That's a whole different can of worms."

Perry had a sour look on his face.

"That's for sure," he offered. "Nosing around in secrets is what *they* do. It's not our job, and they definitely don't like competition."

"Sure, I can see that," Dawson agreed, pushing back his chair and standing up. "Meanwhile, we'll keep watching the hanger and we're running the surveillance views here." Draining the can he tossed it toward a trashcan. Not seeming to notice when it clattered on the floor, he walked away toward the control center. The others watched him go with mildly amused expressions.

"Gee," Perry said, "that was an easy two-pointer and he missed it by a mile."

"Yeah, Tommy's not really into basketball," Bonner observed.

"No," Chris agreed with a grin. "When it comes to sports, he's strictly a spectator." He looked around and spotted Kate working on her laptop at a distant desk. "Kate!"

She looked up and he waved her to come over.

"It's time we start to plan our next steps," he told Bonner and Perry. "Leonard, I'd like you to *parlez-vous* with the others and see what ideas you come up with."

"I love it when you speak French, sir," Perry said, standing up and turning away with a grin on his face.

Kate approached, set her messenger bag on the floor and sat down. "What's up boss?"

"How you coming on the plan for peace?"

She shrugged.

"I wouldn't say that's what I'm actually doing," she said.

"Okay. What are you doing?"

"Documenting what Tommy has discovered, what's happened to us, how we've responded. I'm writing two sets of documents, actually. One is general, intended for the press and the general public. The other's technical, aimed at people in key places. I'm developing news releases, fact sheets, Q&As, a whole packet. And when I have a chance I need to edit the video footage, but that's going to take a lot of time. I've probably got fifteen or twenty hours of raw video to go through."

The door at the far end of the room opened and Stuart Reeves entered. His head was bandaged and he wore a foam neck brace. Chris stood up and walked to meet him. Reeves held up a hand in greeting and tried unsuccessfully to nod.

"Damn," he said. "This thing'll take some getting used to." He held out his hand and Chris gripped it in a comradely two-handed shake. "No real damage except for a minor concussion. The brace is only for insurance, just for a couple of days. But I do have a headache that's of Biblical proportions…" Wincing, he reached up and touched the bandage that wrapped the side of his head. "They say I'll be better soon."

"Great news," Chris said. "We've just started a planning session. Come join us." He led Reeves over to the table and pulled out a chair for him.

The acting director looked around, his eyes lingering on the area in the center of the room that was marked off with yellow tape.

"What happened? After I tried to warn Tommy I don't remember anything."

Bonner and Chris looked at each other and Chris gave a little nod.

"Well, things got pretty nasty," Bonner related. "The official story is that Chris here shot the guard just before he was about to kill Tommy..."

Reeves gasped and turned to gaze appreciatively at Chris.

"...but the real story is that Kate here was the one who saved the day," Bonner continued, touching Kate on the shoulder.

"What? Kate?" Confused, Reeves looked at her wonderingly. "What did you do? You didn't shoot him...?"

Kate nodded and Bonner cut in quickly to advise Reeves of Kate's sensitive situation. While Bonner was talking she reached into her bag and presented the badge case for Reeves to examine.

"Jesus," he said at last. "Thanks. I think you saved us all."

Kate nodded soberly. "Can I get you something to drink?"

"Yeah, thanks, coffee please." Kate jumped up to get him a cup. Bringing it back to the table, she sat down and picked up the discussion.

"Stuart, I was just filling them in on the reports I'm preparing to help get us through this mess. They're calling it a plan for peace, but I told them it's really just documentation. It's certainly not a plan, only justification for what we've done already."

Reeves nodded. "Do we have any ideas about how the plan for peace might work?"

There was silence then Chris spoke.

"No, not really. It's too big to have easy answers. This thing will turn the world upside down. Hell, Tommy's already itching to use space-time viewing to see if there's anything to the rumors of secret flying saucer programs at Area 51."

"Holy mackerel! I hope you told him to lay off of that."

"Yes, we did. But that's just an example of what his space-time viewing system is capable of. There can be no more secrets if it becomes public, none whatsoever. Not as long as there are inquiring minds."

Reeves leaned back in his chair and gazed at the ceiling for a moment.

"That would be different all right: no secrets. Hard to imagine isn't it?"

"It could result in a nearly perfect world," Kate mused. "It might really make peace a reality. We can't even imagine what it could be like."

"Sir Thomas More imagined it nearly five hundred years ago," interjected Dawson, walking up in time to hear Kate's last statement. "He called it Utopia, an Earthly Paradise. It was always considered to be impossible, but now…" Dawson looked around at them with eyebrows raised.

No one said anything.

"I've been thinking about this some more. In a world without secrets, we could finally achieve that Utopia," Dawson added, turning to the coffee urn and pouring a cup.

"But," Kate responded after a moment, "it could also create chaos. We've talked about that."

"Yes, I understand that now," Dawson replied, turning back and taking a sip from the cup. He pulled out a chair and sat down. "As you've pointed out to me, the trouble is that the world as we know it is built on a foundation of secrets, lies, myths and half-truths. That will have to be swept away before a better world could emerge."

"It might mean the collapse of civilization," mused Chris. "It would almost require all social and political institutions to fall, and be replaced by…what?"

No one said anything.

"Faith of every kind could be shattered," Chris continued. "Faith in governments, faith in leaders, in financial institutions. Religious belief would be shaken, certainly."

He paused to think for a moment and frown lines appeared on his forehead.

"On the other hand most humans are incapable of radically changing their beliefs," he mused. "I once read a book about cults and one of the interesting things is that it's almost impossible to shake the convictions of true believers, no matter how bizarre their beliefs."

"That's true," Kate agreed. "A couple of years ago I interviewed a psychologist who was studying cult followers. The hundreds of men, women and children who died when Jim Jones told them to drank cyanide-laced Kool-Aid. David Koresh and his Branch Davidians who burned to death resisting the ATF. It goes all the way back through history, including the Nazi true-believers, Islamic extremists over the last fourteen hundred years and even the defenders of Masada nearly two thousand years ago.

"I remember one anecdote in particular. The psychologist described a California cult. The leader announced that on a certain date a flying saucer would descend and take them to a better place." She paused and looked around the table. "Remember, this was California. Anyway, on that day the followers all assembled to await the arrival of the saucer. Nothing happened. Now, you'd think that their faith in the leader would have been destroyed, right? But no, actually the opposite occurred. They became even more fanatic in their belief and their leader announced that she'd made a mistake in the dates and named another time some years in the future. The cult emerged from the experience stronger than ever."

Kate looked around. Each pair of eyes met hers. No one said anything. She shrugged.

"So, I guess my point is that even if every secret could be revealed, every lie uncovered—that wouldn't mean that Utopia was at hand."

There was another period of silence. Dawson was staring into his coffee cup, idly swirling the contents to create a miniature caffeinated whirlpool. The others were watching him curiously.

"You know," Dawson mused, "I'm a person who lacks empathy." He looked embarrassed. "My condition—it's a form of autism called Asperger's Syndrome—makes it impossible for me to relate to people very well. It's my one great fault..."

He jumped as the coffee whirlpool spun out of control and splashed across the table. Bonner grabbed some paper napkins to wipe up the mess. Dawson set down the cup and murmured an apology before continuing.

"I hear what you're saying Kate, and I believe it. But I can't really understand that people could be like that. The way my mind works, when you see firm evidence that X is true, there's no alternative to accepting it. And unless you can prove without doubt that Y is true, there is no alternative to doubting it. What you're describing just couldn't happen in the world according to Tommy."

Nobody said anything.

Dawson stood up and refilled his coffee cup. He sat back down.

"Now, that doesn't mean that I don't feel any emotions," he continued. "I can be angry, I can have my feelings hurt. I know I can be insensitive and hurt the feelings of others, but somehow when that happens it surprises me. I expect everyone else to think the way I do, which is logically, rationally and with an open mind." He paused and his eyes wandered away into the distance.

"Sometimes I think that somehow I can even be happy, although that's a condition I find hard to relate to," he said in a wistful voice.

Nobody said anything for a while. Finally Chris spoke.

"Tommy, you're a unique, one-of-a-kind individual. You're gifted beyond our ability to imagine it. There are trade-offs and you shouldn't dwell on your shortcomings..."

Dawson laughed. "Oh, I'm not dwelling on them, not at all. Hell I'm incapable of feeling guilt. I'm just trying to explain why I can't understand what Kate said about how NTs think—that's what we Aspies call you 'normal' people, stands for Neurologically Typical. I just don't have a clue how anyone could accept the kind of irrational world-view of those cultists. It just doesn't make sense to me."

"Well, it doesn't make sense to most of us NTs, either," Bonner interjected with a tone of irritation. "We're not all whackos."

Startled, Dawson looked up from his coffee cup.

"Oh, no, I apologize," he said, patting Bonner's forearm. "But don't you see, what just happened is an example of what I'm talking about. I had no intention to hurt your feelings."

"Oh, sure, I understand," Bonner said, relaxing and trying to smile. "In that case it was me who was reacting when I shouldn't have."

There was another period of silence. At last Chris spoke.

"All this is very interesting but it doesn't get us any closer to an answer to our dilemma. I'm starting to suspect that we're dealing with an impossible concept, like an irresistible force meeting an immovable object. I've read some philosophy and always try to solve problems by the Socratic principle of dialectic reasoning, looking at any question from both sides to see which point of view is correct. In this case, I'm afraid that both sides have some pretty strong arguments.

"For one thing, I think we're being awfully cynical about our fellow human beings. Yes, there are bad people, and they do bad things. But most of us are decent, moral, and have nothing to hide. Admiral Chase mentioned something like that just before he left."

Again, nobody spoke for a long time. Dawson drained his coffee cup. Reeves reached up as if to scratch his wound but recoiled when his fingers encountered the bandages. Kate shifted uncomfortably in her chair and appeared to be examining her nails. Bonner looked questioningly from one person to another. It was Dawson who finally broke the silence.

"Well, as I said a minute ago I'm a logical guy too," he almost whispered. "And you're right, absolutely right. All the time I was looking at this from the bright side, imagining a better world, a Utopia. And all the time I was ignoring the very real problems that my discovery might create. Sure, most people probably are decent, but it really could shake our civilization, couldn't it?" He looked around imploringly.

Chris shrugged noncommittally. Kate reached out to lay her hand on Dawson's.

"Tommy, we just don't know yet. Right now we're making the mistake of assuming too much. Space-time viewing could be a positive force in the world, if it's managed correctly. I think for that to happen we'll need to develop completely new laws, regulations, ethics guidelines, international agreements—everything that impacts on this. If we could do that, perhaps space-time viewing can be a very positive thing."

Dawson stared bleakly at her for a moment. Then he looked down at her hand resting on his.

"No, I don't think so," he murmured. "It's probably too much for humanity to accept. I see that other side of the issue now. We're genetically adapted to function in what we perceive as four-dimensional space-time. It's hard-wired in the human brain, or at least, in the NT brain…"

"Well what do you think will stop it?" Chris demanded. "If you don't release your theory, someone else will discover it eventually. That's especially true now, with rumors circulating that it's possible. Everybody and their uncle will be racing to solve the mystery."

Stricken, Dawson recoiled. He withdrew his hand from beneath Kate's and regarded it as if he had never seen it before. Then he raised it to his face and covered his eyes. He bowed his head and his body began to shake slightly. The others realized that he was silently crying.

"And he believes he has no empathy or sense of guilt," Kate told the others quietly.

Faced with the realization that the fruits of his discovery might bear deadly poison, Dawson continued to mourn for nearly a minute.

At last he lowered his hand then used the back of it to wipe his eyes.

"It's the story of the apple," he said with a tone of hopelessness.

The others looked at him blankly, then Bonner inquired: "What apple is that?"

"Oh," Dawson said vaguely, "you know, the primal apple. The one that Eve picked in the Garden of Eden. The fruit of the knowledge of good and evil. The apple that got us humans into the mess we're in." His voice hardened. "And now I've brought us another damned apple…"

Stunned by the force of his statement, for a moment no one could think of anything to say. At last Kate spoke.

"The apple in that creation myth wasn't really evil," she told him gently. "It made it possible for Adam and Eve to take control of their own destiny. It was the symbol of mankind's emergence from the animal kingdom to a higher plane."

Dawson stared at her in astonishment. He broke into a grin.

"You think?"

Chapter 49

Quantum Center
Friday, 2:32 p.m. MST

"The press is here and they're thirsting for blood!"

Bonner strode into the Quantum Center with his radio glued to one ear. Chris and Kate looked up from her laptop where they'd been examining a document.

"There are video trucks from every network parked outside the gates, including the BBC, French, German and Japanese television crews and even al-Jazeera. We've got at least two hundred journos milling around out there, interviewing each other as they usually do."

"Well, we expected that," Chris said. "The question is what to do about it?"

"Yeah, well we can't shoot them all—much as I'd like to." Bonner grinned at the thought.

"No, probably not the smartest move," Kate remarked wryly. "But they're not going to get inside. This is one of the most secure facilities in America. What harm can they really do?"

"No, they're not getting in, but you wouldn't believe some of the stories they're putting out. I've been monitoring CNN, Fox and the major networks and so far I've heard about UFOs, a terrorist takeover, mad scientists run amok, a plot to overthrow the government, an outbreak of Ebola, a black hole that's going to swallow the entire Earth and a military coup by maniacs that are threatening to use nuclear weapons against Washington if they aren't allowed to take over the government."

Chris chuckled and Kate couldn't suppress a grin.

"Nothing wrong with that, is there?" Chris asked with a smirk. "At least they don't suspect anything serious."

Bonner shrugged and waved a hand aimlessly in the air.

"Well, it's going to make them all look like perfect asses," he said.

"Nobody's perfect," Chris quipped. He glanced at Kate. "Present company excepted, of course." Pursing her lips she rolled her eyes and sighed dramatically.

"How long can you keep them at bay without throwing them some red meat?" she inquired.

Bonner pulled up a chair and sat down. He pondered for a moment.

"Not very much longer. After the helicopter attack this morning and the rumors that're floating around town, the press are on the scent and they're in full bay. Jeesh, even some big name reporters are showing up. I saw that dreadful woman from CNN right outside the gate trying to interview the guy that delivers food to the cafeteria, poor bastard. She stepped right in front of his truck and forced him to stop. I had to send three patrolmen out to get him free, and then a whole gang of reporters surrounded *them* and started trying to out-shout each other with questions."

"The usual news feeding frenzy," Kate said. "When I was with the *Plain Dealer* I saw enough of that at national media events. It's one of the reasons I left so-called mainstream journalism."

"Smart girl," Chris remarked. "Paul, I suggest we let speculation continue for a while. It'll be dark soon and if we keep the lab buttoned up there's not much they can do. Meanwhile I'm helping Kate to flesh out her report and we're brainstorming how to handle this thing."

"Okay. One thing that may be in our favor—as long as that pack of wolves is baying outside our door it makes it harder for the enemy to organize another attack. But we need to get a lid on this soon. I suggest we announce a press conference for tomorrow morning. That should defuse things for a while."

"Good idea," Kate told him. "But definitely not here. See if you can arrange the conference somewhere on neutral ground. Is there a community center or something that we can use?"

Bonner thought for a minute.

"The high school has an auditorium big enough to seat several hundred. That should work."

"Perfect. Let's do it. Schedule it for 9 a.m. Chris and I will have our story ready by then, and you and Stuart can handle the presentation and questions."

"Sure, we can do that."

"I think we should keep Tommy out of the picture for now," Kate continued. "Those sharks would tear him to pieces. Anyway, we probably won't be going public with his theory quite yet. We need to come up with a relatively innocent explanation for what's happened, not veering too far from the truth but without creating panic in the streets."

Chris smiled at the way Kate expressed her intent to deceive the press.

"You know," he said, "it's really bothering me that we haven't heard anything from Washington, not a peep. It's been more than a day and a half since the agents were pulled out, and 12 hours since the helicopter attack. Now we've got the press outside *en masse* and everybody in the world knows that something's going on."

"Well, we've been cut off," Bonner observed.

"Yes, by the intelligence agencies. But what about the President, the Secretary of Defense, people in top places? Even the director of the CIA—he's a political appointee, not a professional spook. Surely they're aware that something serious is happening here. Everyone in control of our nation can't be determined to destroy Tommy's discovery…" His voice trailed off.

"No, of course not," Kate asserted. "But we don't know who to trust. We don't know what, if anything, is already in motion that will favor us. If we did, we could start to reach out to build support from above. It's actually we who've cut ourselves off because we don't know who's on our side and who's against us."

"Okay," Chris said, "that should be our prime goal then, to gain support. The first three unofficial rules of engagement are communicate, communicate, and communicate. We've got to clear the air on this one way or another, and quickly."

"Jesus, yes," Bonner breathed. "We're really between the rock and the hard place here, aren't we?" He got up and walked away, holding his radio to his ear.

Chris and Kate watched him go then turned to each other, Chris's brown eyes meeting her green ones. They were showing signs of stress and exhaustion yet each was drawing strength from the other. Chris reached out and laid his hand on one of Kate's.

"Well, lieutenant, we better get back to work."

She continued to meet his gaze for a moment. She reached up to touch his Coyote pendant.

"Remember what Eaglefeather told us when he blessed Coyote?" she asked quietly.

"Yes."

"He was binding the spirit of Coyote to protect you. He said that Coyote could safely lead you to other worlds."

Chris looked uncomfortable. He nodded. Kate continued to gaze into his eyes.

"Neither of us are superstitious," she pointed out, "and yet I can't help but feel…something. I don't know how to explain it."

Chris took his hand away from hers and reached up to touch her Thunderbird pendant.

"Yeah," he admitted. "I understand what you're saying. It's uncanny…" He dropped his hand again and sat back in his chair.

"I never mentioned it, but that night when my SEAL team took down the assassin in Tommy's house I was listening on the tactical net. I reached up and touched the Coyote pendant, just sort of unconsciously, and just then I heard an unfamiliar sound over Della's radio out there in the forest. I asked him what it was, and he told me it was coyotes howling. It was kind of spooky and stunned me for a moment, but had to be a coincidence."

Kate smiled. "There are so many mysteries like that," she said. "Remember that Eaglefeather told us there is no such thing as coincidence We all experience moments that seem impossible to explain except by invoking some extrasensory power or unknown sense. Scientists tell us there can be nothing to it. Still…"

Chris looked off into the distance for a while. After a moment he began to speak in a flat, distanced tone.

"When I was about eleven years old I went with my parents on a fishing trip in Canada, up north of Lake Superior. We stayed for ten days in a little cabin on a lake. There was a Cree Indian family living nearby and the father was our fishing guide. They had a son who was near my age, maybe a couple of years older. His name was Little Fox and he taught me some of the lore of the lakes and forest."

Chris brought his eyes back from the faraway place.

"I didn't understand it then, nor can I to this day," he continued, "but Little Fox could do things that were, well, impossible. I remember one time in particular. We were walking along the edge of a stream. There was a rocky ledge where the water had undercut the bank. Little Fox paused beside the ledge and told me quietly, 'there's a fish down there.'

"I looked at the water but there was no fish to be seen. I asked him, 'Where?' and he told me, 'Under the ledge.' I looked again, but could see no fish. The light was coming from an angle that could cast no shadow. Barring x-ray vision there was no way a fish could be detected.

"Challenging him I demanded 'How do you know?' He shrugged and had no answer. 'Can you see it?' I asked and he shook his head. 'I just know it's there,' he told me.' I didn't believe him.

"Then he said, 'Look.' He knelt down quickly. His hand darted into the water and emerged with a fish struggling in its grip. 'See, it's here,' he told me, holding it up as irrefutable evidence."

Kate was nodding as he spoke.

"It has to be something that we don't understand," Kate said.

"That's the only explanation I've been able to come up with," Chris admitted. "I saw Little Fox do other things that were equally inexplicable by normal senses. I've often thought that perhaps our ancient ancestors had some additional sense that most of us have lost, but that a few who live close to nature like Little Fox have retained."

"He could sense the fish somehow, even though it was hidden from view," Kate said. "And there wouldn't be words to describe what he did because the ability isn't recognized."

"Certainly there are no words in English," Chris agreed. "I wonder though whether ancient hunter-gatherers may have spoken of this around their campfires…"

Kate smiled.

"I've heard a lot of other stories like that," she told him. "I've often wondered why scientists so stridently deny the possibility of special senses…"

"Yeah, if lots of people have had experiences like I saw with Little Fox, you'd think there would've been some serious research on the subject." Chris sat for a moment then appeared to make a decision.

"I've experienced it myself," he said quietly, "in Afghanistan. There were times when I sensed danger and there was no way I could have known. Once we were probing a Taliban position, a patrol convoying in three Humvees. I was in the lead vehicle and suddenly I shouted for the driver to halt. The guy behind nearly rear-ended us, but it was a good thing we stopped because there was an IED hidden along the road just a few yards ahead, a bomb. It took a while to spot it, but we finally did and detonated it from a distance with rifle fire."

He paused and his eyes again went far away.

"Then my men were asking me the same questions I'd asked Little Fox so many years before. 'How did you know?' And of course I had no answer…

"It's well-known among men who have seen and survived combat," he continued, "that some individuals have an uncanny kind of luck. In dangerous situations they always seem to know how to stay safe. It tends to grow stronger with time, as if experience brings out this mysterious quality. By the end of our tour in Afghanistan nearly all my SEAL team shared some of that uncanny power, to one degree or another..." He paused, his eyes still focused in the distance. "There were others who never seemed to get it. Almost without exception they went home on stretchers or in body bags."

"I can see that in the men you brought here," Kate commented. "Look at what Shooter did with the helicopter this morning. It was absolutely surreal. Any reasonable person would say what he did was impossible."

They were silent for a moment, Chris's eyes still focused far away. After a moment he came back from his memory trip. He smiled and continued his story.

"I heard some of the men talking about it later. They said I had the 'sixth sense'. Old-timers like Perry understood it best. Damned if they didn't think I was something. And you know what? I could never explain to it myself, any more than I know how Little Fox 'saw' that fish under that ledge and caught it in his hand." He grinned uncomfortably.

Kate put her hand on his forearm and smiled back.

"I hear you, boss," she said. "I never experienced anything quite that dramatic, but there've been those little moments of, well, my Mom called it 'woman's intuition,' when thoughts passed through my mind that later appeared to be rooted in reality."

She thought for a moment before continuing. Her face formed into a little frown as she, too, plumbed the misty depths of distant memories.

"One night when I was about fifteen I woke up from a dream," she related. "I was convinced that my aunt and uncle were somehow in danger. It took me a long time to go back to sleep. In the morning we learned that they'd been killed in an auto crash during the night, hit by a drunk driver more than a thousand miles away..."

"Jesus," breathed Chris. "That's a real stretch for coincidence isn't it?"

"Yeah."

"I've experienced other things, too," Chris mused. "Often without really questioning them. For example, now that I think about it, something happened just the other day when Tommy was making his presentation over at the New Mexico Institute. I was watching the people in the crowd and somehow I focused on that Indian physicist, Singh. I shook it off, telling myself I was paranoid, racial profiling. But as it turned out he *was* up to something."

Kate's eyes lit up with insight.

"Oh, and it helps explain how you took down the two thugs in the cafe. I couldn't figure out how you were able to do that. It was like you had eyes in the back of your head, able to watch both men at the same time."

Chris thought about that for a moment.

"Well, the mirrors on the wall were some help," he said with a smirk. "But what about you? You sensed my signal and spilled your tea just at the right instant. Something to that maybe?"

Kate pondered that idea for a moment.

"You know, you told me about Tommy's special abilities, how he thought he could 'see' an object in hyperspace."

"Yes, that's true," Chris responded, "and I get where you're going with this."

"Well, maybe not quite. Obviously it would appear that Tommy has well-developed special senses—but what if there's even more to it? What if his theory of six-dimensional space-time provides an explanation? What if human minds really are capable somehow of traveling beyond the mundane, visible world, just as Eaglefeather described?"

Chris looked at her for a long moment. He shrugged.

"Could be," he admitted. "It kind of makes sense."

He sat back in his chair, stretched his arms above his head and yawned.

"It's interesting food for thought, but right now we better concentrate on how to deal with the present situation."

The screen saver had appeared on Kate's laptop. He reached over to tap a key and the Mac's desktop popped into view.

Chapter 50

Quantum Center
Friday, 4:12 p.m. MST

Chris and Kate were hunched over her laptop when a technician came by to deliver a cold snack of sandwiches and a carafe of coffee. They accepted gratefully and pushed the computer back to make room for their Spartan meal.

There had been no warnings from Dawson's space-time views but as night approached they planned to keep up their guard. Chris's SEALs were again taking turns on the roof, manning the remaining 25mm gun which had been field stripped and thoroughly cleaned and lubricated for the second time. Tuttle had examined each cartridge and discarded any that looked suspicious. Martin Sharp was sticking with his Barrett fifty.

Reports from the front gate indicated that the crowd of journalists was beginning to disperse, apparently satisfied that no news would be forthcoming until the announced press conference in the morning. The center was quiet now. Crime scene tape had been removed and the floor scrubbed. Some of the staffers were already taking turns catching some sleep.

Chris took a bite from a turkey and Swiss on rye and followed it with a sip of black coffee. He looked around appraisingly.

"I think your reports are great," he told Kate. "Not perfect, but certainly good enough. We could polish the words forever and it wouldn't make very much difference."

She nodded agreement. "I'd like to edit the video, but we just don't have time. When it comes to the press, we've got to keep away from the sensitive stuff. We can't really release much if anything about these details yet."

Chris grinned. "Yeah, like the space-time viewing, the fact that somebody is trying to kill us, the Iranian infiltrator, murders and intrigue. All that. So what are we gonna tell them tomorrow, that we've been having a costume party here?"

Kate chuckled and took a bite of her roast beef sandwich. She chewed slowly as she thought about how to answer. Swallowing, she met Chris's eyes and shrugged.

"You know, I haven't the least idea what we can tell them."

Chris looked uncomfortable. He said nothing.

"The real way to resolve this is to kick the problem upstairs," Kate mused. "Like you told Paul a while ago, we really need to try to open channels of communication with the powers that be."

"Yeah." Chris took another bite of his sandwich and laid it down. "I wish we hadn't lost the Admiral. I felt he was someone we could trust. Maybe he's out there trying to help…"

"If only there was a safe way to contact him."

Chris looked up sharply.

"You know, there may be."

"What do you mean?"

"When he left, he handed me his sat phone. Said he was going to report it lost. He mentioned that there might be some numbers in the phone's address book that would be useful."

"Oh, crap, you never told me that!"

"Well, I've had a few things on my mind…"

Kate patted him on the shoulder encouragingly.

"Yeah, I'd say we all have. But that could be just what we need. Let's see the phone."

Chris reached in his pocket, pulled it out and handed it to her. She examined it closely.

"It's encrypted to military specs," she said. "Let's see what's in the address book." She began to scroll through the menu. "Here it is." For a long moment she studied the screen as names and numbers rolled across it. She frowned.

"The names are coded," she murmured. "Like this one, 'Chicken Dinner.' What the Hell could that mean?"

Chris shrugged. "Got me. Maybe that's the number for Colonel Sanders."

Kate punched him gently on the arm. "Cut it out."

Chris affected a pained look and rubbed his bicep as if it had been injured.

"Okay, so it's not that. But there must be some connection. For example, maybe the words are code for a name that starts with the same letters... Charles Dempster, Carl Downey, something like that."

Kate frowned more deeply. "If that's the case, we'll never figure out any of these. It's coded so that only Chase knows what they mean."

Chris reached over to take the phone from her and began to scroll through the list.

"Could they be anagrams?"

Kate shook her head. "Try making a name out of 'Chicken Dinner'," she said. "Even if we could, what would it mean?"

Grabbing a pen Chris wrote out the letters of the code name in block letters. Scowling and putting an index finger to the corner of his mouth he stared for a moment.

"Well, I don't think it's an anagram," he admitted. "It could be just a personal memory cue, like someone he once shared a meal with, or a nickname, something like that."

"Your guess is as good as mine. But if we can't figure out even who these people are, what good would it do us? Are you sure he told you there might be something helpful here?"

"Definitely." Chris resumed scrolling through the address book. "Some of these are really bizarre. Here's one named 'Mad Liar'. Doesn't sound like someone you could trust, but who, and why would he have it on his phone?"

"Here, let me see that." Kate grabbed the phone back and studied it for a moment.

"You know," she said hesitantly, "that one kind of stands out from the rest. You might have been right about anagrams, at least in some cases. For example, the letters in 'Mad Liar' can also spell 'Admiral.' Might be something."

Chris studied the letters, nodding.

"Well, yeah—but which admiral?" Could be his boss?"

"No, I don't think so. More likely a self-reference. Maybe his personal home number or the direct line to his office."

"Well, nothing lost, nothing gained. Dial it."

"You think?"

"Why not? This is a secure phone so caller ID is blocked...I know that from experience. You can always hang up and nobody will know the difference."

"Right." Kate selected the 'Mad Liar' entry and pressed the call button. She put the phone on speaker and they listened as it rang three times. There was a click followed by a strange electronic sound. A voice came on the line.

"Hello."

Kate pressed the mute button and turned to Chris. "I think it's his voice, but I can't be sure. Do we dare say anything?"

Chris nodded and reached for the phone. He turned off the muting.

"Good evening, Mad Liar," he said and winked at Kate. There was a moment of silence, then:

"Fisher, you son of a bitch, what took you so long?"

Chris broke into a smile.

"Been pretty busy, sir. Trying to stay alive. You know how it is."

There was some more silence then Chase sighed.

"What's been happening Commander? I can't figure out anything from the press reports. And, I have to say I'm worried. Not only didn't I hear from you, I had another agent in place and..." he paused briefly, "he hasn't checked in either."

"Yes, sir. Would that be 'she' by any chance?"

Silence, then: "What do you know Fisher?"

Chris nodded at Kate who picked up the ball.

"Reporting in now, sir," she announced. "Thanks for the gender disinformation, but it's no longer necessary to keep Lieutenant-Commander Fisher in the dark."

"You've blown your cover?" Chase sounded disappointed but not really angry.

"No sir," Kate responded and began to relate how she had used deadly force to save the others. She assured him that her cover had been protected. The Admiral listened in silence. When she was through, he cleared his throat huskily.

"Good work Lieutenant," he granted her. "Jesus, I never meant to put you in harm's way, either of you."

Chris spoke again: "We've all been in harm's way ever since this FUBAR started. We still are. Early today an air attack nearly destroyed the center and would have killed us all."

"Oh, Christ!" Chase exclaimed. "That helicopter crash was in the news, but I had no idea it was a deliberate attack. What do you know about it?"

Chris described the time viewing of the secret base in Nevada. He continued to describe everything that had occurred, including the capture of Jalaly, the attacks on the center and their responses. When he finished he asked Kate to discuss the plans they were making to document the events for the record.

Chase asked a few questions to be certain he understood the details. Finally he asked if they could provide him with the documentation. Chris and Kate looked at each other and Chris shook his head.

"It may be premature to let that information out, sir," she hedged. "What did you have in mind?"

"Are you questioning my motives, Lieutenant?" Chase's voice turned chilly.

"Not at all, sir. I'm merely following the strictest protocols to preserve the integrity of sensitive information." Kate winked at Chris. "I remember being trained to do that, sir."

Recognizing that Kate had paraphrased from his agency's operations manual, Chase was silent for a moment. When he spoke his voice had a placating tone.

"Okay, fair enough. You need confirmation. I can see that. What about you, Fisher? Where do you stand on this?"

Chris laughed without humor.

"With all due respect, sir, you're the one who Shanghaied me into this mess without a full briefing. You failed to inform me of Kate's position, which left me in the dark. You bailed out with the others the minute the word came down from on high. I think you can understand my, shall we say hesitation…sir."

There was another long pause, merely the hum and crackle of the satellite phone carrier signal.

"You're right, Lieutenant-Commander," Chase allowed. "I fucked up. I put you in a bad situation without giving you all the facts. I admit it and I apologize."

"Apology accepted sir—but it still doesn't entirely satisfy my concerns."

"All right, let me tell you what I've been doing since you last saw me."

As he had told Chris at the time, Chase reaffirmed that he had left the center when ordered because he felt he could do more good outside. He left Kate in place because to do otherwise would have compromised her cover, although he was surprised that she had not tried to make contact with him.

"Sir, we were under a lockdown here and we had to be careful," Kate interjected. "Nothing personal, sir, but like I said a moment ago we felt we needed to follow the strictest protocols."

"I understand, Lieutenant. You acted correctly.

"Thank you, sir. Where are you now?"

"I'm in the study at my house in Maryland. I've stayed away from the office since I got back yesterday, but I've been keeping one ear to the rail. I've got to say there's a lot of confusion. Someone inside the Company seems to be the source of the trouble. It was a call to my boss from the Deputy Director for Operations at Langley that got me recalled. I know a guy at the FBI and he told me the same thing happened there."

"So it didn't come from the top level at CIA," Chris observed. "The deputy is a career officer I presume?"

"Yes he is. As you probably know the director himself, former Senator Higgins, is a political appointee, not a professional spook. Whether we can trust him or not is the real question. My feeling is that he's a weak link. Probably being manipulated by the pros, or a mushroom, being kept in the dark. It's the long-time spooks who have the most to lose, particularly any that have dirt on their hands."

"Why haven't we heard anything or received instructions from higher authority?" Chris asked.

"The silence from the White House makes me think the President has been advised this is a national security issue that needs to be taken care of quietly. If so, the arrival of the press in Los Alamos is going to change that PDQ."

"As I understand it the CIA isn't even authorized to act inside the United States," Kate pointed out.

"Lieutenant, there are people within the intelligence services who act outside of the law, where and when it suits their interests. In this case, I think someone's gotten wind of Doctor Dawson's discovery. Whoever it is apparently has learned enough to realize his ass would be in the wringer if the truth about past actions starts to come out. Lord only knows what they may have to hide. Black ops, illegal renditions, off the books dealing for personal gain, probably even political assassinations." He paused for a moment. "Dallas, Los Angeles, Memphis, catch my drift?"

"Yes sir," Kate said. "The Kennedy brothers and King. There've always been suspicions about that."

"We'd never have known the truth," Chris mused, "but Dawson's discovery could make it all crystal clear. And reveal so much more. But what about the other intelligence agencies?"

"Nobody has dirt on their hands like the Company does." Chase spat out the words. "NSA just handles signals. We at Navy Intelligence are information gatherers, monitors, not an action agency except in unusual circumstances. Same with Army and Air Force intelligence."

"What about the Bureau?"

"Oh, the Fibbies probably have dirt left over from Hoover's era but it's pretty outdated now. Most of the key players from that time are dead. And, I have to say the Bureau's not very competent when it comes to anything but crime investigation and SWAT ops. Most of their agents are lawyers and forensic accountants, not James Bond wanna-be's."

"So it's something to do with the Company that we really need to worry about. But obviously it can wield considerable power over the other agencies."

"Oh, yes they can. I'm convinced they have spies everywhere, right up to the Justice and State Departments and even the White House. That's the problem we face. Right now I don't dare trust anyone. Even my secretary at the office might be a CIA mole for all I know."

"Roberta?" Kate laughed. "You're not serious?"

"No, no, just an example." Chase sounded embarrassed. "It's damned unsettling to have to go around suspecting everyone."

"Yeah," Chris said, "that's the same way we've been thinking. But where do we go from here? We're pretty sure we've stopped the Iranians and we've defended the center against one attack. But they could be coming back. What do you suggest we do now?"

"Keep your heads up," Chase responded immediately. "Go ahead with the press conference tomorrow, but play your cards close to your vest. By all means don't give out the real story about time viewing. There are some people you can trust. I'm pretty sure about Bonner, he's a good guy and so is Reeves. But keep your inner circle tight or there'll be more leaks. Hell, a handful of FBI agents have already gotten a glimpse of what you're doing and we know it got back to Langley. If the press gets the story now it'll be the biggest balls-up in history."

"Okay, roger that. But what about our situation here at the center? Bonner's security force is getting pretty well stretched and my little SEAL team is too."

"I understand. I'm taking personal initiative and I've arranged for a Marine unit to reinforce you. It'll be Company B of the Marine Anti-Terrorism Battalion, based in Amarillo, Texas. I already put them on standby earlier today. They'll arrive tomorrow morning."

Chris and Kate knew the unit of more than a hundred Gyrenes would come with heavy weapons and armored vehicles. Air support by an Apache attack helicopter unit would be on call as needed.

"Thank you, sir. That'll be a big help. Now, about the report you requested…"

"Yes?"

"Can we be assured it will be transmitted safely? I'm not suggesting you're going to leak it, just concerned it could be picked up by NSA and go straight to Langley."

"Yes, I understand but I really need to know everything you've got. I'd like to try to get it into the hands of the President. The sat phone you have is encrypted to very high standards. It would take months or even years for NSA to crack it. I have the same level of encryption at my end. You can squirt the files through the ether without any worries."

Chris looked at Kate and she nodded.

"Okay, we'll do that, sir," she agreed. "Give me some time to organize and format the files. I'll be calling 'Mad Liar' in about an hour. Have your phone ready to connect to a secure hard drive."

"Thank you Lieutenant. Talk to you then."

"Just a second," Chris spoke up. "Just curious sir: Who is 'Chicken Dinner'?"

Chase chuckled. "That's the number for Colonel Sanders," he said, and paused for effect. "Colonel Blake Sanders, Royal Marines. One of our British military spook friends." He clicked off as Chris and Kate broke out laughing.

He reached over and took her hand in his with an affectionate squeeze. She responded by patting him on the cheek then looked around. Seeing that nobody was paying attention she raised her face to his and gave him a quick kiss.

"We're going to make it through this," she breathed.

Chris returned the kiss, gave her hand another squeeze and smiled.

"Bring 'em on," he said.

"The CIA?"

"Actually, I was thinking of the press corps tomorrow," he said with a shudder. "They scare me more than anything. Maybe we should take a couple of the SEALs with us when we meet the press. Mouse would be a good choice, don't you think?"

They both laughed and turned their attention back to the laptop.

Chapter 51

Quantum Center
Friday, 5:32 p.m. MST

As Kate arranged to transmit her files to Admiral Chase, Chris and Bonner retired to a private corner to consider their position. For the first time since the murders took place in the quantum center, the two had a chance to discuss what had happened and analyze the situation.

Bonner briefed Chris about on-going investigations. He reported no change in the status of the two men who had attempted to kidnap Chris and Kate. The man Chris had struck with a plate was still in a coma and the second was refusing to talk. Attempts to ID them so far failed. No fingerprints for either man had been found at IAFIS, the FBI's automated fingerprint identification system. Further inquiries to Interpol were awaiting a reply.

"Of course, now that the Bureau's cut us off, there's been no further cooperation," Bonner noted with a scowl. "So that's a dead end for now."

He went on to discuss the investigation of Jalaly, the Iranian spy who had been captured, then killed just as they began to interrogate him. There was firm proof from the retinal scan that it was Jalaly who had entered the center and killed two guards. He had admitted as much during his interrupted questioning.

"We've guessed that Jalaly was Eminence Grise, either that or his handler who was killed when Jalaly was captured," Bonner told Chris. "We know from the text messages we decoded that Eminence Grise was behind the attack on Tommy, so it fits."

Chris demurred. "It's possible, but we don't know that for a fact. The attempt on Tommy could have been unrelated."

"We know there's a CIA connection and we presume they got word about Tommy's ability to view the past," Bonner responded. "That precipitated the pull-out of FBI support and Admiral Chase. The attack on the center came shortly after that. My take is that someone there was monitoring the situation and panic set in when they realized their fabric of lies could come unraveled."

"We don't even know if it was the CIA that attacked us."

Bonner sat back in his chair and crossed his arms.

"Frankly, absolute proof is thin on the ground in all of this," he admitted. "We know the Black Hawk came from a secret base within Area 51, and that it appeared to be separate from the military presence there. As far as seeing guys wearing jackets with CIA printed on the back, no way. Covert CIA personnel never openly identify themselves with the Company."

"Yeah, some old-time Navy veteran once told me that during the Laos operation some of them carried business cards for a supposed religious organization."

Bonner chuckled. "Yeah, I heard that too. It was called Christian Independent Aid or something like that. Interesting first letters. But that's just one example. The Company guys that worked in Laos and 'Nam also pretended to be affiliated with the Army, Navy, Marine Corps, Air Force, even the Coast Guard. And of course there were the CIA-owned front organizations such as Air America and Southern Air Transport."

"I remember you said yesterday that if they attacked us it would probably be through a front like that. Of course, they had no idea we could track that copter back to its source. That's a valuable piece of information, thanks to Tommy."

"Yes, it is. If the past viewing system were more advanced, and if we had time, we could learn a lot more. For example, the Hummer attack. The so-called terrorists that showed up outside our gate were probably merely supposed to fire RPGs into the guardhouse then skedaddle. Their job was just to create chaos and set us up. All we know so far is that they came off of Interstate 25 coming from the south. Given enough time, Tommy could trace them to the source, but we haven't made that a priority because it probably wouldn't lead us anywhere."

Bonner paused, opened a folder and riffled through some reports.

"We've identified two sources of threat to the center. We think we've taken out the Iranian threat with the death of Jalaly and his handler. Whoever is behind the helicopter attack remains the viable danger."

Chris glanced at the security director and shook his head.

"What?"

"I think we're missing something. We need to go over the whole thing again."

They started from the beginning, passing reports back and forth and sharing observations. After half an hour they sat back and looked at each other. Chris raised one eyebrow inquiringly but said nothing.

"You were right," Bonner said after a while. "Our suppositions have more holes than that shot-up Hummer. There are a number of anomalies here."

"Yeah. Let's make a list. Sometimes it helps to get things down on paper." Chris pulled a yellow legal pad in front of them and took a pen out of his shirt pocket.

He wrote:

> **Rahman Singh**
> —Observed secretly recording Dawson.
> —Probably acted under duress; family held hostage.
> —Presumed source of pass codes used by Jalaly. Strong connection.
> —Disappeared; whereabouts unknown; possibly (probably?) dead.

—Other connections unknown.

"So we do have a positive link between Singh and the Iranians," Chris pointed out. He began to write again:

New Orleans Connection
—Unknown individual held Singh family hostage.
—Did not kill victims.
—No physical evidence to identify this person.
—Direct connection with Iranians through Jalaly.
—Indicates organized effort by Iranians.

"That's what we call an 'unsub,' an unknown subject. A blind alley if I ever saw one," Bonner commented.

Chris nodded and began to write again:

Harvey Pearson / Sanjar Jalaly / Jim Rowland
—Long-time secret agent working for Iran. Freely admitted same during interrogation.
—Dual legends with Jim Rowland as second cover.
—Penetrated Quantum Center. Proven through iris scan recognition.
—Killed two guards. This was "witnessed" through time view.
—Admitted working for Iranian secret intelligence agency.
—Unidentified handler caught and killed at remote cabin.

Chris laid down his pen and sat back. He poured some coffee from an insulated carafe and took a sip.

"We have a pretty good idea about what Jalaly did, and we can connect him with Singh because of the passcodes. He's also connected with the handler, but we can't connect the Iranians to Grise."

Bonner sighed. "That's so often the way it is when you're trying to put together a case. What looks obvious at first can't be proven. It's the bane of the Bureau and law enforcement in general."

Chris wrote again:

Presumed Iranian Agent
—Unknown subject killed during capture of Jalaly.
—Identified by Jalaly as his contact or "handler".
—Jalaly presumed him to be an Iranian secret agent.
—No identifying papers or marks found on body.
—Possibly Grise? No direct connection.
—Possibly unsub who intimidated and recruited Singh.

Bonner studied the pad. He grunted with impatience. "Lots of nothing there," he remarked. "Jalaly did suggest he was the one who put the squeeze on Singh, but we may never know for sure."

"Okay, next we have the CIA." Chris flipped the pad over to a fresh page, picked up the pen and began to write again:

Central Intelligence Agency
—Suspected to be behind attacks on Quantum Center Speculation, but backed by knowledge of secret base at Area 51.
—DDO ordered FBI and Navy Intel agents out of Los Alamos. Verified by Chase.
—Further involvement? Unknown. Speculation.
—Probable infiltration of LANL. Possible connection with assassination of Jalaly and attempted killing of Dawson.

Chris stopped writing, thought for a moment, frowned then laid down the pen.

Bonner slid the pad over in front of him and read the latest entry.

"That's pretty thin beer," he commented. "We really don't have anything to definitely pin on the Company, do we?"

"No. All we know is that the Deputy Director ordered the other intelligence agencies to pull out. He could have been acting in good faith or under orders from somewhere else for all we know."

Bonner rubbed his forehead with the palm of his hand.

"What else do we have?"

"Okay, there are some other links in this chain." Chris began to write again:

"Red Falcon" Assassin
—Attempted to kill Dawson.
—Fingerprints eradicated, no ID.
—Probable professional assassin.
—Origin unknown.
—Phone intercept indicates working under orders from "Eminence Grise".

Again Chris stopped and studied the list.

"As we saw, we've assumed a connection between Red Falcon and the Iranians but we don't know that." He added two more lines under the Red Falcon listing:

—No demonstrated connection to Iranians.
—Possible CIA connection? No evidence of that.

"And that leads us to another subject." He continued to write:

"Eminence Grise"
—Code name for unknown individual.
—Recently in Los Alamos area, per phone intercept.
—Known connection with Red Falcon.
—Ordered attack on Dawson per phone intercept.
—Speculation: Possibly Jalaly's handler, which makes an Iranian connection for both Grise and Red Falcon. No evidence of that. Speculation.
—Possible CIA connection? Speculation.
—No further information.

Chris stopped writing, laid down his pen and sat back, sipping coffee. Bonner looked at the pad, running his finger down the items Chris had written.

"The more we look at this, the less it turns out we know," he said glumly. "What else is there?"

Chris didn't reply. He flipped the pad over to a fresh page and wrote:

Kidnappers
—Two perps attempted to kidnap C. Fisher
  and K. Elliott.
—Unable to ID. One man in coma, second man
  refuses to talk.
—Subjective evaluation by Fisher: individuals
  appeared to be common thugs. Mafia hit men?
—No known connection with Iranians.
—No known connection with "Eminence Grise".
—No known connection with CIA.

He laid the pen down and took another sip of coffee, then looked inquiringly at Bonner.

"Shit!" The security director poured coffee for himself and they sat gazing at the yellow pad. "That's what we've got: shit," he added after a moment.

"There's more." Chris leaned forward and began to write:

Jalaly Assassin
—LANL security patrolman volunteered
  to guard Jalaly.
—Killed Jalaly after Iranian agreed to talk.
—Also killed second security guard.

Chris looked over at Bonner.
"What have you learned about this guy?"

Bonner looked depressed. He shuffled the papers and pulled out a document. He put the others together and replaced them in the file folder.

"I have to say I'm puzzled about that one," he admitted. "He was definitely a ringer, but how he got inside my organization is beyond me. Supposedly, his name was Joe Ricardo and he was an ex-cop from Philly. His jacket shows he came on board just a few weeks ago. But I checked and the personnel office with the Philly cops had no record of him, which is bizarre because we always run a full security check on incoming employees. This is disturbing because it indicates I've got someone inside my organization who's fudging records."

"A bit more than fudging, I would say."

"Well, yes, but please don't rub it in."

Chris added some more lines:

—Fake ID, origin unknown.
—Ricardo name probably alias.
—Recent hire.
—Possible inside involvement by security staff.
—Motive for killing Jalaly unknown.
—Expressed intention to kill Dawson.
—Assumed intention to kill Fisher, Bonner, Reeves and Elliott.
—No known connection with any other actor.

"Okay, once again there's not much there." Bonner wiped his forehead. "What else do we have?"

"Here's another." Chris turned the page and continued to write:

Attack Helicopter
—Suspected CIA connection; unconfirmed.
—Originated in Area 51, a highly secret facility.
—Intended to attack with deadly force, Hellfire missiles, with intention to destroy Quantum Center.

He drew a line beneath that entry and made another one:

Humvee RPG Attackers
—Attacked front gate of Los Alamos National Lab.
—Apparent coordination with Black Hawk helicopter
   attack. Connection highly likely but unproven.
—Probably intended as a diversion. Again
   likely but speculation.
—No identification of bodies.

"We're running out of perps, but we can add a couple more." Chris continued to write:

Inside Connection for Jalaly Assassin
—Presumed inside help for "Ricardo" alias.
—Individual(s) had access to personnel files and
   ability to alter content.
—Possibly someone tasked with security
   checks. Speculation.
—Probably requires fairly high-level position
   in LANL security.

"That hurts," Bonner said, scowling. "But I have to believe it's true. Any more?"

"Here's another." Chris wrote again:

Inside Informant
—Unknown party inside project informed outside
   party of time viewing discovery. Speculation.
—Could have been a project employee, FBI, security
   officer, scientist, engineer or technician;
   all these witnessed past viewing.
—No evidence of connection with any actors
   other than CIA.

"There's just one more individual we can put on the list," Chris said. He wrote:

Deputy Director Operations, CIA
—Per Adm. Chase, known to have been source of order to pull-back FBI and Naval presence at center.
—May have inside informant.
—Career intelligence agent. May have secrets to hide or be working with those who do. Speculation.
—May be controlling the DCI, a political appointee. Speculation.
—Also possible DDO is himself being manipulated, blackmailed, deceived or otherwise indirectly involved.

Chris read what he had written, then added one final entry:

—DDO could also be taking orders from higher up, perhaps from highest levels.

"That's all I can think of," Chris said, laying down the pen. "Not a lot there, but at least it tells us some of what we don't know."

"Lot of help that is."

Chris shrugged. "Yeah, but knowing what we don't know could be the first step to figuring out what we're really dealing with."

Bonner grunted in disgust. He poured some more coffee.

"Damn, at this rate I'll be a caffeine addict like Tommy," he muttered.

Chris had written on three pages of the yellow pad. He tore out the sheets and walked over to a nearby copy machine. He made three sets of copies, came back and handed one to Bonner, setting aside another for Kate. He folded and tucked the third copy into the inside pocket of his sports coat. He spread out the yellow originals on the desktop, leaned back and closed his eyes for a moment.

Bonner stared bleakly at the pages for a while before speaking.

"Since we have no evidence at all to tie Eminence Grise and Red Falcon to either the Iranians or any other connection, we need to reconsider that. If Jalaly was Grise, as we once suspected, we've found nothing to indicate it. For example, no encrypted phone such as Grise used."

"Right," Chris agreed. "But you know the old maxim, absence of evidence is not evidence of absence."

"Yeah, I know. They used to beat that into our heads all the time at the Bureau. The way Jalaly was assassinated raises questions, such as why and who ordered the hit. If it was Grise, that means Jalaly definitely wasn't our dark eminence. But we can always speculate. Let's run through some scenarios."

Chris picked up the yellow sheets and scanned them again.

"All right, here's a possibility. Let's assume that Jalaly and his handler had no connection with Grise or Red Falcon. That's step one. Let's further assume that they weren't connected with the Company either."

"They were engaged through a third party, perhaps the Iranian secret service or another party."

"Exactly. Now let's assume that the thugs that tried to kidnap Kate and me were equally not associated either with Grise, Jalaly, or the Company."

"Jesus! Now you're proposing a fourth entity." Bonner grimaced and swiped his forehead.

"These are just suppositions, but we have no evidence to the contrary so as speculation they're perfectly valid. This is dialectic reasoning, dissecting the facts to find the truth."

"Oh, yeah, that Plato thing you mentioned the other day. Looking at every side of the question, I believe you said."

"Right. Now let's carry it even further: Who was the Jalaly Assassin? We don't know of any connection there, either. He could be coming at us from a fifth entity."

"Christ, Fisher, stop it before you make my head hurt." Bonner stared glumly at the pages in front of him.

"Now here's a thought," Chris said, sitting back in his chair and looking into the distance. "We've been assuming that the Jalaly Assassin was tasked to kill the Iranian, to keep him from talking. What if that wasn't his real assignment at all?"

"What do you mean?"

"Maybe he was there to kill Tommy, and us. To finish the job Red Falcon failed to complete. Jalaly just happened to be his entré to get to us, and killing him created a convenient cover for his actions, pointing suspicion at the Iranians."

"Oh, shit, that makes sense. We know that Grise wanted Tommy dead, so that may be a connection. Jalaly may have only been collateral damage."

"Yeah. And if that's the case, it could also mean a connection between Grise and the inside connection."

Both men thought for a moment.

"Okay, let's go with that. What about the kidnappers? If they weren't connected to the known perps, who were they working for?"

Chris thought for a moment. He doodled on the blank page of the yellow pad. He drew a series of boxes and labeled them with the names on the list. Then he drew lines to connect the boxes they knew had a relationship. He drew a solid line between Grise and Red Falcon. He connected Singh and the New Orleans Unsub with the boxes for Jalaly and the unidentified handler. A third line was drawn between the CIA and its deputy director. He drew a final line between the "Jalaly Assassin" and "Inside Connection" boxes.

"Those are connections we know for sure," he said. "Now here are a few more connections that we can only guess at."

He drew dotted lines between the CIA, DDO, and Black Hawk helicopter boxes, paused for a moment then added dots between those boxes and the one for the Hummer attackers.

"The reason for suspecting those connections is that only a major agency would have the assets to launch operations on that scale. We can't be sure of the CIA connection, but it seems fairly likely. Plus, even though it appears that the Hummer and Black Hawk attacks were coordinated, we don't really have any evidence of that other than the timing."

"Circumstantial," Bonner agreed. "Could have been coincidence and wouldn't hold up in court."

Next Chris drew dotted lines to connect the boxes for Grise, the Inside Informant and the Jalaly Assassin. That left only the "Kidnappers" box unconnected. Chris stared at the chart, lightly tapping the pad with his pen.

"There have to be more connections, but we have no basis for them."

"Crap, this means there could be several more operations going on against us." Bonner slumped despondently in his chair. He was haggard and Chris noticed that dark bags had appeared beneath his eyes.

"Maybe not. Let's make some assumptions." Chris picked up his pen and drew another, larger box. At the top he placed two question marks. Inside the box he wrote "Mastermind".

"What if we connect this hypothetical unknown party to the kidnappers, Grise and Red Falcon? It's possible that this hypothetical Mastermind might have more than one game in play." He drew more dotted lines to connect the boxes.

"Okay, now let's try to guess who the Mastermind is. It could be Grise himself, which is a possibility, but it could be someone else further up the chain."

"Ah," Bonner breathed. "Very interesting."

"And let's add in one more connection." Chris drew lines to connect the fake security officer who had killed Jalaly and the presumed "Inside Connection" to the "Mastermind" box.

"Now we've tied together most of it except for Jalaly, his handler, Singh and the New Orleans Unsub into one ball of wax— except that there's still no firm connection with the CIA." Chris pushed the pad over so that Bonner could study it. "It also raises the possibility that some of the boxes might be different glimpses of the same individual. Grise could also be the inside informant and possibly even the inside connection for the guy who shot the Iranian. Or, he could be the Mastermind."

"I like it, but it's still full of holes."

"Yeah, we still have only a speculative link between any of these and whoever organized the overt attacks. There's a lot we still don't know. For example, we need to consider the possibility that there's another foreign power involved. What about the Russians? From what I've read they're past masters at infiltrating our intelligence agencies, and when you think about it they'd likely have a presence here."

Bonner looked up sharply.

"Jesus, Fisher, we've been fighting those rats forever. The Cold War never ended, you know. Hell, the Soviets placed hundreds of moles inside our country, maybe thousands of them, starting even before the Second World War. This 'Mastermind' of yours could be someone the KGB put in place decades ago."

He sank back in his chair and swiped at his face with a damp palm. He looked exhausted and his eyes were bloodshot.

"Paul," Chris said gently, placing a hand on Bonner's forearm, "let's set this aside for now. We both need some rest. Go check on your guys then hit the sack for a few hours. I'll do the same."

The security chief nodded and stood up slowly, stiff, moving like a much older man.

"Jesus, you're right. I'm beat." He stretched and twisted his torso slowly back and forth from the waist then turned toward the entrance and walked away, giving Chris a vague wave.

Chris watched him go for a moment then took one last look at the diagram. He folded up the yellow sheets including the chart and put them in his pocket along with the copies.

Chapter 52

Los Alamos
Friday, 7:42 p.m. MST

The courier was late and Arthur "Tim" Timmons was getting nervous. Ever since learning that his encrypted cell phone messages could be decoded he'd been forced to communicate through unreliable runners and outdated methods from Cold War era. That meant significant delays and the tedious inconvenience of manually encoding and decoding each message using a one-time pad. Beyond that there was even more work for him to do, creating microdots to further conceal and protect the flow of information.

Today he'd laboriously coded his latest report, looking up each letter on the one-time pad and substituting it for the original. When finished and having double-checked everything he'd burned both his original text and the one-time code sheet. He disposed of the ashes in the toilet bowl, flushing twice. Then he undertook the tedious job of making a microscopic photograph of the encoded message.

Now Timmons was getting pissed. He'd worked under pressure to be ready when the courier was expected. Now the gofer was more than an hour late.

He got up and went to the heavily curtained front window of his house. He pulled the curtain aside an inch or two and peered out. The porch and lawn were vacant. No strange cars were parked nearby. There were no signs of any human presence except for the flickering glow of a television set in a window several houses down and across the street. Somewhere in the far distance he heard the sudden bark of a police siren, then silence.

He let the drapery fall back and walked into the kitchen. A bottle of 15-year-old single-malt Scotch sat on the island. Selecting an elegant crystal highball glass from a cabinet he poured two fingers of the amber liquor. Holding the glass to his nose he inhaled, enjoying the peaty fragrance. He sipped and let the whisky roll around in his mouth before swallowing. Sighing impatiently he carried the drink back into the living room and sat down in a leather recliner.

Timmons had known plenty of pressure during his career, but as he grew older he knew that the razor edge that is the special gift of youth was inexorably slipping away from him. More than three decades before, as a rookie agent in the jungles of Laos, he'd gone for days suffering leeches, eating native chow and sleeping with only a rubberized ground cloth on the dirt floors of Hmong huts.

Better times had come. During his long career he'd enjoyed postings in Hong Kong, Paris, Mexico City and after the fall of the USSR, for four long, dreary years in St. Petersburg.

Timmons was tired of the game, tired of the stress, tired of the danger. He was burned out and he knew it. Although still in his 50's he was ready to hang up his cloak and dagger for good.

This latest assignment was the worst. For two years he'd been stuck in the backwater of Los Alamos, far from the comforts of deluxe hotels and elegant restaurants boasting stars in the Guide Michelin. Perhaps in part because of the rough conditions of his early days as a field agent he'd developed a taste for the pleasures of the *haute monde*.

He took a bigger drink from the crystal glass and set it on the side table. He looked at the mantle clock over the fireplace. It was after eight. *Damn!*

He rose and went into the bathroom. Standing over the toilet bowl he gazed at the image of his face in the bathroom mirror. He was tall and his skin had the rich tones of Southern Europe. His face was drawn in bold strokes with dark eyes brooding beneath heavy brows. His nose was patrician but bent from a long-ago injury, a parachute jump into a remote jungle village. The jaw was decisive but flanked by jowls that were starting to sag. There were even traces of a double chin!

Disgusted, he flushed the toilet and returned to the living room, barely pausing to swoop up the bottle of Scotch as he passed through the kitchen. He went to the window and pushed the curtains aside. Still no sign of the courier. He topped up the drink and sat down, sipping and staring into some distant place.

It had been damned inconvenient being caught in the lockdown at the National Lab for the last two days. *Damn Bonner! That fat clown!* That and the need to stop using his encrypted phone. He'd been cut off, out of touch. Now the lockdown had been relieved on a rotating basis. He had to report back to the lab at 7 a.m., but for now he was free.

And where is the damned courier? Impatiently, he began to tap on the chair arm with his index and middle fingers, a drumming beat, cycling paradiddles on the padded leather.

Timmons was a hardened agent, working undercover. His real identity was something else, but for two years he had been an administrative Captain on the security staff at the National Lab, with special responsibility for security clearances and personnel. Until the last few weeks it had been the ultimate dead-end assignment. Now things had gone beyond anything he had ever experienced.

Earlier he had prepared his report, now transformed into unreadable code. Using telegraphic sentence fragments he'd written the following:

> Confirming earlier report: Quantum Center capable viewing time, both past, future. Totally unforeseen. Future viewing resulted failure of Hummer diversion, helicopter attack this a.m. All friendly personnel killed.
>
> Agent assigned to silence Center principals also failed. Iranian spy and second guard killed but Dawson, associates survived. Agent "Ricardo" dead.

"Ricardo" cover revealed. Likely suspicion of me. May need to skip. Awaiting orders. — Eminence Grise

At last Timmons heard a sound, a car engine slowing to a stop outside then shutting down. Setting aside the drink he turned off the table lamp beside his chair and stood to peer out the window. A nondescript tan car was parked on the opposite side of the street. There was a magnetic sign on the roof advertising a pizza restaurant. The driver remained inside for a moment, then the door opened and he watched as a gangly young man stepped out. The youth opened the back door of the car, pulled out a pizza box and looked around to get his bearings. Then he came straight toward Timmons's front door.

The agent sighed and got up from the chair. When the doorbell rang he was standing right inside. He opened the door. The youth was wearing baggy jeans with a hole in the right knee, a T-shirt with an ugly design promoting a rock band that Timmons had never heard of, and a cap with the logo of the pizza company. There were metal rings in the boy's left nostril and eyebrow, tattoos on his upper arms and a gold ring in his right earlobe. Grinning, he extended the pizza box but said nothing.

"How much?" Timmons asked in a flat tone, struggling to contain his revulsion.

"Twelve fifty-seven with tax," the youth told him.

Timmons handed him a ten and a five.

"Keep the change." He took the pizza box.

"Thanks, man." The courier folded the money and stuffed it casually in his pocket. Hidden in the image of Alexander Hamilton on the front of the ten-dollar bill was the microdot containing Timmons's coded message.

The courier started to turn away then glanced back and winked. Timmons almost flinched. *Outrageous! What's the Company coming to?* The agent watched as the courier sauntered back to his car, got in and drove away. Only when the car had disappeared did he close the door and carry the box to the kitchen.

Inside were a Chicago-style deep-dish pepperoni pizza with onions and black olives plus a grease-stained copy of the bill from the pizza restaurant. It took Timmons a several minutes to find the microdot stuck onto the invoice, inadvertently camouflaged by a smear of tomato paste. *Idiot!*

He picked up a slice of the pizza and took a bite. Chewing thoughtfully, he bleakly pondered the tasks that lay ahead. He would have to use an eye loupe and surgical tweezers to remove the microdot, carefully read and transcribe its message using a 100x microscope, then decode the message using the next sheet from his one-time pad. *Shit!* It would take him at least a couple of hours before he could read the instructions that had come in the guise of a pizza delivery.

He wolfed down the slice and reached for another, carrying it back to the recliner in the living room where his half-finished glass stood on the side table. Pizza tasted best with Italian red wine, but single-malt would do.

Chapter 53

Great Falls Park, Virginia
Friday, 11:33 p.m. EST

The light green 2005 Chevy Impala was moving slowly to avoid hitting one of the many whitetail deer that were the bane of night driving along the Potomac. The car headed north on Old Dominion Drive. The windows were down, night sounds dominated by the rushing sound of nearby Great Falls.

The driver slowed as he approached the turnoff to a parking area on the right. Most days there were dozens of tourists thronging to walk the scant few dozen yards to the overlook that provided a view of the falls from upriver. The Maryland state line followed the opposite bank.

Admiral Chase accelerated and continued for several hundred yards to where a service road led into the forest to the left. He turned into the narrow lane and followed it until the car was out of sight of the road. He parked, turned off the lights and shut down the engine.

He sat quietly for a full ten minutes, listening and looking. At last he opened the door. The courtesy lamp switch had been disabled with duct tape so the car remained dark. Stepping out of the car, he leaned across to grab a small brief bag that had been riding beside him on the passenger seat. He quietly closed the door. Slinging the bag by its strap over his left shoulder he walked cautiously back to the main road, paused to reconnoiter again, then crossed into the trees on the Potomac side.

He continued east to the far northern end of the long parking lot, which stretched for almost two hundred yards along the river's edge. There were no cars or signs of human presence. He crossed the parking area and walked quietly through the last of the trees to the riverbank. He was about 400 feet north of the public viewpoint.

Chase stepped to the edge of the river and looked out over the water. Here, less than a quarter mile upstream from the falls, the sound of the foaming water overwhelmed the stirring of wind in the upper leaves of hemlock, birch and maple trees.

He checked his watch. Nearly midnight. Shivering in the night wind and cool, damp air rising from the river he turned up the collar of his suit jacket and put his hands in his pockets for warmth.

Headlights appeared to the south, their beams flickering through the forest trees. The vehicle was slowing. The lights swung to the east as the car turned into the parking lot near the far end, then turned toward where Chase was waiting. He remained concealed among the trees until the car stopped.

Keeping his cover, Chase moved closer. The car was parked at an angle facing northeast. It was a canary yellow Miata convertible. *Jesus!* The driver turned off the lights then shut down the engine. Chase watched intently. As the car door opened he saw a single figure outlined against the courtesy lights. *Amateur!* The door slammed, loud even above the rumble of the falls.

Alert to the possibility that the man had been followed, Chase remained in the shadows of the forest. By the light of the rising Moon, he saw the driver was a stocky man wearing a three-piece suit. The pale light gleamed on a head of silver hair. Chase continued to wait for a minute, listening and watching. The other man looked around impatiently then walked in a slow circle around the sports car. He looked at his watch. Another minute passed, then:

"Chase, damn it, where are you?" The driver leaned inside the car and punched on the horn button. *Christ!* Chase quickly moved forward, approaching the parking lot.

"There you are," the man cried. "What are you doing, hiding from me?"

Chase kept coming, waving his arms. "Quiet!" he hissed.

Walking up to the man in the three-piece suit Chase gripped him by his upper arm and leaned close.

"Senator, please, I warned you we have to be careful." He attempted to draw him toward the trees.

Indignant, the man reached up to remove Chase's hand from his arm. He put his hands on his hips and glared haughtily.

"Admiral, I did as you said. I even borrowed my daughter's car to avoid attention."

"Yes, Senator. Nothing like a yellow sports car to keep people from noticing."

"Are you making fun of me?" The man was beginning to lose his temper.

"Frankly, yes," Chase told him. "But I apologize. This is a very delicate situation which is why I got word to you through outside channels."

The Director of Central Intelligence and former Senator from Wisconsin continued to glare. His name was Brian Higgins and he was six-two and weighed about 230 pounds, hardly more than when he played linebacker for the UW Badgers.

"All right, Admiral," he said coldly. "Apology accepted. But this is ridiculous. You're acting like this is some kind of silly game, something out of a cheap spy movie."

Chase stared for a moment, then laughed.

"Senator, I'm the deputy director of Naval Intelligence and you are the head of the CIA. What kind of game do you *think* we're engaged in?"

For an instant Higgins was at a loss, then he nodded. "Well, yes, I'll give you that. But this isn't the way we do things. Sending me that message at home by some kid pretending to be a Mormon missionary. Ridiculous! Insulting, even. Why didn't you just call me at the office like everyone else does?"

Higgins paused to take a breath before resuming his tirade.

"And then to absolutely demand that I meet you here in this God-forsaken place, at midnight no less. Ordering me, yes ordering me not to tell my staff. An outrage, sir!" He paused to take another breath. "And telling me not to use my official car, then having the audacity to make sarcastic remarks when I drive the only other car available."

He gave Chase another glare. "Well, what's so all-fired important that you need to play these childish games?"

Chase glared back. He spoke harshly. "Senator, Mister Director, I'm trying to protect your life and mine, and maybe preserve the entire nation."

Higgins recoiled.

"What?"

"As we Navy people say, we're in perilous waters, Senator. I needed to see if you were part of the problem, or could be part of the solution."

"Part of the problem? What'dya mean by that?"

Sighing, Chase led the DCI into the forest, angling southeast to reach the river at the overlook. Here the sound of the falls was a constant roar of white noise. Higgins followed without objection.

Chase approached the safety railing and beckoned the DCI to stand beside him, facing out over the river toward the Maryland side. The roar of the falls would cover any possibility of them being overheard.

"Senator," he began, "the Department of the Navy has funded a research project that will have important implications for the security of our country…"

Higgins cut him off.

"Oh, I know what you're talking about. It's those Loonies out in Los Alamos. I was briefed on that. We have it under control."

Chase gazed at him for a long moment before speaking.

"Do you?" he asked quietly.

"Yes, some of my best men are on top of the situation." Higgins began to appear slightly uncertain. "It's a real cock-up," he added. "I'm surprised the Navy allowed itself to be drawn into such a hoax."

Chase regarded him for a moment. Higgins was leaning forward with his elbows on the railing, looking out at the rushing waters.

"Maybe you better tell me about it," Chase suggested. "You first, then I'll tell you what I know."

Higgins turned away, shaking his head.

"You're not going to trick me that easily, Admiral."

"Trick you?"

"Yes, I know what you're trying to do. I was briefed, remember? I know you were part of this thing and now you're trying to cover it up. My knowledge is confidential, even from you."

Higgins remained standing with his back to Chase, his body language a clear signal of rejection.

Chase pondered. *Is this guy really that big an ass, or is he just being clever?* He shook his head despondently. He knew that Higgins had a spotless record during his twelve years in the Senate, rising to a senior position on the Senate Intelligence Committee and spearheading innovative legislation concerning national security. But Chase reminded himself that appearances are often deceiving. Now as DCI Higgins wielded a different kind of power, with a direct line to the President. Yes, Homeland Security was now part of the picture, but Central Intelligence remained a world apart.

The Company was noted for its entrenched bureaucrats and had a history of mobilizing resistance against political appointees from outside the agency. Chances are Higgins had been fed a steady diet of baloney while those who considered themselves the true masters of the nation's largest intelligence organization continued to do as they wished.

"Okay," Chase said at last. "I'll go first then. You just listen and wait 'til I'm done. Then we'll see if what you've been told matches anything I tell you."

Higgins turned back, a tight little grin on his face. "Agreed, Admiral." He resumed his position leaning on the railing.

Chase began to describe the quantum research, the goal of creating a computer capable of making immensely fast calculations. He briefly described the potential advantages, both to America and to the benefit of its enemies should they get their hands on the technology.

He carefully avoided discussing the time dimensions or the ability to view the past and future, waiting to see if Higgins would respond. He could see no indication that the DCI was aware of those facts.

"Then the killing started," Chase continued.

Startled, Higgins, turned to stare, wide-eyed.

"What killings?" He was agitated now.

If he's acting, he's really good, Chase thought, continuing to observe the DCI. It was in situations such as this that his training in psychology paid dividends.

"Murders, sir. An attempt to steal the secrets of the quantum computer in which two guards were strangled. The assassination of a prisoner and another guard by a security officer who turned out to be a ringer using a fake identity. Attempted kidnappings of two individuals associated with the project. The disappearance of another who is presumed dead."

"Jesus!" Higgins exclaimed, seeming genuinely surprised. "I…I didn't know about that…"

"Then you probably didn't know about the attack on the Quantum Center by a Black Hawk helicopter that tried to destroy it with Hellfire missiles, do you? Or that we have a pretty good idea that it was a Company operation."

"What!" Now Higgins was outraged. He stepped back from the railing and faced Chase. "That's a pretty nasty accusation, Admiral. What proof do you have?"

"We know the copter came out of a secret base in Area 51. It appears to be a civilian operation. Do you know of any other agency that might have that kind of capability?"

Higgins stuttered for a moment, shaking his head wildly.

"The Company absolutely does not have any such secret base," he declared.

Chase regarded him calmly.

"As far as you know," he informed the DCI.

"What do you mean by that?" Higgins sputtered. "I *am* the director, they tell me everything…"

"As far as you know," Chase repeated. "Do you have any inkling of what your agency has really done over the years? Have they briefed you on absolutely everything?"

"You're talking about those old lies about the Company? Nasty rumors? The discredited products of whacko conspiracy theorists and hack journalists? Yes, for your information, I've been thoroughly vetted on those idiotic examples of misinformation.

"And furthermore, as I'm sure you're quite aware, we are an intelligence gathering agency, not a paramilitary organization. And we're specifically prohibited from acting within the borders of the United States of America."

Chase sighed. He had hoped to determine whether the DCI was involved in the apparent plot within his agency to destroy the Quantum Center and murder its core personnel. If he believed Higgins was being kept out of the loop, he'd hoped to recruit him to help counter the plot. But he'd failed to consider a third possibility—that the DCI was not only being manipulated but was also dumb as a truckload of rocks.

"Okay, Senator, let's give it a rest. I suggest you keep this meeting confidential. I'm sorry I couldn't convince you. Maybe you're right. Anyway, thanks for meeting me and I'm sorry for the inconvenience."

Higgins was taken aback.

"Admiral, I'm offended. You drag me out here to listen to a cock-and-bull story, you imply that I'm out of touch and that my staff are engaged in illegal acts, even murder—and then expect to just walk away and tell me to forget about it?"

"Senator…"

"This is an outrage and it is you, sir, who are a danger to our national security. Your superiors are going to learn of this."

Higgins spun on his heel and walked quickly up the path that led to the parking lot. Chase watched him disappear from sight. In a few moments he heard the throaty roar of the Miata, distinctive even over the rumble of the waterfall. The engine surged as the DCI spun the rear wheels in some loose gravel, then settled down as the car tore back toward the main road, turned south and ran up through the gears.

As the sound faded into the distance, Chase sighed again. He reached into his shoulder bag and touched the package containing copies of the reports Kate had sent him. He had hoped to give them to Higgins, to be passed on to the President. *Idiot!*

He shivered, chilled by the damp air blowing from across the Potomac, and perhaps something else... He turned and began to walk back to his car, keeping to the cover of the trees.

Several miles away Higgins was driving southeast on Georgetown Pike, headed toward the CIA headquarters at Langley, about six miles from Great Falls Park. Agitated, he was eager to share what he had learned with his trusted deputies. They were the pros, after all. He was merely an administrator, a political figurehead for the Agency. They needed to know what he'd learned tonight. Wrapping his mind around that idea, he began to feel smug.

He approached a dark stretch of road with open land on left and right and only the Moon and the Miata's headlights to illuminate the scene. Smiling he dropped the transmission down a gear and punched the accelerator. Passing 80 mph he up-shifted and the engine whined as he hurtled down a long straightaway. The windows were open and a gush of cool air was clearing his head, wiping away the tension of the last hours since he received Chase's enigmatic request for a clandestine meeting. He broke into a smile then laughed out loud.

Behind him a cluster of red and blue flashing lights suddenly came to life. *Shit!* Higgins let up on the gas and slowed down, watching in the rear view mirror as the patrol car caught up with him. Coasting to the side of the road he stopped and set the parking brake. He reached into his coat pocket for his credentials. Because he wasn't a covert agent he carried official identification, and anyway almost everyone—even a stupid cop—would recognize his patrician face and silver hair from many appearances on Sunday morning news programs.

The car stopped behind the Miata's rear bumper. In his side mirrors Higgins watched as two men stepped out of the vehicle. The uniformed driver walked forward, stopping just behind the open window of the Miata. Higgins craned his neck to see.

"Sorry, officer," he said, holding out his ID.

"No need to apologize Senator," said a low voice. Startled, Higgins twisted his head further around. The policeman was aiming a pistol at him!

"What the hell…"

The last thing Higgins saw was a brilliant flash of light from the muzzle of the Glock.

The 9mm Parabellum hollow-point bullet entered the forehead of the Director of Central Intelligence at a slight downward angle. Fragmenting as it penetrated his skull it spattered bits of hot lead and bone fragments through his brain. Much of the metal and macerated gray matter blew out the back of his head—but by that time Higgins was far too dead to know or care.

Chapter 54

Near Ashburn, Virginia
Saturday, 1:46 a.m. EST

The house was a historic building on a remote farm that had long ago gone to seed. Former fields and pastures were grown up in scrub brush and trash trees. An old barn that may not have seen fresh paint for half a century was leaning precariously to one side, held up by sagging baulks of timber. Built in 1843, the main house's exterior clearly showed its age. Inside was something else.

A clandestine group within the Central Intelligence Agency had purchased the country estate through a front some thirty-five years previously. Carefully, so as to avoid drawing attention, the building was renovated to provide comfort and security. A full-time staff maintained the safe house that had been expanded by the construction of a large underground bunker featuring quarters for six permanent staff. A cook and housekeeper shared the facility with four armed agents. Everything was off-the-books, black ops money, known only to a few members of an inside clique. In fact, the safe house appeared on no records at Langley. It had been made to disappear from all but a secret clique within the agency.

On the top floor were five comfortable bedrooms. The ground floor featured a spacious reception hall dominated by a seven-foot grandfather's clock. French doors led into a dining room furnished with an elegant Chippendale table and twelve chairs. To the left lay a large living room with native stone fireplace and rough-hewn beamed ceiling. In the rear was a comfortable eat-in country kitchen with modern appliances but the rustic look of the past.

In the left rear corner of the country mansion was a very private office featuring an antique walnut desk with swivel chair, three conference chairs, a sitting area with couch and two easy chairs, and a bookcase wall flanking a 52-inch flat panel television.

This was one of Jamie Delancy's favorite places. Located less than 45 minutes from his house in McLean and the nearby CIA headquarters at Langley, he considered it his own private fiefdom. In the bad old days of the Cold War the safe house had been used to warehouse "politicals," key individuals from the Soviet Bloc who had claimed asylum and were willing to trade information for their safety in the West.

Now the clandestine facility had been usurped as a private meeting place and control center for what Delancy liked to call "sensitive ops." In fact, everything that took place here was unauthorized and usually illegal.

In the early morning hours when even the more sensible owls are asleep, Delancy was sitting behind the desk. He was very erect. His face was suffused with anger.

"Could you have missed something?" he asked for the fifth time.

The man sitting across from him shifted nervously in his chair. He was wearing a military-style shirt that might be mistaken for part of a policeman's uniform. He looked appealingly at Delancy and shook his head. A second man stood to the side, gazing at his shoes. He was similarly dressed.

"Damn it!" Delancy pounded the desk with his fist. A pen set jumped in the air and loose papers shifted away from the point of impact as if trying to scuttle to safety.

"Sir, he didn't have anything. We thoroughly searched him and the car. Nothing. We even stripped off his clothes before we disposed of the body. There was nothing in them, nothing. Cell phone, Blackberry, you've seen them. Clean."

"Christ, I can't believe this. You're telling me you killed the Director with nothing to show for it?"

The faux cop sank back in his chair. His expression hardened. The second man looked away.

"We followed directions, sir." He spoke in a flat, toneless voice.

Delancy relaxed and nodded. "Yes, you did," he admitted. "What I don't get is why? Why didn't he have the documents?"

"Sir, that's way above my pay grade." The second man nodded his agreement.

"Yes, yes, of course. You may go. And take off those damned shirts. You got rid of the magnetic light bar of course?"

"Yes sir."

"Good. And the weapon?"

"Wiped clean and deep-sixed in the Potomac."

"All right. Sorry I took off on you."

"That's okay sir." The men left the room, closing the door carefully behind them.

Delancy rubbed the tense cords in his neck. *How did this happen? It was so clear, so plain.* He stood and walked to a drinks cabinet. Opening the lid he pulled out a two-ounce whisky glass and poured a shot of Jameson's Irish. He raised it to his lips, took a sip. He briefly rolled the smoky liquor around his tongue then tipped his head back and emptied the glass at one go. He filled it again and sank down on the leather couch.

Just hours earlier Delancy had been elated in the knowledge that the end was in sight. The bugs his group had planted in the Director's house had finally yielded pay dirt. A video recording made earlier that night had given him the opportunity he needed. There was to be a document transfer and the DCI would deliver them right into Delancy's hands. It was too bad the old fool would have to die. *Shit! Why didn't Higgins have the file?*

He picked up a remote control and activated the television. He scrolled through a menu and pressed a button to start a video sequence. It showed the family living room of the Higgins mansion. A young woman, Higgins's 17-year-old daughter, was sitting in an easy chair watching a sitcom. The door opened and the Director came into the room. He looked around then moved some magazines from a chair and sat down opposite his daughter. She looked up.

"Hi Daddy."

"Hello darling."

She turned back to the TV screen. A group of young people with strange hair styles were sitting on a couch insulting each other. Canned laughter punctuated each lame remark.

"Honey?"

"Yes Daddy?"

"I'd like to ask you a favor. I need to use your car tonight."

That drew her attention.

"No, you can't," she said. "I'm going to meet some friends later."

"I have to," Higgins said.

"Daddy!" the teen implored. "You have a perfectly good Lincoln Town Car. Why would you need my little Miata?"

"Can't tell you, sweetheart."

He really had her interest now. She muted the TV and focused her full attention on him.

"Is it something to do with the secret stuff?" she asked. She was intrigued by her father's new job as head of the Central Intelligence Agency. "I bet you're meeting a spy, aren't you?"

Higgins flinched and she noticed his discomfort.

"It is, isn't it? You're finally taking your job for real, getting down with the spooks, right?"

"Certainly not." Higgins squirmed in his chair and didn't meet her eyes.

"Okay, tell me what's going on and you can use my car," she said, bargaining like a rug merchant. "This is so cool."

"No, I can't tell you."

A gleam of inspiration came into her eyes.

"It's something to do with that Mormon boy who came to the door right after we finished dinner," she concluded. "I thought he looked fishy. Wondered why you even talked with him." Higgins flinched again.

"No, no, nothing like that," he denied, still refusing to meet her eyes. She knew she had him.

"Puh-leeze, Daddy, just tell me what you're doing. I'll never tell anyone, not even my friend Mandy."

"Honey, you're making things very difficult for me. I need to use your car, that's all. Please don't resist."

"Oh," she crowed. "Now you sound like one of those Borgs on Star Trek. 'Resistance is futile.' I love it. Come on, Daddy, spill."

Higgins shifted again in the chair.

"Well, all right, but you have to swear you'll tell no one."

"Of course not. You know me."

"Unfortunately, I do, but this time I mean it. You cannot breathe a word about my official duties. Ever. Is that understood?"

"Okay, cross my heart and hope to die." She made a zipping motion across her lips.

"You're going to be disappointed," Higgins said. "It's just routine. I have to meet a man from another government agency. He let me know he has some information for me."

"Oh, that sounds exciting," the girl squealed. "Where?"

"Up at Great Falls Park around midnight," Higgins blurted, then looked as if he wished he hadn't mentioned those details.

"Oh, Daddy, can I go with you? Now I see why you want my Miata, in case you get into one of those big car chase things. Right? I'll be your driver; I'm really good. Please?"

"Absolutely not." On the big screen Delancy watched as Higgins gazed at her speculatively, no doubt wondering what unreported adventures had taken place in the little yellow car.

Her face fell and she seemed to lose interest. She glanced at the muted sitcom, then back to Higgins.

"Who's this guy you're meeting?"

"Just somebody from the Navy, honey. It's nothing important, really it isn't."

"It's just like one of those Spy vs. Spy stories from *Mad Magazine*, right?" She'd lost her enthusiasm and reverted to the normal state of sarcasm. "Which one are you, the black one or the white one?"

"That's quite enough young lady," Higgins admonished her. "You aren't getting anything more out of me. Now give me the keys."

The girl raised the remote and turned the sound back on. She looked at the TV.

"They're in the kitchen, Daddy. On the counter next to the door to the garage. Get them yourself."

Sighing, Higgins rose and left the room.

Delancy turned off the video recording and stared at the blank screen. They'd gotten almost the full story, right there. It hadn't been hard to connect the dots. Clearly, Higgins had been asked to meet someone from Naval Intelligence. It was highly clandestine. The only issue involving the Navy was the quantum project, and that meant it was Admiral Chase. Had to be. Attempting to contact the Director could mean only one thing—Chase intended to pass over secret documents about events at the quantum time project. Probably wanted Higgins to take them to the Oval Office. It was as plain as day.

And that would have been the end. This time-viewing trick would have uncovered everything, a massive organization that had taken decades to build. It would all have come toppling down. And there, thanks to a miniature camera concealed behind an air duct in the Director's house, was the key to preventing that disaster.

And yet... And yet when his agents had searched Higgins's body and the car there was no sign of anything. They took apart his clothing, every stitch. The Miata had been searched from one end to the other. Nothing.

Delancy bent forward with elbows on knees and covered his face with the palms of his hands. *Shit! Shit! Shit!*

He got up and refilled the neat whisky glass. Holding it up in an imaginary toast to the empty room, he intoned: "Point to you, Admiral." He downed the shot, set the glass down and reached for the phone.

Chapter 55

Los Alamos
Saturday, 2:20 a.m. MST

Tim Timmons, the spy styling himself as Eminence Grise, had completed the tedious task of decoding the message brought by the courier posing as a pizza deliveryman. Following strict procedure, as soon as he was finished he tore off the top sheet of the one-time pad and carried it into the bathroom. Holding the sheet over the toilet bowl he ignited one corner with a Zippo lighter. He turned the sheet slowly from side to side, letting the flames consume it. As the edge of fire neared his fingertips he dropped the sheet into the bowl and flushed twice to make sure the code sheet was entirely destroyed.

Satisfied, he returned to the home office in the house's small third bedroom. He gathered up the code pad and other tools of spycraft and locked them away in a floor safe hidden beneath a Navajo rug.

Only then did he return to his desk chair and read the message a second time. His mind was churning.

Timmons began to review his options. Like any smart actor in the world of black ops—and no other kind lasted for long—he had made provisions for his future. In safe deposit boxes scattered around the world were several sets of credentials for different "legends," each complete with passport, driver's license, credit cards, even copies of birth certificates. Each set of documents, while flawless, was phony as North Korean dollars. His employer had provided many such legends, but these were "unofficial," front identities he had set up personally, unknown to his superiors.

Now he faced the possibility of having to take his retirement a bit early, and off-the-books.

There was money, too, cash in safe deposit boxes or numbered accounts in tax havens across Alpine Europe and the Caribbean. Not really enough, though, especially if he had to go dark and forego his pension. He'd managed to sequester only a few hundred thousand dollars in his "rainy day" emergency fund. Always Timmons had waited for the chance to make a big haul, and always his taste for expensive things and high living had drained his resources.

No, it would be not nearly enough, not if he planned to live the lifestyle to which he had become addicted.

He stood up and walked back into the living room, carrying the decoded message with him. He poured another two fingers of single-malt and sank into the leather easy chair. Sipping, he held the sheet of paper up and read the words once more, than laid it on his lap and stared across the room.

Timmons had done many things during his career with the spy agency, some of which he was not particularly proud. Even now, in what was supposed to be an easy assignment, a reward for years of wet work, he'd personally ordered the professional killer code-named Red Falcon to attempt to assassinate Tommy Dawson. More than that, he'd provided information for the Hummer and Blackhawk attacks on the Center. He'd infiltrated "Ricardo," the fake security officer, into the Quantum Center with orders to kill the central players in the project. All that he'd done for his masters in Langley, just during the last few days.

And he had done one other thing, supposedly as a bluff, a bargaining chip only, never to be actually used.

And now this? It's outrageous. How could they possibly expect me to do something so...

He had to pause to think of the right word. The one that came to him was "monstrous." *Yes, monstrous.* He rolled the word around in his mind and it fitted like a glove.

Timmons took another sip of Scotch. He got up and walked into the home office, carrying the glass with him. At one end of the desk was a rack of well-used pipes and a cylindrical teakwood humidor decorated with a carved dragon that wound around the outside. He chose an English-made flame-grained briar and gently packed it with a large pinch of aromatic tobacco from the humidor. Striking a wooden match he carefully stoked the pipe alight with short puffs.

Timmons reflected that in truth he'd done many "monstrous" things during his career. The differences were only matters of scale. Yes, he could refuse his orders, but to do so meant to live out his life as a fugitive. It wouldn't do.

He drew a deep draught of smoke and exhaled it in a stream.

No, he decided, *it definitely wouldn't do.*

He began to plan how to achieve the task he'd been assigned.

Chapter 56

Suitland, Maryland
Saturday, 2:42 a.m. EST

Admiral Chase had just laid his head down on the pillow when the phone rang. Grunting he heaved himself upright and reached for it. The caller ID was blocked. He hesitated, then shrugged and pressed the "talk" button.

"Hello."

A woman's voice came on the line. She sounded worried.

"Is this James Chase? Admiral Chase?"

"Yes."

"Oh, Admiral, this is Marilyn Higgins? Senator Higgins' wife?"

Oh shit. What's this about? Chase composed his voice.

"Why, Mrs. Higgins, whatever could cause you to call me, especially at this hour?"

She remained silent for a moment. He heard a sniffling sound.

"I'm worried about my husband," she said at last. "He told me he was going to meet you, and he never came home..." She seemed to be holding back tears.

Oh my God, the idiot named me! Chase was wide-awake now. *Got to focus.* He affected a tone of incredulity.

"Me, Mrs. Higgins? The director of the CIA told you he was going to meet me? That's ridiculous. Why ever would he want to meet with a Navy desk commander?" Chase's association with intelligence was not widely advertised.

The woman was silent again for a long moment. He could hear her breathing.

"You didn't see him?" Disappointment verging on despair.

"No, of course not. When was this supposed to have happened?"

"Midnight, at Great Falls Park? He said he needed to borrow our daughter's car. Her yellow Miata? And now he's not home. I called the office and they had no word from him." She tried unsuccessfully to suppress another sob. "He said you're the deputy director of Naval Intelligence..."

Christ! Was there nothing the idiot had held back?

"Mrs. Higgins, please, I don't know anything about this," Chase told her, desperate to quell the sudden breach in his security. "I'm sure he had something important that he couldn't tell you about, so he made up a story."

"No, he'd never do that!" The woman was outraged. "Brian always tells me everything. He never hides anything!"

Bad quality for someone in the spy business. Chase let the silence ride on for a few moments.

"Ma'am, I surely hope that doesn't extend to state secrets?"

He heard a sharp intake of breath, followed by a quick denial.

"Well, Mrs. Higgins, I'm afraid I can't help you. I'm sure he'll turn up. Probably something of a highly sensitive nature that he needed to keep under his hat."

"Brian never wears a hat..." the woman said, a desperate non-sequitur.

"Yes, ma'am," Chase said and switched off the phone.

He stared at the glowing screen until it timed out and went dark. He knew he was in trouble and needed to act. But for the moment he was stunned at the ignorance and carelessness of the Director. Had Higgins even been briefed on security protocol when he assumed the post? Probably not, Chase admitted to himself. He was being led around like a prize bull with a ring in its nose. And the stakes were high enough that mass murder had already been attempted to stop

Tommy's discovery from becoming known. Now the CIA director was missing...and he'd connected himself with Chase!

He snatched up the phone and dialed a number from memory. On the third ring a voice came on the line. It had the hard-steel quality that only military professionals seem to be able to master.

"Sir?"

"Chief, we've got trouble. Meet me in twenty minutes. I'm at my house, but come to that wooded area about a quarter mile west on the back road. Pronto."

"Aye aye." The party at the other end hung up and Chase went into action.

Stepping into his walk-in closet he hurriedly began to dress. First he pulled on a black T-shirt and topped it with a Navy-issue turtleneck. He strapped on a quick-draw shoulder holster, followed by a pair of dark gray Dockers. Next he shrugged into a black silk sports jacket. He returned to the bedroom and sat down to put on a bulky pair of dark blue socks and low-cut hiking shoes. He picked up a black Kangol driving cap and settled it onto his head.

Chase reached into the drawer beside his bed and pulled out his Glock 19 and two extra magazines loaded with 9mm Parabellum hollow-points. He safed the pistol, racked the slide to load a round into the chamber and seated the gun in the shoulder holster. He stuffed one extra magazine in each jacket pocket. The drawer also contained a black Smith & Wesson spring knife. It went into his left pants pocket. He snatched his new sat phone from its nearby charger and put it into the right-hand pocket. His keys and wallet went into the appropriate places along with a pair of compact binoculars and a small LED flashlight.

Chase took a look around, then sighed and returned to the walk-in closet. A small wall safe was hidden behind a stack of shirts. Quickly he opened it and drew out several items. First came a banded stack of hundred-dollar bills, ten grand in cash. He stuffed it into his inside jacket pocket.

Next came a small pistol in an ankle holster. The tiny Kahr automatic had a lightweight polymer frame and held seven 9mm rounds. Kneeling he strapped the holster below his left calf. He checked the magazine, snapped it back into place and shoved the pistol into the holster, shaking down his trouser cuff to cover it.

Last to come out of the safe was a thin billfold, designed to be hidden inside the waistband of a pair of trousers. Marketed for travelers to deter pickpockets, it contained a set of phony credentials including a Texas driver's license, passport, Visa and American Express cards, and several other pieces of identification. Loosening his belt he threaded the stainless retaining wire of the billfold around it and tucked the wallet inside his waistband.

Returning to the bedroom he snatched up the thin portfolio case that contained Kate Elliott's report from the quantum center. He slung it over his shoulder and checked the time. Just eight minutes had passed since he had called for a pick-up. He had twelve minutes to make it to the rendezvous.

Leaving the bedroom lights on he slipped down the hallway and descended to the dark foyer. Carefully he peered through the sidelight window next to the front door.

Nothing.

Feeling his way he passed through the house, pausing at the kitchen window to check the view into the back yard.

Again nothing.

Quietly he slipped into the mudroom between the kitchen and the garage. Leaning in the corner with some brooms and mops was a 12-gauge Mossberg stainless steel automatic shotgun. A box of double-aught shells sat on a shelf above. Working by feel in the dark he tore open the box and filled the weapon's tubular magazine to capacity. He dropped a handful of shells into his pocket. The shotgun was fitted with a sling, making it easy to carry while leaving both hands free.

Finally, he used the keypad inside the garage door to activate the house's security system. As the alarm device beeped through its countdown he checked his watch again. Ten minutes. He unlatched the door to the garage and stepped around a riding lawnmower to the back door leading into the yard. He opened it a crack and paused to look and listen. All was quiet. Swinging it further open he stepped outside, eased the door closed behind him, and crouched in the cover of a lilac bush. Again he listened and watched before moving.

Behind the house was a long, sloping yard leading to a band of forest. The residence was located in a semi-rural area with no other houses nearer than a hundred yards in any direction. Chase moved quickly through the yard and took cover among the trees. He paused to take another long, slow recon.

Somewhere in the distance he heard the squeal of tires as a car took a corner too fast. A few seconds later there was a second protest of rubber against road. Two cars, coming his way, fast. Chase melted back into the forest and waited. About a half-mile away the vehicles began to slow and he knew he was right. They were coming for him.

He drifted further into the forest, keeping the house in sight and getting a view of the road. He pulled out a pair of compact binoculars and raised them to his dark-adapted eyes.

As they passed beneath a street light, he saw two black Chevy Suburbans with blacked-out windows and oversized wheels. They were driving with the lights off. Chase knew the drill. These were specially equipped armored SWAT-style trucks, killer trucks, death on wheels. The kind of vehicles a special ops "wet work" team would choose.

"Thank you Mrs. Higgins," he whispered to himself.

The intruders slowed to a crawl. The first Suburban eased to the curb about fifty yards from his front sidewalk and stopped. The second drove slowly past and pulled over an equal distance beyond the front of his house. They sat still for a long moment, then as if on a signal all the front doors swung open and four dark-clad figures emerged. Chase saw the faint glint of light from tactical goggles, making the killers look like spacemen or aliens from some Sci-Fi movie. The figures rapidly deployed, surrounding the house.

Time to go. He turned and walked quickly back through the woods toward the distant road. *Four minutes.* He stepped out briskly, confident now that no one would be coming to his house from the rear.

Behind him he heard almost simultaneous crashes and knew that both his front and rear doors had just been breached. The siren on the roof began to scream and through the trees he could see security lights flashing as the alarm system responded to the sudden intrusion. Chase grinned. The sheriff's police would be on their way in seconds. He savored the thought of the spooks getting caught red-handed in an illegal break-and-enter. But in reality, he knew, there was no chance of that. These men, unlike their director, were pros. They would be in and out fast and would cover their trail at all costs.

Hurrying he reached the far edge of the forest and took cover behind some brush along the edge of the country road. A car was coming his way from the left, about a quarter-mile distant. He raised the binoculars and glassed the oncoming vehicle. *Yes!*

As the dark green Lincoln Town Car neared, Chase flicked the flashlight once and stepped out to the side of the road just as the vehicle pulled over and stopped in front of him. He opened the passenger door and leaned in to position the shotgun between the front seats. He grimaced at the driver as he sat down and closed the door.

"Get out of here," he said, reaching to pull down and buckle his seat belt. "Keep to the back roads."

"Aye aye," came the terse reply.

The car eased away from the verge and accelerated into the night. Chase turned to look back but in the starlight there was nothing to see but the dim shapes of trees and the pale ribbon of road.

Chapter 57

Los Alamos
Saturday, 1:14 a.m. MST

Chris Fisher had lain down on one of the folding cots but sleep would not come. He and Kate had spent several hours talking about the upcoming press conference. They were still uncertain what to do. Finally, deciding to try to grab some shut-eye, they left word for Reeves and Bonner to meet for a last-minute planning session at 6 a.m.

Although he was deeply tired, Chris couldn't stop thinking about their situation. Several times he turned on the hard cot, pounding the pillow into shape. After a half hour of that he decided to get up and check on the SEAL team.

Leonard Perry was on watch at the eastern pillbox along with Carlos de la Vega. Chris chatted with them for several minutes but they had nothing to report. As he started toward the other end of the center he noticed Floyd Garrett sleeping. The Army surplus cot was not configured for a man of Mouse's size and his feet were sticking off the end. Chris had to chuckle at the comical sight. Nearby, Jim Chen was snoring lightly.

He climbed the ladder stair to the western pillbox where Martin Sharp was stretched out with his Barrett rifle at his side, one arm around it. Joseph Tuttle was sitting on the opposite side, leaning against the sandbags with a droll smile on his face.

"I ask you sir, look at Shooter there," Tuttle remarked. "He looks like a teenager with a bad case of puppy love. Shameless, and while on duty, too."

Sharp grinned and patted the rifle suggestively. "You're just jealous my friend."

Chris and Tuttle laughed.

"Just so long as you don't get 'er preggers," quipped Tuttle. Now even Sharp joined in the laughter.

"You never know," he replied with a mysterious look.

After a brief situation review Chris started down the ladder stairs to check on the time viewing project. Tommy Dawson, ever the insomniac, was still wide-awake and working with two assistants. As Chris approached the control center the physicist jumped out of his chair.

"There's no sign of any trouble," he reported. "We're keeping watch 15 minutes into the future. Everything's quiet. The two helicopters are still in their hangers."

"Any luck tracking that Hummer?"

"We followed it back to a used car lot in Albuquerque. Looks like it was stolen. Now we're trying to back-trace the two men that took it, but it's tough. They flew in from different parts of the country. We've tracked one of them back to Seattle, and the other came from somewhere out East."

Dawson was obviously thrilled by the excitement of the chase. He bounced up and down on the balls of his feet and for a moment Chris was afraid he was going to start one of his manic dances.

The two men stood watching the changing images on the big ten-foot display. The scenes cycled from various views of the lab to the dark interior of the secret helicopter base to crowds in an airport terminal.

Dawson excused himself and scurried back to his workstation. Chris turned toward the break area to get a cup of coffee. As he was walking through the center the encrypted cell phone rang. He pulled it out of his pocket and read the screen. Caller ID was blocked. He pushed the talk button. Before he could say anything a voice came on the line.

"Fisher, this is Chase." The Admiral's gruff voice was tense, commanding.

"Aye, sir."

"Listen Fisher, we've got trouble." Chase filled Chris in on what had happened since earlier that night when Kate sent her file to him.

"I'm afraid they've probably iced the director, and they're hot after me," he concluded. "I've gone dark—got a few things left in my bag of tricks—but our idea of getting the facts to the President is a bust."

Chris was silent for a moment. This was truly bad news. They'd been hoping to get word to the very top and that the power of the Oval Office would come down on their side. Now things had spun out of control once again.

"What are you going to do?" Chris asked. Through the ether he heard Chase sigh.

"Damned little I can do. I'm on the run…"

Holding the phone to his ear, Chris stared into the distance for long seconds. *Now what?* Dejected, he walked over to a chair and sank down.

"You still there, Commander?"

"Yes, Admiral, I'm here. I was just thinking…" He paused and both men remained silent for long seconds.

"What if…" Chris began then stopped again.

"Yes?" Chase coaxed.

"Well, it's a long shot, but what if you could find out where the SWAT team came from? Could that help?"

"I don't know," Chase admitted. "It could. All depends. What did you have in mind?"

"Well, it occurred to me that we could use past viewing to track them from your house to where they came from."

"Oh, yes!"

"Would you be able to do anything with that information?"

"Again, it depends. I might be able to. I've got some guys I can trust so I could put together a small strike team. But if the trail leads to the Pentagon or CIA headquarters, something like that, then no. Too big to touch."

"Why don't we give it a try," Chris suggested. "Give me the location and time when they showed up at your house and I'll have Tommy add it to his workload."

"Sure. This phone keeps a continuous GPS and time record. As soon as we end our conversation I'll email you the coordinates for my house."

"Great."

"So bring me up to date from that end. Have you figured out how to handle the press? That's gonna be a real fur ball."

"Well, yes sir. Lieutenant Elliott and I spent a couple of hours brainstorming and we have some ideas, but frankly the whole thing is daunting. The less we tell the press the better, but anything we do reveal only invites more questions. Either that or speculation, and there's already plenty of that."

"I can imagine," Chase commented.

"We're going to get together with Reeves and Bonner at six to plan our strategy. The two of them will handle the presser, with me as a support. Kate, of course, will stay in the background. She's going to video the event for our record."

"Jesus, I don't envy you guys. Look, I've got to get moving. I'll line up my team and hope you can give me something useful. The coordinates will be coming in a moment."

"All right, sir. I'll be in touch as soon as we know something."

Chris turned off the phone. He got up from the chair, walked over to the coffee bar and poured himself a cup. As he took the first sip the phone binged to announce the arrival of the text message. Smiling grimly he headed for the control center to get Dawson's team to start tracing the men who had attacked Chase's house.

An hour later Kate was awake and she and Chris were going over their plans one more time. They were huddled at a desk in a remote corner of the center. Kate had her hands resting on the keyboard of her laptop but hadn't typed anything for several minutes. Chris was absently doodling on a legal pad. For several minutes neither had said anything. Suddenly they both started to speak at once.

"You know…"

"What about..."

They stopped in confusion.

"You go ahead."

"No, you."

They stopped again and looked at each other, then broke out laughing.

"I was just thinking out loud," Kate said. "You go."

"Oh, well, I guess I was too." Chris frowned. "We just seem to get in deeper and deeper. I'm half inclined to cancel the press conference. We're just going to make damn fools of ourselves."

With a thoughtful look on her face, Kate sat back and put her hands in her lap.

"Yeah," she said quietly. "I've been thinking the same thing." She reached over and patted Chris on the cheek. He reached up to cover her hand with his.

"Well," Chris mused, "I've said that SEALs don't run, so I guess it's not an option." He twisted his face into a wry grin. "And I don't think Navy spooks do, either."

Kate chuckled and pulled back her hand. "No, we're stuck with this situation...but that doesn't mean we have to throw ourselves into a tank full of sharks and piranhas. Maybe we need to think outside of the box on this. Who says the press has a right to know what's happening here? It's a top-secret project, after all. We aren't authorized to reveal the details. Plus, we don't really know what's going on anyway."

"We've gone over that before, and you're right as far as it goes." Chris picked up a pen and began doodling on his pad. "But what about the helicopter we shot down? The public knows about that, not to mention the cover story about a radiation leak. In hindsight I think that was a mistake. It raised too many questions and concerns, and started the rumor mill going."

"Yeah. There's even a wave of speculation about the kidnap attempt at the cafe. I know the cops tried to keep it quiet, but I've heard it's been blown up to sound like the battle of the century. Do you know that the mysterious hero is alleged to be a former Mister

Universe, Medal of Honor winner, or world martial arts champion depending on who's telling the story?"

Chris broke out laughing.

"Well, since that's all true it reaffirms my faith in unsubstantiated rumors and wild speculation," he quipped.

"Well, of course it's *true*," Kate said with a snicker and punched him on the bicep. "Everything's true, in its own way."

Chris laid down his pen and looked off into the distance.

"You know, that's a funny thing you said…"

"What?"

"About everything being true. There's something to that, isn't there? Truth is an elusive quality. It can be twisted, spun, turned on its head, wrapped in webs of doubt until it can't be recognized. There's probably no such thing as real, objective truth. They say that history is always written by the victors, and what kind of 'truth' is that?"

"Well, I was only making a sarcastic comment…"

"But, what you said is true…in its own way." Chris brought his gaze back from the distance and looked into her eyes. They both laughed.

There was a distant shout and they looked up to see Dawson making his way toward them. He was obviously excited. They stood up to meet him as he arrived, puffing. He took off his Quantum Cowboy cap and wiped his forehead with the sleeve of his T-shirt, exposing a hairy armpit.

"What is it Tommy?"

"We tracked down the guys that attacked Chase's house," he told them proudly. "They came from an old farmhouse, about forty miles west of Washington."

That got Chris's attention. "Does it look like they went back there?"

"Oh, yes, we skipped ahead and sure enough they came right back. It seems to be a secret hideout or something. There's an underground bunker they use as their base."

"Ah." Chris's mind was working. This sounded like a safe house, the kind of thing that Chase might be able to penetrate. "Tommy, let me have the coordinates."

Dawson handed over a printout. Chris reached for the sat phone and auto-dialed Mad Liar's number.

Two rings, then. "Chase here." The Admiral sounded tired.

"Admiral, we've tracked down your rats. They're in an old farmhouse. I've got the coordinates. Are you ready to copy."

Chase seemed to be invigorated by this news. "Yes, go ahead." Chris read off the GPS location of the farmhouse.

"We can give you details of the layout. Tommy says there's an underground bunker where the rats hang out."

"It's a Company safe house," Chase said with certainty. He sounded upbeat. "We can do them. Thanks, Fisher."

"You're welcome, sir," Chris said, but the Admiral had already clicked off.

Chapter 58

Near Ashburn, Virginia
Saturday, 5:30 a.m. EST

Dawn was beginning to stain the eastern sky, swallowing the skyglow of the nearby Washington, DC metro area. Here, less than forty miles from the world's most powerful seat of government it seemed that time had frozen in a previous century.

Concealed behind a clump of tall grass Admiral Chase studied the old farmhouse through a break in the trees. He was located to the left side, giving him a three-quarter view. The building appeared almost, but not quite deserted. A faint glow appeared in several windows, including one at the far left corner. Tommy Dawson's quantum computer had provided him with a detailed layout of the building, including both the two visible stories and the hidden bunker. Chase knew the room where the light showed was an office, and he knew that the CIA's deputy director of operations was there, pacing uneasily and sipping occasionally from a glass of Irish whisky. Such was the power of time viewing.

The four men of the strike force were sleeping in private rooms below ground. There were two others, one a cook and the other apparently a gofer and maintenance man, responsible for the grounds and security systems. The DDO was the only other person on the premises.

Much as Chris had done with his SEALs, Chase had brought together a team of warriors to mount a surprise attack on the safe house. Master Chief Art Lewis, the man who had picked up Chase earlier, was at his side. Lewis had worked with Chase for several years. Far more than a driver, he was an experienced intelligence agent himself.

Lewis had brought together seven other men from their circle of friends and comrades, including three active Marine Recon sergeants, two Army Stryker troops, an undercover agent working for the Navy in Washington and a recently-retired Army Delta Force senior noncom.

The little band was poorly equipped by modern military standards. There had not been time to draw weapons and gear, nor was Chase certain his authority would still have made it possible. He had to assume that he was a "person of interest" in the fate of the CIA director and because of his own sudden disappearance. Each man came armed with personal weapons and equipment.

Having no tactical comm systems, they were relying on cell phones, each set on silent vibrate. Chase had programmed his sat phone to allow him to dial up all eight men with a single button push. Lewis had found an all-night store and obtained headsets for each man. Also on the jerry-rigged communications net was Chris Fisher back in New Mexico. Chase could also speak to any individual by pushing a single pre-set number.

The selection of weapons was eclectic but each man had at least one serious handgun, rifle or shotgun. Chase had his personal pistols and 12-gauge shotgun. Lewis was better prepared with an AR-15 semi-automatic rifle, the civilian version of the military's AR-16 combat rifle. Lewis had also managed to scrounge three flash-bangs and two thermite grenades. "Don't ask," he told the Admiral as he showed him the grenades. Chase smiled.

One of the Marines had a vintage Colt .45 automatic in a belt holster and a Remington model 700 hunting rifle slung over his left shoulder. Another had an M-1 carbine of the World War II era. A third was armed with an imposing .44 magnum Smith & Wesson revolver with an eight-inch barrel in a cross-chest holster. Two others had brought hunting rifles, one had an AK-47 assault rifle, and the last was armed with a Browning 12-gauge over-and-under shotgun augmented with a snub-nosed .38 special revolver. All wore various items of camouflage clothing, including coveralls, field boots and hats.

Now the members of the band were spaced along the rural road around the outskirts of the little farm. Ill-equipped though they may be, they had one huge advantage: The information being provided by Tommy Dawson and his team back in New Mexico.

The safe house was surrounded by an electronic security system that would alert its occupants the moment the inner perimeter was breached. From the outside the place had the appearance of a rundown farm. It was ringed with a sagging barbed wire fence with faded "No Trespassing" signs posted every ten yards. Brush and weeds had been allowed to grow up around the fence. A close examination would reveal that the barbed wire was of a particularly nasty kind normally used in high-security prisons, and that what appeared to be rust was actually polymer paint.

At the entrance was a standard farm gate with a combination padlock. There was no alarm set on this outer perimeter, Chase had learned. Inside the outer gate a lane wound through some second growth timber and brush. Beyond sight of the road the real security perimeter appeared. This second gate was also camouflaged to appear rustic but in fact was connected to an alarm system that could be overridden by a keypad hidden behind a concealed panel in the massive fencepost to which the gate was connected.

Thanks to Dawson, Chase had both the padlock combination and the keypad code. Dawson had simply scanned back to the time when the SWAT team returned to the safe house, then zeroed in as the locks were opened.

Similarly the remote time viewing system had probed the security system further inside the house. There was a second keypad to disarm the alarm system inside the front door. This code too had been provided. Armed with these keys the rag-tag team could penetrate the no-longer secure safe house without triggering an alarm.

There were three entrances to the building. Besides the main front door and a rear door leading from the kitchen to the back yard, there was a hidden back entrance to the underground bunker. A tunnel led to the foot of a concealed stairway in the old barn that was used as a garage. At the bottom of the stairs was a door made of thick carbon steel in a heavy metal frame. Chase did not intend to allow that door to be opened. In the first phase of the attack two men would secure the steel door with thermite grenades, thus sealing off the escape route. They would then take positions at the rear of the house.

Chase looked at his watch. It was time to move. He keyed his phone.

"Fisher here," came the reply.

"We're ready. What's the latest intel?"

"Sir, we just completed a five minute look-back and all's quiet. The four goons are still asleep. So's the handyman. The DDO is in the office and the cook is working in the kitchen."

"Splendid. We're go."

"Aye sir. Good luck."

Chase punched another number connecting him with the entire team.

"Heads up boys. We're going to go on a countdown of five minutes from …" he paused as the sweep second hand of his watch approached the 12 o'clock position, "mark!" He started the stopwatch function.

As planned the men began to move from their surveillance positions toward the first gate. The last man arrived as the timer counted down to one minute. The motley group silently gathered around Chase. He gestured at his watch and held up one finger.

He pointed at Lewis and then at the combination padlock. Lewis walked over and turned the dial, opening the lock. He swung the gate open about three feet and stepped aside.

"And…go!" Chase commanded in a low voice, gesturing with his arm. The men filed quickly through the gate with Chase. Lewis followed, wheeling around to close the gate and reset the padlock. When he turned back the other men seemed to have disappeared. He grinned and stepped into the cover of brush and scrub trees lining the lane. Like the others, his camouflage clothing caused him to melt into the background.

About a hundred yards further on they reached the second, alarmed gate. Chase nodded at Lewis and the master chief opened the hidden panel to reveal the keypad. A light began to blink. Carefully he entered a five-digit number and pressed the "Enter" key. The pad gave a little double beep and the light stopped blinking.

"We're in, sir," he whispered. Chase nodded and gestured to two men who were standing ready. Designated as the A-team their assignment was to disable the rear exit from the bunker. Lewis opened the gate and the two disappeared in the direction of the farmhouse. The rest of the team filed through and gathered around while Lewis secured the gate behind them.

Chase pointed at the ground then held up three fingers. They knew this meant to stay in place for three minutes, giving the two advance attackers a head start to be ready to seal off the steel door.

Speaking quietly into the phone, Chase filled in Chris on their progress. So far there was no sign of alarm. Chris confirmed that another time snapshot showed the men in the bunker were still sleeping.

Three minutes passed and Chase signaled the others to begin to advance. He took the middle position with three men spread out on each side of him. Except for the serious expressions on their faces they could have been mistaken for a band of deer hunters.

"Sir, we're approaching the barn," came a voice through Chase's headset. The A-team leader was a former Delta Force top sergeant.

"Roger. Hold." Chase looked around then focused his attention on the house about 75 yards ahead. His men were spread out in the tall weeds and grass. Chase himself had taken cover behind a tree. He judged the distance then made his decision.

"A-team, burn the door now," he ordered. "The rest of you, wait for my command."

In place at the bottom of the hidden stairs the advance force positioned a thermite grenade against the bottom of the door. A second was taped over the lockset area to fuse the deadbolt. A wad of pizza dough was slapped over it to focus the heat.

"Pulling pins," the leader said into the microphone stalk of his headset. On a count of three they pulled the pins, turned, and ran up the short stairs to the main floor of the barn. A few seconds later a bright, sustained flash lit up the stairs and the smell of hot oil, metal and burned flour wafted up to mingle with musty barn odors. The steel door had been welded shut.

"Door sealed," the leader reported. "Moving to house."

Chase waved his team forward to the front door. One man was carrying an eight-pound urethane coated sledgehammer. Stepping to the hinge side of the front door he took a roundhouse swing at the lockset. With a crash the deadbolt shattered and the door swung open. Lewis surged inside, his AR-15 pointing the way. Chase was right behind him with the Mossberg at the ready.

The foyer was empty except for the stately grandfather's clock, slowly ticking off the seconds. Chase stepped left and Lewis right and two more men surged in behind them. One turned to sweep the barrel of his gun around the dining room, the other covered the living room. Lewis snapped open the cover of the alarm system touchpad and punched in a quick series of numbers.

There was nobody in the side rooms. A second pair of men continued straight ahead toward the kitchen, just as the cook appeared with an astonished expression on his face. His sleeves were rolled up and his hands and lower arms were covered with flour.

"What the fu…"

The barrel of a deer rifle poked him in the solar plexus and he doubled up as the air was driven from his lungs. One man swung him around, looped two daisy-chained plastic wire ties around his wrists and pulled them snug. He shoved the man into the dining room and pressed him down on his face.

The seven members of the front assault group were now inside the house. Two men took stations at the top of the stairs leading down to the underground bunker. Two more cleared the kitchen and opened the rear door to admit the A-team.

While that action was taking place, all in the space of about ten seconds, Chase and Lewis were concentrating on the door to the ground floor office. There was a crash from inside the room as if somebody had knocked something over. They waited. The door began to open and Lewis was ready. He raised his right leg and aimed a powerful kick. The combat boot struck the door just below the handle, causing it to slam back against the person on the other side. They heard a howl of rage and pain.

Lewis followed through, swinging the door wide and rushing the man behind it like an NFL lineman, causing him to sprawl backwards, arms and legs akimbo. A handgun clattered away to one side. Lewis stepped between the man and the pistol and kicked the gun. It flew under a couch and hit the wall with a thump.

Chase was right behind him with the Mossberg pointed at the man's chest.

"Don't tempt me, Jamie," he gritted. "Hands on top of your head. You know the drill."

Jamie Delancy, the CIA's Deputy Director for Operations, stared into the mouth of the shotgun. Then, he slowly laced his fingers together on his expensively styled hair.

"Jesus, Chase, I think your boy broke my wrist," he said.

"You're lucky you dropped that gun, or you would have been in a whole world of hurt," Lewis informed him, stepping back to one side and keeping the AR-15 trained on Delancy.

Somewhere from beneath the house there was a sharp explosion.

The three men stayed still, the CIA executive spreadeagled on a Persian rug, Chase and Lewis watching him with guns pointed. After about a minute one of the volunteers tapped at the open door and stepped inside.

"The bunker's secured, sir," he told Chase. "A couple of them opened the door at the bottom of the stairs to see what was going on so we let them have a taste of a flash-bang. They didn't have the stomach for any more. Two others tried to escape through the tunnel to the garage, but the door seemed to be locked from the outside." The trooper smiled ironically. "And not only that, in their haste they forgot to pick up their weapons. A couple of the guys went down and brought them up like good little lambs."

Chase nodded.

"Good work. Put them all together in the dining room and keep a guard on them. Shut the door when you go out. We're going to have a little private chin wag with our friend Mr. Delancy here."

"Yes, sir." Glancing with distaste at the sprawled CIA agent the man backed out of the room and shut the door quietly behind him.

"Mister Lewis, please search our prisoner," Chase instructed. Lewis patted Delancy down roughly while Chase made a quick call to let Chris know the raid had been a success. The search yielded an empty shoulder holster and a spring knife from a pants pocket. Lewis tossed the knife into a corner.

"He's clean sir."

"Okay, Jamie, I want you to sit up very slowly while keeping your hands where they are." Scowling, Delancy complied. "Now get up on your knees and duck walk over to that chair." Chase indicated a leather lounger in the corner. Lewis had already checked the room for weapons.

Fingers still laced atop his head the angry CIA man clumsily clambered into the chair and sat back, glaring at Chase.

"Admiral, you're dead," he spat. "You know that, don't you?" His voice rose in tempo with his rage. "How dare you attack me, attack a CIA facility?"

Chase met his gaze with implacable calm.

"Jamie, you're wrong. It's you that's just run into a meat grinder."

"You've got nothing on me!"

"Oh?" Chase contemplated his prisoner for a moment.

"Yeah, nothing." Delancy was becoming belligerent. "I'll have your ass for this. It's treason. You'll be lucky to get life."

Chase continued to gaze into the DDO's flaring eyes for a moment, then turned and walked over to the desk. Lewis remained standing about ten feet away from their prisoner, AR-15 poised.

"I see you've got some nice whisky there," the Admiral observed. "Mind if I have a wee taste?"

Delancy didn't reply. Chase poured two fingers and swirled the oily, golden liquid around the Waterford crystal glass. "The water of life," he intoned. He tipped the glass at Delancy and took a sip.

"Fuck you, Chase!"

"Now, now, mind your manners. We're gonna have us a little talk now and it won't do to get us all riled up. My friend here," he nodded at Lewis, "tends to get excitable at times."

Reacting to this wild mischaracterization, Lewis smirked and rocked the rifle up and down in a suggestive way. Delancy flinched.

Chase sat down in a chair facing the prisoner, crossed his legs and laid the shotgun across his lap. He took another appreciative sip of the Jameson's Irish.

"Let me ask you something," he said, looking up from the glass. "Why did you have Director Higgins killed?"

Delancy struggled to hide his reaction.

"Why, that's outrageous," he shouted. "I've never been so insulted…"

"Shut up Jamie," Chase cut him off. "You're going to tell me some cock and bull story about how innocent you are, cover your shame with pretended anger. Right?" Chase pointed an accusatory finger. "It won't wash with me, Jamie. I know your kind. I know you, and you're guilty to the bone."

Delancy scowled and started to sit forward in the chair. As Lewis raised the AR-15 to center on his face he thought better of it. He took several deep breaths and struggled to settle himself.

"Okay, Chase, we'll have our little talk. May I please lower my hands now? It's pretty uncomfortable."

Chase nodded. "Slowly," he said, "Very slowly." Lewis grunted a warning.

Delancy placed his hands in his lap and crossed his legs. Appearing calmer now he forced a thin smile.

"Admiral, you've got no proof of anything. I don't even have a clue what you think I've done. I'm a senior intelligence officer, as are you, and I've done nothing beyond my duty to my country."

Chase regarded him as a cat might observe a fat mouse. He said nothing. Delancy's bravado began to crack and he squirmed slightly in the chair. He glanced nervously at Lewis. The rifle was still pointed directly at his face.

"Okay, Chase, why don't you tell me what you think?" he challenged.

"Jamie, I don't *think* anything—I *know* what you've done. All of it."

The spy chief relaxed again. He smirked.

"Oh, really?"

"Yes, really."

Gaining confidence, Delancy chuckled.

"That's impossible, Admiral," he stated.

"In your dreams, Jamie."

Chase continued to stare at his prisoner, waiting for the next retort.

"Well, why don't you show me your proof?" Delancy challenged. "You've got a big mouth, Admiral. You're bluffing, you asshole. You've got nothing."

Chase allowed himself a small, grim smile and watched the chilling effect it had on Delancy.

"Okay, let's start with something recent. As you very well know, by your order Director Higgins was shot in the head with a 9 millimeter Beretta pistol at about 12:24 this morning. He was sitting in the driver's seat of his daughter's yellow Miata sports car. The shooter was dressed to look like a cop. He is in fact one of your men. We have him in custody in this very house, along with an accomplice."

Delancy's face drained. He gasped.

"That's a lie!" he said at last, his voice strangled. "You can't prove that."

"No?" Chase reached into his inside jacket pocket and pulled out some folded papers. Slowly he began to unfold each sheet and smooth it flat. Delancy's eyes were riveted on the documents.

"What kind of bullshit is this, Chase?" he demanded. "You can't pull this psychological shit on me."

"Oh, no, this isn't psychology," Chase said matter-of-factly. "This is *photography*. Quite a different thing."

"What!" Delancy was becoming agitated again. "What the hell kind of evidentiary photographs could you have?"

Chase fanned the sheets out like a hand of poker. He ruffled them, taking note of each image. Finally he looked up at Delancy and began speaking in a quiet voice.

"Well, Jamie, you want proof? I've got it right here. Here's a picture of your goon standing beside the car shooting Director Higgins in the head. You can see the blood splattering. Quite a revealing photo, actually. You can see the killer's face quite clearly. I've got another shot where you can read the serial number of the gun."

He rose and stepped forward to hand Delancy the picture, downloaded and printed out at a 24-hour Kinko's about an hour earlier from Tommy Dawson's computer.

Delancy goggled at the picture. Chase could almost see his eyes trying to pop out of his head, like some cartoon character.

"Wha... What..." the CIA man stuttered. "How..."

"Oh, just shut the Hell up Jamie," Chase advised him. "We have you dead to rights. Here's the next photo. It shows the killer with you here in this very room, date stamped about two hours before Higgins died."

He handed it over and Delancy took it with a limp hand. He stared incredulously for a moment before letting it drop into his lap with the first picture. He raised his eyes to look at Chase with a forlorn expression.

"I've got some more, Jamie," Chase told him, riffling through the spread of photos. "Here's one you might like. It shows you handing your goon the Beretta pistol he used to shoot Higgins. Again, the picture caught the serial number. It matches. And, oh yes, we know where your man threw the gun into the Potomac. We'll be able to recover it pretty quickly. We know where he dumped the Director's body, too and, oh, all kinds of stuff that implicates you in this and a whole lot of other things."

He stopped and resumed his contemplation of the CIA agent.

"Is that proof enough?" he asked in a soft voice. "We've got a lot more." He riffled the photographs.

Delancy seemed to shrink into himself, taking on the appearance of some dangerous, cornered animal. His expression turned feral, beginning to verge on madness. His eyes glittered with loathing. A violent tic started to pulse in his right cheek.

"Damn you, Chase," he intoned. "It's that time thing isn't it? That's what all of this is about. You think you're smart but you're not, sailor-boy, you're not." Delancy smiled in manic delight. He laughed. "You, my friend are well and truly fucked because the source of all this so-called 'proof' of yours is going to disappear in a puff of radioactive dust."

"What!" Alarmed, Chase leaned forward intently. "Jamie, what in the name of Christ did you do?"

Delancy grinned broadly now. There was a trace of spittle in the corner of his mouth. His eyes had a wild look. "Not gonna tell," he chanted in a passable imitation of a twelve-year-old. He smirked and returned to a normal voice. "You'll find out soon enough, asshole."

Chapter 59

Quantum Center
Saturday, 4:48 a.m. MST

Chris and Kate were waiting in the small conference room for Reeves and Bonner to arrive for a 5 a.m. planning session. Neither had gotten any sleep and fatigue was beginning to wear on them. They each had a steaming cup of coffee and had helped themselves to the ubiquitous donuts from a tray in the middle of the table. Kate's had pink sprinkles and Chris chose one with chocolate frosting. He took a bite, wincing as the high-octane sugar hit his palate.

Kate had set up her video camera to record the session and her documents were spread out on the table. She rummaged in her bag, pulled out her steno pad and a pen and laid them on the table. Idly, she carefully aligned the stack of papers.

Neither said anything.

There was a tap on the door and Bonner stepped inside, closing the door behind him. He was still haggard. He tried to smile but it was a futile effort.

In his hand was Kate's Kimber .45-caliber pistol. Bonner handed it to Chris.

"Here's your gun, Fisher," he said with a wink. "Forensics didn't think they needed it any longer and you may as well have it."

Kate's eyes lit up at the sight of her personal sidearm. Wordlessly

Chris laid the gun on the table and pushed it across to her. She snatched it up and ejected the empty magazine. Bonner reached in a pocket and silently handed her seven shells. Deftly she filled the magazine, snapped it back into place and tucked the weapon into her capacious messenger bag. Chris and Bonner looked on approvingly. Grinning, Kate pulled Chris's Beretta out of the bag and slid it across to him.

"Thanks," she said, smiling at Bonner. He nodded and winked again.

Reeves arrived a moment later. Like the others, the acting director looked as if he had not slept in days, which was almost literally true. He and Bonner poured cups of coffee and grabbed donuts. They sat. They looked expectantly at Chris and Kate. Nobody spoke.

Kate was first to break the eye contact. She lowered her eyes to the table and shuffled her papers. She picked up her pen then set it down again. Chris took another bite of his donut and followed it with a sip of coffee.

"So, I gather we don't have much of a plan?" Reeves deduced. Bonner grunted and bit off a third of his donut.

Chris shrugged. "No, we really don't," he admitted. "We've discussed canceling the press conference on the theory that it might do us more harm than good. But, of course, that would just cause more speculation. The press would go into an even bigger frenzy."

Reeves looked glum. "I don't know…" he began.

"Wait, there's something else." Quickly Chris told them about Admiral Chase's impromptu raid on the CIA safe house that had just been completed.

"He's captured the DDO, Jamie Delancy. He's interrogating him now," he concluded.

"Jesus, you've been busy!" Bonner exclaimed. "And Tommy, too. He actually got the proof that Delancy was behind the killing of Director Higgins? That's incredible."

"Yeah," Chris replied. "It was another good demonstration of just how powerful the time viewing system is."

"But how is this going to help our situation?" Reeves asked, glancing around the table. "We obviously can't tell the press that Chase has raided a CIA facility and imprisoned a deputy director."

"Definitely not. But we have intel that we can use. Chase believes Delancy was probably behind the attacks on the project. We're waiting to hear what his interrogation turns up."

"Ah." Bonner grinned. "An off-the-books interrogation in a CIA safe house can be pretty productive."

"Yes, although I'm sure that Admiral Chase would never violate the rules against human torture or anything like that," Chris said dryly. He paused as Bonner chuckled in appreciation of this obfuscation. "But remember that his specialty is psychology. He might get something out of the rat."

As if on cue the encrypted cell phone buzzed. Chris picked it up and pressed a key. The faint hum of a carrier wave came from the phone's speaker. "Hello Admiral. I have you on speaker with Stuart, Paul and Kate."

Chase's voice sounded strained. "Hello, Fisher, everyone. I wish I had better news..." Bonner groaned and Reeves rolled his eyes toward the ceiling.

"What is it?"

"It's Delancy. He's clammed up, but only after telling me that the source of our proof against him is, quote, 'going to disappear in a puff of radioactive dust' end of quote."

For a long moment nobody said anything, then Bonner swore under his breath.

"You couldn't get anything more out of him?" Chris asked.

"No, like I say, closed up like a clam."

"Did you, uh, try anything to encourage him?"

"Of course, but nothing that would leave marks. No luck. He's had training in resisting interrogation, of course."

Chris drummed his fingers on the table.

"Okay," he said, looking around at the others. "I assume we all agree what he meant?"

Bonner and Reeves both nodded. Kate fondled her Thunderbird pendant for a moment then met Chris's eyes.

"This is where they invented the damn things," she said calmly.

"They're going to blow up the center with a thermonuclear device."

Chase's voice came over the phone speaker.

"I agree, Lieutenant. That's my conclusion also."

Bonner's face had turned cherry red. "Jesus Christ! Why does everything always go from bad to worse?" He jumped up and began to pace around the room, pounding his right fist into the palm of his left hand.

The others were silent. Reeves had focused his eyes somewhere outside of the room. Chris was leaning forward over the phone on the table in front of him. Kate picked up her pen and began doodling on the steno pad, drawing little circles and squares.

"I don't know what to tell you." Chase's voice was hollow, flat, with an overtone of suppressed rage. "It's in your hands I'm afraid."

"What are you going to do?"

"I'm staying here in the safe house. We have seven prisoners and there's plenty of food. I'll be available on my cell phone anytime, but there's damned little more that I can do. Of course, we'll keep trying to get more information out of Delancy, but as I said he doesn't seem inclined to cooperate."

Chase stopped, cleared his throat then continued in a formal voice.

"Lieutenant-Commander Fisher, in Navy parlance, you have the conn. There are dangerous waters ahead, so steer your ship well and fight her hard."

"Aye aye, sir. Thank you sir."

Chris turned off the phone and sat staring at it. Bonner was still pacing around the room, swearing under his breath. Kate was drawing little mushroom clouds along the side of her pad.

Reeves reached out like a robot and picked up his coffee cup, raising it to his lips. With his eyes still far away he began to speak over the rim of his cup.

"About a month ago three W-87 thermonuclear warheads were brought here to the lab for testing. Those are 300 kiloton devices." His eyes returned from the distance and he glanced at Chris and Kate. "Little Boy, the bomb we dropped on Hiroshima, was about one-tenth that size," he added.

Bonner stopped pacing and sat down. His face was pale.

"Those bombs have been under strict security protocols," he declared. "They're locked in vaults and under twenty-four hour armed guard."

"Or so we thought," Reeves said. "We already know that our security staff has been penetrated. Who knows how big a presence they have here?"

Bonner let his head drop, his chin nearly touching his chest. "Yeah, you're right..."

He looked up resolutely, wiping his forehead with the back of his hand.

"I spent all night running through personnel records," he informed the others. "There are some I red flagged, others I just put a question mark beside. All told, I found seventeen possibly false records and more than thirty that look suspicious."

"Christ, Paul, how can that be?" Reeves was outraged.

Bonner held up his hands, palms out.

"Stuart, I sure as Hell don't know," he declared. "My top security officer vetted all of those files, so I've put the biggest red flag of all on his file. My problem right now is that I can't be sure of anything and I don't know how to begin verification because we can't trust anybody. I've made a few calls to some old friends in the Bureau, but nobody wants to cooperate. They know we've been fenced in."

He pounded his fist on the table. "We're like lepers," he declared. "Nobody wants to touch us."

There was silence for a long time. Kate began to draw question marks in the circles and squares she had sketched.

Chris took a deep breath and sat up straight in his chair. The others could sense a change, as if they could see him assume the mantle of command. "Paul, first let's have someone check those vaults," he said. "If the weapons are still there, we need to secure them. If not, well..."

"Right." Bonner reached for his radio, thumbed a button and waited for a response. "Major, would you please ask a security team to check the three special packages in the vaults?" He listened for a moment, then: "Do it now, and get back to me ASAP." He signed off.

"That was my second-in-command. I can trust him as much as I can trust anyone. He's going to personally look into it."

"Good," Chris told him. "Next we need to begin to identify and isolate those names you red-flagged or questioned. You say you don't know who you can trust, but we can start by not trusting any of those." Bonner grunted agreement. "Most of them will turn out to be innocent, but we must quarantine them until we know for sure."

"You're right," Bonner agreed. "I'll get on that. As you recall, I already selected a key group of officers that I thought were clean. None of them turned up on my review last night."

"That's a good indication your gut instincts are sound," Chris told him. "Give each of them the list of suspects and ask them to draw in others who aren't on the list. That way you can build a network around the suspect group, isolate them from the start."

"Okay."

"How many of those suspects are part of your security force and how many are civilians?" Chris asked.

Bonner thought for a moment. "I'd say about a third are security, the rest are civilians. That includes several fairly high-profile engineers and physicists. One problem we have is that most of the civilians are not on site."

"I realize that," Chris said. "I suggest you get some of your plain-clothes officers outside to check up on them. And, see if any of them have come into the lab. In our present lock-down, that alone could be cause for suspicion."

"Right." Bonner stepped away from the table and keyed his radio.

Kate looked up from her pad. "May I make a suggestion?"

"Of course."

"Let's not forget what Tommy can do," she said. "Can't we check out the future to see what we're facing?"

Reeves slapped his thigh and stood up.

"Crap, what are we doing sitting here?" he asked, striding to the door. Chris and Kate followed him. Bonner tagged along behind with his radio pressed to his left ear.

When they arrived at the control center a moment later Dawson had just gotten up from a nap and was wiping the cobwebs out of his eyes. A large cup of coffee was precariously balanced next to his keyboard as he pondered a plate of donuts, apparently trying to decide which one to eat first. As they approached he turned and his elbow tipped the cup, slopping coffee on the keyboard.

"Tommy, be careful!" Kate exclaimed.

Dawson turned back and moved the cup to a safer place. He pulled a handful of tissues from a box and wiped at the keyboard. "Don't worry," he mumbled. "Every keyboard in the center is the spill-proof kind. Something to do with me, I understand…"

"Tommy, we've got trouble. We need your help, right now."

Startled by the tone in Chris's voice, Dawson stopped wiping at the keyboard and looked up with wide eyes.

"We want to take a look-ahead, longer than the fifteen minutes you've been using," Chris told him. "We need to view the center."

"Sure, no problem."

Grimly Dawson turned back to his workstation. He shouted for an assistant who was napping on a nearby cot. Together they fired up the quantum computer.

"How far ahead?" he asked over his shoulder.

"Let's try twelve hours," Reeves suggested.

"That'll take longer to process, you know. About ten minutes, I'd guess. The further ahead we look, the more data to handle."

"I know, Tommy. Just do it."

Dawson nodded and began to type code as his assistant monitored the interface between the quantum computer and the racks of daisy-chained servers. In a moment they heard the computer array begin to spin up. The center's automatic cooling system kicked on.

"Now we wait," Dawson said, reaching for a donut. It was halfway to his mouth when he suddenly froze. The hum of the servers was winding down to a stop. "What's this…?"

On the screen in front of him appeared two words:

INSUFFICIENT DATA

Dawson goggled at the message for a moment.

"That's impossible," he muttered. He scowled at his assistant. "Let's try again, looking just three hours ahead."

In a moment the server array whined into life again. This time it continued to work. Time passed too slowly for anyone's satisfaction, but at last the process was complete. An image of the center appeared on the large overhead screen. It was fairly stable, but with the fuzzy appearance caused by the uncertainty of events that had not yet occurred.

"Hmm," Dawson mused. "I don't get it." He spun around in his chair to address the four hovering behind him. "A few days ago, when we first launched the system, we did a two-day look-ahead just to see what would happen. It took a while to process, but it did fine. This is really strange."

"Try going ahead an hour at a time," Chris suggested. Dawson set to work at his keyboard again.

At present time plus four hours there was a clear, if fuzzy image.

Present time plus five hours yielded a similar result, as did present time plus six hours.

At seven hours into the future the servers ground to a halt and once again the "insufficient data" message flashed on the screen.

Agitated, Dawson sat back in his chair. "I just don't get it," he muttered.

Chris stepped back, pulled a swivel chair away from a nearby workstation and rolled it over next to Dawson's. He sat down and contemplated the message spelled out by the cold, green letters.

"Tommy," he asked quietly, "what would you see if at some point the quantum computer and this entire center might have been utterly destroyed?"

"Destroyed!" Dawson squeaked. "Why, why..." He lapsed into thought for a moment. A grave expression of realization spread over his face.

"There would be a time disruption at the instant of destruction," he said in a wooden voice. "The very possibility of such an event would shatter the time dimensions. That would overwhelm the system, and..." He pointed at the words on the screen. "It would yield that message."

Quickly, Chris explained the possible danger. Dawson's face turned white. He stood up from his chair and began to rock gently back and forth from one foot to the other, staring at the green letters. "Oh, this is terrible. Terrible!"

Reeves stepped over and placed a hand on Dawson's shoulder.

"Calm down, Tommy. We're not licked yet. There's a lot we can do. Remember, we've already changed the future twice, with the Hummer and the helicopter. We can do it again."

Dawson stopped rocking and sat back down, glancing gratefully at Reeves. "Yeah," he mused. "Yeah, maybe we can...but from somewhere beyond six hours we'll be blind."

"That's okay, Tommy, it gives us plenty of time. We need to focus on what to do about this."

"You're right. First, we'll try to narrow down the time of the future event." Dawson turned to glare at his assistant. "Get the other techies up here and get to work," he ordered. The man nodded and scurried away.

He turned to point at Chris. "Meanwhile, you're the military genius here. I don't know shit about that so you better get those reputed brain cells of yours on overdrive. You, too," he added, sweeping his arm to indicate Kate, Reeves and Bonner. Then, as if they had ceased to exist, he turned to his workstation, picked up a donut and started to work.

Chapter 60

Quantum Center
Saturday, 7:12 a.m. MST

The new threat to the quantum center put paid to the press conference. Instead of canceling it altogether they decided to postpone until that afternoon. That would gain them some time, but would assuredly create another media reaction. Reeves agreed and Kate left to make the arrangements with the lab's public information officer.

Bonner had already gone to the security headquarters to coordinate investigation of the warheads and to brief his trusted officers. That left Chris and Reeves to ponder their next steps.

At Chris's request, Dawson had posted a large count-down clock in the upper right corner of the big overhead monitor. He was working in five-minute increments from six hours ahead to determine the approximate time at which future viewing cut off, the presumed time of the detonation. Every ten minutes or so the gap closed as a new future snapshot was confirmed, each step narrowing the event time.

Now the decision weighing on Chris and Reeves was whether to initiate an evacuation of the Los Alamos lab and the surrounding town. It was a tough call, since panic could result in chaos and deaths. What if people died and the threat turned out to be false?

Reeves reviewed their alternatives. If they did not announce an evacuation and their fears were realized, they could be responsible for thousands of deaths. On the other hand, if they started a panic for what turned out to be no reason and people died or were seriously injured in the chaos, that would also be on their heads.

"It's a classic Catch 22," Reeves concluded somberly. "We're damned either way."

"Okay, fair enough," Chris agreed. "I vote for punting. Let's hold off on that option until we know more."

"I can agree with that, at least for a while."

"But we must start right now to plan for how to facilitate the evacuation if we need to go ahead. For example, we can assign Bonner's security force to aid the local police and fire departments. The school district can be put on standby to make its busses available."

"Good," Reeves said, making some notes in his PDA. "Also we should make special arrangements for the residents in nursing homes, hospitals, homebound individuals."

"Good. We've got the outline of a plan so let's move on to the real problem. Damn, I wish Bonner would get back to us on the status of the warheads."

"Give him a call. Meanwhile, I'm going to start detailing an evacuation plan."

"Okay." As Reeves walked away Chris picked up the encrypted cell phone and punched the autodial code for the security chief.

"Bonner." The voice was brittle. He was obviously on edge.

"Paul, can you give me an update?"

"Oh, hi Chris. Yes, dammit, we're waiting for a passkey to open the vaults. They're located in the pit production facility, where they used to machine plutonium to make bomb cores. As you can imagine this is a pretty high-security area. The department chief has to personally enter the code by thumbprint scan. He's not here and isn't answering at his house. I've sent a car to check on him."

"Jesus."

"Yeah."

Chris thought for a moment.

"Listen, if the vaults can only be opened by his authority, doesn't that make him a major suspect in case the warheads are missing?"

"Well, yeah, maybe. I thought about that. But then I realized that since our security files have been breached, someone could have used the thumbprint pattern from the man's personnel file to make a fake thumb. I asked one of my technicians if that would be possible, and he said it was."

"So we're back again to someone with access to the private files. Who's your top suspect?"

"Captain Timmons. He's in charge of the records and verification process. He's been here two years, from before my time. His file looks clean as a whistle—I wouldn't have even flagged it except for his unique position with the access and authority to manipulate personnel records."

"If he's working under a professional legend you'd expect it to be convincing," Chris offered. "Where is he now."

"He was off-duty last night but checked into the lab a few minutes ago through the main gate. He hasn't shown up and isn't answering a page."

"Not good. You have people looking for him?"

"Yeah, about a dozen officers are on his case."

"Okay, good. What about blowing open the vaults?"

Bonner said nothing for a long moment, then in a quiet voice:

"You would actually propose to set off high explosives near a thermonuclear warhead?"

"Oh! No, I guess not." Chris shook his head as if checking to see if his brain was still inside. "Can you use a diamond saw?"

"Sure, if we had all month. Those vaults are serious shit. The doors are eight inches of tool-grade tungsten steel."

"Okay, so what about doing what you suspect the perp may have done and fake a thumbprint?"

"I've got my chem lab guy working on that, using a scan from the file print and working it into latex. He says it's going to take some time. Plus, we're not even sure the file hasn't been altered. Anything's possible I guess."

"Time's the one thing we don't have..." Chris paused. "Wait a minute, we do have time, of a sort. Hell, Tommy could look into those vaults through past viewing!"

"Christ, why didn't I think of that? Sure, let's do it. I'm standing right outside the vaults now. I'll email a GPS reading to Tommy. Get him set up and I'll send it along in a few seconds."

"Right." Chris clicked off and headed for Dawson's station. In moments the coordinates were being set up and entered for the first of the three vaults.

"This won't take long," Dawson told them. In a moment a plan view of the vault facility appeared. The physicist toggled the positioning arrow until it was just inside the door of the first vault, set the time for a half hour previously and hit the enter button.

In a few seconds a dim infrared image appeared. It revealed the ugly shape of a nuclear warhead resting in a padded cradle on a heavy steel pallet.

"That one's there," Chris said sharply. "Go to the next vault."

Dawson complied and in a moment a second image appeared. Again, it showed a warhead on its cradle.

Could Delancy have been bluffing? Chris shook his head in frustration. "Go to the next one Tommy."

A few more keystrokes, a turn of a trackball and the past view began to process. An image appeared.

The third vault was empty.

Chris stared momentarily then reached for his cell phone. He punched redial and waited impatiently through two rings before the security director picked up. Chris didn't wait for Bonner to speak.

"The third one's gone," he informed him. "Start a search. Could it have been removed from the lab grounds?"

"Oh Jesus." Bonner thought for a moment. "I don't know for sure. We check things pretty closely—generally not even a paper clip leaves the lab without our knowing about it. But since our security's been breached..." His voice trailed off.

"Oh, that's just great," Chris declared. "So we not only have a loose warhead, but it could be anywhere in the whole town."

"Yeah." Bonner sounded cowed. "It could, and now that I think about it that would make sense because it would be a lot easier to hide the warhead outside of the lab."

Chris had reached the boiling point. "Back in the service we used to talk about FUBARs—that's an acronym for 'Fucked Up Beyond All Recognition'—but I've never seen as big a one as this."

"Listen, I agree," Bonner replied stoically. "I want you to know that as soon as this is over, if I'm still alive I'm immediately tendering my resignation."

Chris calmed down as if he'd received a dash of cold water.

"Oh, shit Paul. Don't say that! I'm not blaming you. We're all in this mess through no acts or omissions of our own. You inherited this FUBAR when you took the job. Come on, I don't want to hear any more talk of guilt or blame. You're one of the good guys. How could you know the lab had been infiltrated with sleeper agents? That should have been the FBI's responsibility."

The line was silent for a long moment. Chris could hear Bonner breathing heavily.

"You're right," the former FBI special agent said at last. "Thanks. But I still may decide to take my retirement and go lie on a beach somewhere with a tropical drink in one hand and a pretty woman in the other."

"Well, okay then. Let's get on with it."

"Right. Here's an idea: Can Tommy look back to find out when that warhead was taken out of the vault? If so, we could follow it to wherever it went."

"How long ago was the vault last opened?"

"Not since the warhead was delivered, about a month ago."

"We could try that. I'll get Tommy on it. Meanwhile, start searching the lab for that bomb. Stuart and I decided not to announce an evacuation quite yet, but the clock's running down on that option. If we do decide to go ahead, we thought your staff could help the local authorities."

"Sure. How long until you need to make that decision."

Chris glanced up at the distant screen. The count-down clock read +6 hours 40 minutes."

Looks like Tommy's narrowed the event down. We've got a little over six and a half hours until the balloon goes up."

"Jesus! Okay, I'll get that search underway. Keep me informed."

"Roger. And Paul, please, stop beating yourself up. This isn't your fault. You're doing a great job."

"Thanks." Bonner clicked off the connection.

Chris stood up and stretched. So tired. He yawned and looked around the center. Everything seemed to be in order. He headed back to the control center where Dawson and his technicians were busy narrowing down the exact time of the looming disaster.

"Tommy, can you get us the location of the explosion? I think if you can narrow down the time close enough you could glimpse it just before the center is destroyed."

Dawson shook his head.

"Sorry, no can do," he said with a doleful expression. "You're an atomic physicist, so you should have known that."

"Uh, okay, so remind me. I'm still trying to get my head around this time dimension stuff."

"How long does it take for a thermonuclear warhead to detonate?"

"Oh, I see," Chris responded. "About a microsecond."

"Exactly. One-millionth of a second. And how fast does the radiation front expand?"

"Well, of course, I see it now. The X-ray pulse travels at the speed of light and the neutron wave front is only a little bit slower. Anything within fairly close range would be totally destroyed in nanoseconds, including this center."

"Yes, indeed, give the man a cigar."

"And that means you'd have to narrow down the event to within a very small fraction of a second in order to glimpse the ignition."

"Right. And we couldn't possibly do that."

"Okay, sure. There's something else I want you to work on then. Bonner suggested that you search back in time to discover when the warhead was taken from the vault. Then you can move forward and trace it from there."

"Oh, that's an excellent idea. We can track the bomb just like we did that Iranian spy. We'll get right on it." Dawson turned back to his workstation and began shouting orders to his assistants.

Kate had walked up in time to hear the end of this conversation. She smiled weakly and plopped down in one of the swivel chairs.

"That does sound excellent," she said. "If we can trace the bomb, we can stop it." She looked relieved.

"Yes, it could work, if we have enough time," Chris said without too much enthusiasm. "Actually, it might be our only chance." Quickly he filled her in on what he'd learned from Bonner.

"So you think it may have been taken out of the lab grounds? If that's the case, it could be anywhere within a radius of a mile or more." Her sense of relief faded as quickly as it had come.

Chris pulled a chair over next to hers and sat down. He rubbed his face with his hands and ran his fingers through his hair. He let his hands drop to his side and sagged for a moment, trying to relax tired, tense muscles.

"Here, Honey," Kate said, standing up and moving behind him to massage his neck and shoulders.

"Ahh." Chris squirmed with pleasure. "That feels so-o good! Just a little lower please."

The encrypted cell phone buzzed and Chris pulled it out of his pocket and looked at the screen. It was Bonner. He punched the speaker button.

"What is it Paul?"

"We got Timmons," Bonner said. He sounded more upbeat than he had in days.

"Where?"

"The bastard was trying to leave through the south gate. I had him on a watch list and the gate guards stopped him. We have him in one of the detention cells right now."

"Has he told you anything?"

"Nada, but it's funny. He's as nervous as a cat. Keeps looking at the clock on the wall. And that's not all. He lives nearby so I sent a couple of men over with his house keys. They just checked in and they found something interesting. It seems he was poised to make a run for it. All his luggage was packed and sitting inside the garage, ready to load in his car."

"Sounds like a man with something on his mind. See if you can get him to squeal." Chris thought for a moment. "We need to find out where the bomb is, if he even knows. Here's an idea: Tell him we're going to evacuate the town and that you'll leave him behind in the cell."

Bonner chuckled with appreciation.

"Fisher, you have a devious mind. I'll do it. Talk to you later."

"Oh my," Kate breathed. "This is really getting down to the wire isn't it?" She glanced up at the countdown clock. It now stood at a little over +6 hours 20 minutes.

"Well, that gives us a second possible way to locate the bomb," Chris observed. "Let's hope one of them works. Meanwhile, I need to find out what we need to know about that warhead. Here comes Stuart. He's the one who can tell me."

Reeves walked up with the gait of a zombie. He was still wearing the foam neck brace. His eyes were bloodshot. Chris pointed at a chair and the acting director gratefully sank down. Kate was still administering a neck rub to Chris. Reeves watched enviously and Kate nodded. "You're next," she told him.

Chris briefed Reeves on the latest news. "We have a chance to locate that bomb, and we need to be ready to act when we do. I need to be briefed. What are the specifications of this thing? How big is it, for instance?"

Kate started on Reeves's shoulders and he grimaced as her fingers ground into tense muscles.

"The W-87 warhead was designed for the MX or Peacekeeper ICBMs," Reeves recited. "These were MIRV rockets, multiple independently targeted reentry vehicles. Those delivery systems could carry up to a dozen of the W87 warheads, each in a separate reentry capsule. They would part from the boost rocket on the downward path of the trajectory and be steered to various targets. Now some of them are being retrofitted onto the older Minuteman Three ICBMs. It's in connection with that that we were tasked to test those samples."

"I suppose they can't be very big then?"

"Not compared with the city buster devices designed for delivery from B-52 or B-1 bombers. With their individual reentry capsules and heat shields they're about six feet long and around twenty inches in diameter. They weigh about 500 pounds."

"That's quite a bit to carry around."

"Well, yes, but what we have here are just the bombs, without the re-entry shells. They're small and light enough that a strong man could possibly lift one. They could certainly be moved with a hand cart."

"Okay, so someone could have hidden it in a car and snuck it out of the lab?"

"Only if there was a security breach, but as we know that to be the case the answer is yes."

"How would the device be triggered?"

"That's a good question. They have all the latest safety features including arming and fusing security. No ordinary individual could possibly jerry-rig one of these to explode." Reeves paused. "Trouble is, this place is crawling with individuals who might be able to do exactly that…"

"Yes, I see," Chris replied. "Paul told me he had his eye on several atomic engineers and physicists. How would they do it?"

"First they'd have to over-ride and disable the safety features. That could be done with a few modifications as long as they knew exactly what they were doing, which we presume they would. Then they'd probably wire it up to a laptop computer and program it to initiate the physics event—that's what we call the ignition, when the fission and fusion reactions take place."

"Yes, I'm familiar with that term. Was this warhead designed here at Los Alamos?"

"I see where you're going, wondering if one of the original designers might be involved. But, no, the answer is that it was designed at Lawrence Livermore. However, it's very similar to the W-88 which was designed here."

Chris reached for his phone and pressed the number for Bonner. The security director answered curtly. Chris could hear the frantic chatter of voices in the background.

"Paul, run the personnel records on every engineer and scientist on the staff. See if any of them were ever at Livermore, and if so, what they did there."

"Sure, I'll get back to you." Chris clicked off the phone.

"That's a thought," Reeves said appreciatively. "Hadn't occurred to me. We could be dealing with a deep mole who transferred in."

"Anything's possible."

"Well, yeah, I sure would agree with that after what we've seen these last few days." Reeves squirmed to the side. "Kate, a little to the right if you don't mind."

She chortled, the first time Chris had heard her brilliant, two-toned laugh in several days. "Your private masseuse, at your service," she told Reeves, who grinned appreciatively.

Just then the lights went off and an instant later there was the dull thump of an explosion. All three of them jumped. Emergency lights powered by batteries came to life, casting a dim glow in the spacious center.

"What...?" Reeves began.

"The power went down," Chris shouted. "That can't be the nuclear event, it's still six hours away."

They hurried over to where Dawson and his technicians were frantically saving files and shutting down systems before battery backups were exhausted.

"Tommy, what is it?"

"It must be the turbine. Something happened to it."

Chris and Kate looked at each other then started for the ladder stair leading to the roof. The nine-megawatt General Electric gas turbine generator was located in a small building about a hundred yards away from the quantum center. It provided the power for the Qubit Farm, the massive server array and all the other equipment needed to process the quantum computer's output.

Chris took the stairs three at a time and popped out in the west end pillbox. Perry and Chen were standing at the edge of the sandbag wall staring outward. Chris stepped up beside them. Kate was there two seconds later.

"Oh my," she intoned.

A cloud of black smoke was rising from the generator building. Flames were shooting from a hole in the roof. It was obvious the turbine was out of action, possibly for good.

Chris looked at the two SEALs. They shrugged.

"There was a hell of a bang, sir. That's all we know," Perry said.

Chris nodded and ran back down the stairs. The control center was a whirlwind of activity. He grabbed Dawson by the shoulder to get his attention and shouted to him.

"Can we go to alternate power?"

"Hell no," Dawson shouted back. "Why do you think we had that multi-million-dollar generator in the first place?"

"Then we're shut down? No more time viewing?"

Dawson nodded solemnly.

"Bye bye to that," he groaned, sinking down in his chair. "We're blind now."

Chris saw a tear roll down the physicist's cheek.

"I wasn't keeping watch on the near future," Dawson moaned, beginning to rock back and forth. "I should have seen it coming…"

"It's okay, Tommy," he said consolingly. "We had you doing other things. You've done great. We'll take it from here."

Chapter 61

Los Alamos
Saturday, 8:47 a.m. MST

The quantum center was deserted. After shutting down the system, Dawson's team was dismissed. After some thought Chris released his SEALs from 24 hour surveillance and told them to get a couple hours of rest, but to stay near and on call. Chris, Dawson and Kate had returned to the safe house inside the lab grounds. Reeves had gone to his office in the executive center and Bonner was working at his security headquarters.

Dawson had fallen into a depression. Kate tried to cheer him up, offering a Coke and roast beef sandwich, but the physicist merely looked at them and turned away. Kate glanced at Chris and shook her head.

Chris was sprawled on the couch, hands behind his head, staring at the ceiling. His mind was racing. He beckoned Kate to come closer and took the sandwich that Dawson had refused.

"Well, we're a quiet bunch," Kate declared at last. "You guys aren't going to accomplish anything just moping around."

Chris sat up and took a bite of the sandwich. "Excuse me. I was actually just thinking about what to do next." He chewed another bite then looked wistfully at the Coke. Kate handed it to him.

"And excuse me, too," Dawson muttered, "I was actually just thinking that my life is over."

"Tommy, cut that out!" Kate swatted him on the shoulder causing him to jump. "We can't tolerate negative thinking at a time like this."

Startled, Dawson stared at her, eyes wide. "Negative thinking? Negative thinking! You don't know the half of it."

"Yes I do, Tommy. You're acting like a spoiled child. Now straighten out and act like a grown-up."

The physicist stared at her in astonishment. Slowly a smile began to form on his lips.

"You know, my Mom used to say that to me," he said in a wondering voice. "Haven't heard it in years..." His voice trailed off as his mind seemed to travel back in time.

"Well, I apologize Tommy. I shouldn't have said that."

He looked at her solemnly, the little smile still playing around his mouth.

"No, it's okay. I'm really glad you did. It brought back pleasant memories—and you were right. I needed a swat on the butt."

"Well, all right, then."

Dawson watched enviously as Chris ate the last of the sandwich. His eyes followed the can of Coke as Chris raised it to his lips. He turned back to Kate.

"Do you have any ham and Swiss?"

Kate laughed out loud and went to make another sandwich.

"And a couple of Cokes...please?" he called after her.

"Sure, Tommy, coming right up."

A minute later Chris's phone buzzed and he picked up a call from Bonner. He put it on speaker just as Kate walked out of the kitchen and handed Dawson his sandwich and soda.

"Whatcha got?"

"Not much," Bonner admitted. "Timmons won't talk, but we're sweating him pretty hard. Haven't found the bomb. Nothing much else is going on. Stuart says you decided to postpone pulling the plug and calling for an evacuation. How long do you think we can wait."

"Well, we figured we have until about 2:40 this afternoon to get everyone out. If we allow five hours we should announce the evac at

9:30." Chris glanced at his watch. "That's only forty-five minutes away. I don't know, maybe we should just go ahead and do it now…"

"It might not make much difference. Meanwhile, I've got that information you asked for."

"What information?"

"You asked me to check the personnel records to see if any of the staff had been at Lawrence Livermore lab."

"Oh, yeah. Find anything?"

"I don't know if it means much, but there's one guy here that fits that profile."

His interest piqued, Chris sat up on the couch.

"Yeah?"

"This guy's a heavy hitter, Ph.D. in nuclear engineering from Princeton. He worked at Livermore for about fifteen years before moving down here a year ago."

"What did he do there?"

"Well, that's just it. He was project manager on the W-87 warhead, the same kind that's missing."

Chris felt a surreal sense begin to emerge from somewhere deep within his mind. It was an eerie feeling, but he vaguely recognized it.

"What's his name?" he asked sharply

"Harold Muntz."

"Where is he now?"

"Off site since the lockdown. Called his house and nobody answered."

Chris felt a flash of insight. *That's it. Muntz is the key.* He stood up.

"Paul, I've got a feeling…" He stopped, confused. *This sounds silly. How can I be so sure?* He hesitated, leaving Bonner hanging on the line.

"What feeling?" Bonner urged.

"Well, a feeling about Muntz. Somehow, he's got to be part of this."

"Well, I don't know…" Bonner sounded skeptical. "It's pretty circumstantial. There's nothing to indicate this guy…"

"Humor me, Paul," Chris cut in. He looked imploringly at Kate. She smiled and touched her Thunderbird pendant meaningfully. Chris glanced down at his own fetish, Coyote. The chain was outside his shirt collar and the little figure was dangling on his chest. *This is ridiculous.* But then he thought about the time he'd sensed the roadside bomb in Afghanistan. *Is this any different?*

"I'm calling my team. Get me the address for Muntz. We're going there right now."

Bonner cleared his throat. "Chris…"

"Don't argue with me. I'm pretty sure about this."

Bonner hesitated for a moment before deciding.

"Get your men to the front gate. I'll meet you there in five and lead you myself."

"Right."

Chris snapped off the phone and redialed to reach Perry.

"It's a duty call, Leonard. Get the men together. Front gate, vehicles and weapons, five minutes."

"Aye aye, sir."

Chris turned off the phone and began shrugging into his shoulder holster. Kate jumped up and grabbed her messenger bag.

"What are you doing?" he asked.

"I'm going too."

"No you're not."

Kate gazed at him sternly. He almost flinched at the fire in her eyes.

Dawson spoke up from his workspace. "Want some advice, Fisher?" he asked quietly. "It's free."

"Sure, Tommy, what is it?" Chris asked patiently.

"Don't argue with the woman. I know her type. My Mom was like that, and you didn't ever want to cross her. Not ever."

Chris frowned for a moment then found he couldn't hold back a smile.

"Get your coat and let's go," he told Kate. "Tommy, you're in charge here."

"Sure, but Kate, could I have another sandwich before you go?"

"No, Tommy," she told him firmly. "You have to get it yourself."

"Yes, Ma'am," Dawson said placidly, bowing his head.

"Oh for crying out loud, cut it out!" she told him. He looked up at her slyly and winked.

"Okey dokey, Mom," he said. Kate laughed out loud.

Chris pulled on his sports coat as he and Kate ran out the door. As they jumped into his Jeep, Kate laughed again.

"What?"

"Oh, just Tommy. He's so precious. And, I was thinking about how much this is like the time we made a run for it after the kidnap attempt."

Chris started the engine and slammed the transmission into gear. The rear tires spun gravel. They drove in silence for a moment before he spoke.

"That thing with Thunderbird and Coyote...? I don't get it. I'm a scientist. I can't believe in this stuff."

Kate observed him for a moment as he drove.

"Do you think science has all the answers?" she asked.

"Well, no, not yet anyway."

"What you told me about Little Fox and how he saw the fish...do you think science has an explanation for that?"

"No."

"And that roadside bomb? You knew it was there."

"Yeah." Chris concentrated on his driving. "It's the strangest feeling, but somehow as soon as Bonner told me about Muntz I just knew that was the answer..."

Kate reached over and touched his arm. "Yes, I saw it in your face when you had the inspiration. Chris, you can see the fish!"

He glanced over at her and smirked.

"Yeah. Now all we have to do is catch the damned thing."

Chapter 62

Los Alamos
Saturday, 9:16 a.m. MST

Perry and his team were already waiting at the gate when Chris and Kate arrived. Bonner drove up a moment later, skidded his black Durango to a stop and buzzed down the window.

"Muntz lives about a mile away," he shouted. "Follow me." He threw the truck into gear and took a hard right onto the main street. Chris and the SEALs followed in their four vehicles. Bonner lit up his light-bar, hooted the siren and led the convoy through a red light. In less than a minute they were turning into a residential street. Bonner pulled to the curb in front of a large house, shut down his engine and jumped out.

"This is it," he said as Chris joined him on the sidewalk. "I don't know what you have in mind, but try to avoid mayhem." He eyed the heavily armed SEALs who were gathering around Chris. Kate was standing nearby recording with her video camera.

"Only as necessary," Chris replied noncommittally.

He studied the house for a moment. It was faux Dutch Colonial with two stories and a high hip-style roof. He made a hand signal. Wordlessly Perry and Garrett headed to the front door and took up positions on either side. De la Vega and Chen went around the left side of the house while Tuttle hotfooted it around the right side.

Martin Sharp threw a sandbag on top of his SUV and laid the Barrett 50 across it, covering the house from the street.

"Jesus," Bonner said admiringly. "You've got those guys trained pretty well."

"You have no idea," Chris told him, striding away toward the house.

After watching the front windows for a moment to spot any sign of life, Chris pushed the doorbell. He waited five seconds and pushed it again, then knocked hard and rang the bell a third time.

"Mister Perry, please note that there is no response to my lawful attempt to gain a response from this house."

"Noted, sir."

"Is there any indication to you, Mister Perry, that a civilian inside this house may be in danger?"

Perry glanced at Mouse who was towering at the opposite side of the door, his M-4 carbine poised.

"That would be correct, sir," he said, trying to suppress a grin.

"Then am I also correct, Mister Perry, that it is our duty to force an entry in order to assist that civilian?"

"Aye, sir." Perry gestured to Mouse. The giant SEAL stepped in front of the door. He cocked his right leg and delivered a powerful kick with a size 16 combat boot. The impact would have stunned a mule or given an elephant pause. With a splintering crash the front door spun open on one hinge, pieces of the shattered lockset flying onto the floor. Twisted by its own weight the door pulled the screws out of the last hinge and fell to the floor with a heavy thud.

Watching from the sidewalk, Bonner observed to Kate that this wasn't exactly the way the Bureau did things.

"This isn't the Bureau," she pointed out. "This is a military action."

"Yeah." Bonner looked worried.

Chris was first in the door, stepping aside to let Perry and Garrett past him. Garrett headed straight back through the house to secure the kitchen and open the back door to let the others inside. Responding to hand signals, Garrett, De la Vega, Tuttle and Chen spread quickly through the ground floor, securing the dining room, family room and a small office in a rear corner. The attached garage held a sedan, lawn mower and other tools. As they did that Chris beckoned to Perry and the two thundered up the stairs.

The top floor had a typical layout with a central hallway and doors leading off to bedrooms and baths. They started in the back and worked forward. There were three small bedrooms and a stand-alone bathroom, all empty. At the front of the hallway stood a set of French doors that presumably led to a master suite. Chris approached the door, stood to one side and turned the knob. He pushed the door open with the toe of his boot as Perry covered him from the other side.

Nothing. The room was empty. So were the bathroom and a walk-in closet. Perry looked at Chris but said nothing. Chris stopped in the middle of the room and looked around. *There's something here.*

Just then his cell phone vibrated. He pulled it out, saw it was a call from Reeves and pushed the speaker button.

"Yes Stuart?"

"It's coming up on 9:30, the deadline for the evacuation…"

"Delay it," Chris ordered.

"But…"

"I need a little more time."

"Well…"

"Just trust me, Stuart."

"Okay, we'll go to 9:45, but that's it."

"Thanks." Chris snapped off the phone. As he talked he'd been examining the room. Now he stepped out into the hallway. Glancing at Perry he pointed up. A short rope with a handle dangled from a rectangular door set in the ceiling. Perry grabbed the rope and pulled. A folding stair came down, access to the attic. Chris shot up the stairs with Perry right behind.

To the left was a stack of storage boxes, some luggage and a hanging rack holding winter clothing on hangers. Toward the front the attic had been partitioned off with plywood. There was a standard doorway. Chris kicked it open and stepped aside as Perry swung the barrel of his carbine around the room.

"Sir!" Perry did a quick two-step into the room, lowering the barrel of his weapon. He pointed with his chin.

Following him inside, Chris stared briefly at the sight. The attic space featured two square windows looking out toward the Los Alamos lab like a pair of eyes. In the distance stood the quantum center, like the sweet spot in the center of a bull's-eye target.

Dominating the room was a bulky shape hidden under a blanket.

And in the far left corner a human figure was huddled, wide eyes staring over a patch of duct tape that covered his mouth.

Chris stepped closer. The man was chained and manacled to the wall, hands twisted behind his back. Chris leaned down and stripped off the duct tape.

"Oh, thank God," the man croaked. "I didn't think..."

"Doctor Muntz?" Chris cut him off. The man nodded uncertainly, looking frightened.

Behind him they heard Perry shouting down for a bolt cutter and in a moment Tuttle appeared with the requested tool. He snipped the chains and Muntz collapsed weakly onto the floor. Chris kneeled down in front of him and lifted his chin with one hand to look into his eyes.

"How long have you been here?"

"Two...two days," the engineer croaked.

"Water," Chris shouted over his shoulder and a bottle appeared. Muntz grabbed and drained it without stopping.

Chris looked over at the blanket-shrouded shape.

"A W-87?"

Muntz nodded.

"Armed?"

"Yes..."

"Can you stop it?"

The engineer shrank back. His head dropped to his chest. Again Chris lifted the man's chin and looked him in the eyes.

"I... He...he said he'd kill my daughter," Muntz said. "She's in Santa Cruz. Someone kidnapped her. He put her on a cell phone. Said I had to do this or she'd be killed." Fat tears were rolling down

Muntz's cheeks. "He said it was just a bluff, that it would never be detonated..." He lapsed into silence, staring at the bomb.

"Okay, now you're going to help me disarm this thing," Chris said. He carefully lifted the blanket to reveal the familiar shape of the W-87 nuclear warhead. It was identical to the two he had seen in the vaults through Dawson's past viewer. The bomb was set on a sturdy workbench. Beside it was a compact laptop computer.

Muntz squirmed to bring himself into an upright sitting position with his back in the corner of the room. He tried to raise himself further but fell back. Perry reached into a pocket and tore open an energy bar. Muntz grabbed it and ate hungrily.

Chris stepped in front of him.

"Muntz, did you hear me?"

Chewing ravenously, the engineer nodded. He swallowed and took another bite. Chris waited patiently for a ten second count before snatching the remainder of the bar out of Muntz's hand.

"Later," he said. "First this." He pointed at the warhead.

"I... I don't know..."

"What do you mean you 'don't know'?" Chris demanded, his voice rising. "You made this thing, you fixed it to detonate. Do you realize what you've done?"

Muntz slowly nodded, his eyes squeezed tight shut. Tears were still dribbling down his cheeks.

"What do I need to do?" Chris demanded, grabbing Muntz by the collar and pulling him partway upright. He gestured for Tuttle to bring over a chair that was sitting in an opposite corner, pointing to the warhead. Tuttle put the chair down in front of the thermonuclear bomb and helped Chris lift Muntz and carry him over to sit in it. The engineer stared hopelessly at the machine of death that dominated his attic.

"What do I need to do?" Chris demanded again. "Is it booby trapped?"

Muntz shook his head slowly. Cautiously Chris flipped open the laptop computer. At the top of the screen it displayed a countdown clock. It read -3 hours 52 minutes 14 seconds and was clicking downward with all the seriousness of a heart attack.

A blinking field appeared in the middle of the screen. A legend instructed "ENTER PASSWORD."

"What's the password?"

"I... I don't know. He changed it."

"Shit." Frustrated Chris stepped back and circled the bomb warily.

"Okay, if we can't access the computer we need to disable it some other way. You're the expert—what do we do?"

Beginning to regain self-control, Muntz sat up straighter in the chair, rubbing his wrists where the chains had cut off his circulation.

"There are certain problems," he observed. "May I ask who you are?"

"Lieutenant-Commander Chris Fisher, U.S. Navy."

"Not the Chris Fisher from Stanford? The physicist?"

"Yes, on active duty."

Muntz relaxed slightly. Still rubbing his wrists he looked over at the blinking screen.

"If you disconnect the computer, the bomb detonates," he said.

"Damn it! You said it wasn't booby trapped!"

"Well, all right, maybe I didn't understand the term. There isn't any way to stop it without accessing the computer."

Chris sagged. *This isn't going well.*

"What if we simply begin to disassemble the bomb, physically take it apart?"

"No, that won't work. The safety mechanisms have been corrupted, reversed if you will. They've been set to do the opposite of what was intended, causing detonation rather than disarming the warhead."

Chris felt his cell phone vibrate. He walked over to the windows to take the call from Reeves.

"Chris, time's up."

"Stuart, we've got the bomb!"

"Oh my God. That's wonderful."

"Maybe not. It's armed and set and we need to figure out how to shut it down. Muntz was forced to modify it. I need you here. I'm going to send Bonner to bring you over."

Without being told Perry ordered one of his men to alert the security chief.

"So, is it safe?" Reeves asked. "What about the evacuation plan?"

"Not yet. Just get over here, Stuart." Chris snapped off the phone and turned back to Muntz with a grim expression. The engineer shrank back at the sight.

"Better start thinking, Muntz," Chris told him. "We need to fix this problem, now. We're on the verge of having to call for a total evacuation, with all the chaos that would cause. No matter what happens, if this thing goes off I'll make sure you're not among the survivors."

Frightened, Muntz nodded emphatically. "I'll help you as much as I can," he murmured. "But…"

"No buts," Chris said with finality. "Start thinking."

For a long moment no one said anything, then Perry cleared his throat.

"Begging your pardon, sir, but what about moving the bomb somewhere outside of the city?"

"No!" Muntz almost shouted. "It's grounded in place. If we break that ground it will start the physics event."

Perry looked at Chris.

"He means it'll detonate," Chris explained.

Again there was silence.

"Okay, tell me about this guy. Who did this to you?"

"It was a captain in the lab's security force. He didn't tell me his name, but I've seen him around…"

"Timmons," Chris said.

"Oh, that's right. I heard him use that name on his cell phone."

"All right, and he reset the password just before he left. When was that exactly?"

"Early this morning. He came back around dawn. Didn't even say anything, just walked over to the laptop and started using the keyboard. Then he grinned at me and told me he'd activated the countdown. Next he typed a few more times and said he'd changed the password."

"So you don't have any idea how to reset that computer and safe the bomb."

"No." Muntz frowned in thought. "There is one thing..."

"What?"

"Well, I don't know if it would help at all, probably not. But, you see I'm pretty sharp with numbers. I counted the keystrokes when he entered the new password."

"Oh," Chris sighed. "You're right, probably not much help. What was it?"

"Fourteen characters. Shift on the first, ninth and tenth."

Chris pondered this. Well, it was something, but that pattern could describe literally millions if not billions of combinations of letters, symbols and numbers. *If only we still had the use of Tommy's quantum computer...*

Chris heard the scream of Bonner's siren approaching and in a moment Reeves was pounding up the stairs. Entering the room he stopped as if he'd run into an invisible wall, his eyes drawn to the compact shape in the middle of the room.

"You really found it," he murmured. "Now what?"

Quickly Chris filled him in on the situation.

"What do you think, Stuart? Is Muntz right that the only way to disarm the bomb is through the computer?"

Reeves pondered for a moment, slowly circling the warhead and studying the laptop. Without touching anything he examined the connections. There were three separate lines leading between the computer and the bomb, one Firewire, one USB and one Ethernet. A hole had been drilled through an access cover on the bomb and the three wires threaded inside.

"Is the computer connected to the Internet?" he asked Muntz.

"Yes, through my wireless router downstairs."

"Could that give us a back door into the box?"

"No. Unless the password has been entered, the computer will recognize no instructions."

"Umm, okay." Reeves continued his examination, asking a question from time to time. Meanwhile Chris stepped over to one of the windows and assumed a parade rest position with his hands clasped in the small of his back. He looked out at a typical New Mexico morning. Puffy white clouds resembled ethereal sheep grazing in the deep blue sky. Chris contemplated the surrounding mountains, imagining how they would look after a nuclear blast.

His mind was working. Casually, without even thinking about it, he brought his right hand around and grasped the Coyote pendant. In his mind he traveled back to the time when as a boy he'd watched Little Fox reach into a stream and catch an invisible fish. He thought of the time in Afghanistan when he'd been struck with the uncanny knowledge of a roadside bomb. *They said it was a sixth sense.*

He raised Coyote up in front of his face, turning it to sparkle in the beams of the morning sunlight streaming into the attic room.

What was it that Eaglefeather said? That Coyote could take me to other worlds and bring me safely back? Something like that.

Still clasping the silver pendant, Chris closed his eyes. He felt the warm rays of the Sun on his face. He tried to clear his mind—and discovered a mysterious and unfamiliar consciousness.

Something was happening. He began to feel the same strange certainty that had come over him when Bonner told him about Muntz, an assurance that could not have come from mere knowledge but from some deeper insight.

He stood for a while in the sunlight, eyes closed. The Coyote pendant was grasped firmly in his right hand. He felt a shift. He seemed to open his eyes, but what he saw was not the scene before him. In his mind's eye Chris floated upward.

He looked down at himself, standing rigidly in the warm New Mexico morning, eyes still closed. *I look like some ancient priest, offering up a sacrifice, holding a tiny idol aloft to the Sun.* He realized this was what is called an out-of-body experience.

Shifting his bird's-eye viewpoint he surveyed the room. Reeves was still intently examining the bomb, talking quietly with Muntz. Tuttle and De la Vega were standing uneasily near the door. Kate had come up to the attic and was recording the scene with her video camera. Her attention was focused on the two men huddled beside the warhead.

Perry was on the other side of the room, staring curiously at something. Chris shifted his viewpoint and realized that Perry's eyes were locked on him, the real him, standing at the window as if in a trance. Perry didn't move or say anything, just watched.

There was another change. The room below shrank and fell away. He was falling, but upward, into the sky. Calmly he watched the ground dropping away beneath him. Clouds flew past. The Earth was spinning, the Sun traversing the sky, sinking in the West. Darkness fell and the glowing band of the Milky Way divided the glorious night sky. Defying gravity he continued to fall upward, faster now. The Earth itself became a glowing disc shining against the vast backdrop of the Universe.

Somehow he imagined he was not alone, but accompanied by a spirit creature, the dog-mind of his totem. It was a soothing sensation, casting a protective mantle of peace across him. He clutched the silver emblem more tightly.

I've gone mad, he thought. *Utterly mad.* But in his trance state the spirit of Coyote took on a reality that he could not deny.

It's true, I'm traveling to a world beyond. He gazed in awesome wonder at the grandeur of the stars, sprinkled like confetti across the bowl of the sky.

Now his spirit body was soaring high above the Earth. He passed from the cold shadow of the night into the brilliant warmth of the Sun.

And it all began to fade.

Slowly, Chris became aware that he was back in the attic room, standing at the window, his face bathed in morning light. Cautiously he opened his eyes. The silver pendant was there before him, silhouetted against the brilliant sky. Slowly he dropped his hand, letting the image of Coyote fall to his chest.

Something had changed inside of him, bringing a new sense of freedom, of confidence. He shook his head and turned around.

Perry was gazing at him with a strange expression on his face. Kate looked up from the monitor of her video camera and met his eyes briefly. She smiled uncertainly and he smiled back.

Reeves and Muntz were still talking but with little enthusiasm. Weakened by two days in chains, Muntz was beginning to collapse. Reeves, exhibiting the first hints of an outburst of anger, turned to Chris, his face drawn.

"Fisher, it's no go. I'm going to call the evacuation." He reached for his cell phone.

Saying nothing, Chris shook his head. He walked over to the laptop, pushing Muntz aside. He clicked to activate the cursor in the password field and began to type, very carefully, one-finger style.

"First character shift for capital," he murmured. "Next seven characters lowercase. Shift for underscore. Shift for capital letter. Four more lowercase letters."

He paused and studied the screen for a moment.

"What are you doing?" Muntz cried. "It won't let you guess! If you enter the wrong password…"

Chris glanced at the frightened engineer with disdain. "Who said anything about guessing?" Muntz's jaw dropped. Perry, Tuttle and De la Vega shifted nervously on their feet and looked at one another. Kate resumed recording the scene as Reeves bent down to look at the computer screen.

He read what Chris had typed in the password field:

Eminence_Grise

He looked up. Chris met his gaze with a degree of calm certainty that Reeves had never before seen in any man's eyes. He nodded slightly and Chris hit the "Return" key.

There was a flash. The laptop's screen cleared. He was in.

Chapter 63

Los Alamos
Saturday, 12:55 p.m. MST

After the warhead was successfully defused there was a busy flurry of activity. Guarded by the SEALs, a team of specialists had removed the W-87 and returned it to its vault, thanks to the appearance of the department head with his required thumbprint. Muntz, under guard, had been transported to the infirmary to be examined and treated for exhaustion and dehydration.

About mid-morning the Marine unit arrived from Amarillo in a convoy of Humvees and with four Bradley fighting vehicles on flatbed trucks. As ordered, they accepted Chris's command and took up positions around the national lab. Bonner was glad to let his security officers stand down from full alert and return to regular schedules.

Chris, Kate, Dawson, Bonner and Reeves had gathered at the guest house that was their impromptu headquarters. The cafeteria had promised a much-needed hot lunch of fried chicken, mashed potatoes with gravy, green beans, salad, and apple pie. Dawson's computer and monitor had been set aside to let them sit around the dining table.

Although they were still exhausted, their spirits were high. The crisis had been dealt with successfully and the project seemed safe now. Earlier Bonner had reported how Timmons had broken down and confessed his role in the attacks on the lab.

"You were correct that Timmons was our Gray Eminence," Bonner told Chris. "He admitted he was working for deputy director Delancy. He coordinated the assassination attempt on Tommy and the kidnapping of you and Kate."

Chris listened attentively, a small, knowing smile on his face.

"And of course, the thing with the warhead was completely his doing," Bonner continued. "He named three other mutts who helped him with that. One of them sabotaged the turbine generator. Those three have apparently skipped but we have an APB out for them." He grimaced with embarrassment. "All three were high-ranking members of my staff, two lieutenants and a captain."

"It makes sense," Kate had noted. "The more rank, the more valuable. What about the helicopter attack?"

"He didn't know about that, or the Hummer diversion," Chris informed her. "I talked to Chase a few minutes ago and he's sweating Delancy."

"Hope he makes him do more than sweat," Bonner muttered under his breath.

"And of course the officer who killed Jalaly was another one of Timmons's moves?" Kate inquired.

"Yes. He was following orders from Delancy. The real goal was to assassinate Tommy and the rest of us. Thanks to you, he failed." Bonner turned to Chris. "Timmons was Grise, and Delancy was the unknown mastermind, just as you guessed."

At that point the caterers arrived with steaming stainless steel serving trays and they set the conversation aside to eat. After surviving for several days mostly on sugar and caffeine they were ravenous. Even Kate helped herself to seconds and Dawson was just beginning on a third plate when Chris's cell phone buzzed.

"It's Chase," he announced, pushing the speaker button. "Hello Admiral. You're on speaker with Lieutenant Elliott, Dawson, Bonner and Reeves."

"Good," rumbled Chase. "My congratulations to all of you."

"Where do things stand with you, sir?" Kate inquired.

"I'm out of the doghouse, Lieutenant. After we confronted Delancy with Timmons's confession he started to sing like a soprano on steroids. He admitted he was handling Timmons. As we suspected it was he who ordered the helicopter and Hummer attack not to mention the murder of Director Higgins.

"Delancy was just the tip of the iceberg, though. He's admitted being part of a long-term rogue element inside the Company, going way back to the beginning of the Cold War. It started as a Soviet initiative, then took on a life of its own. There are dozens of people involved, including a few regular case officers and a lot of contract agents, all off-the-books and funded with black ops money."

"So it was all about nothing more than personal power and wealth," Chris interjected.

"Yeah, that's pretty much it. They sold their services to enemy states, wealthy individuals and even criminal enterprises. They had a lot at stake. We've got his confession on video with four witnesses. About an hour ago I emailed the confession to about a dozen top players in Washington, including the FBI director, the NSA chief, my own boss at Naval Intelligence, key members of the Senate Intelligence Committee and the National Science Director."

"What's been the reaction?" Chris asked.

Chase laughed. "Seems I set some tails on fire. The White House has already been on the phone with me. I sent the President your report, Kate. He's read it and is on board with us."

"Great! Where are you now?"

"I'm still at the safe house, but now there are about twenty FBI agents with me, including a SWAT team. They've decided this is as good a place as any to keep working on Delancy. That secret underground bunker makes a dandy holding facility. The Bureau's got several of their top interrogation specialists up from Quantico to milk him for everything they can squeeze out. Arrest orders are already starting to go out."

"That's excellent news, sir," Chris exclaimed. "What do you think will happen now?"

"Well, Lieutenant-Commander, that's kind of up to you folks. You've got to decide how to go forward. The President told me he's ready to back you up whichever way you go." Chase paused for effect before continuing. "I believe the world is waiting."

Chris and Kate looked at each other. Bonner cleared his throat and shifted in his chair. Reeves picked at his salad. Dawson reached for another drumstick.

"All right, sir. We'll need to make some hard decisions. We've committed to a press conference at four o'clock, so that gives us about three hours to prepare."

"Then I better let you get on it," Chase responded. "Oh, one more thing. We've been invited to dinner at the White House, including your SEALs, Commander. The President wants to thank all of you in person."

"Aye, sir. We'll keep you informed from this end." Chris clicked off the phone and pushed his empty plate away.

"I'm not going to the White House," Dawson announced, his mouth full of fried chicken.

"Tommy, why ever not?" Kate asked with surprise.

Dawson looked embarrassed.

"Don't have any decent clothes," he mumbled. "Made a damned fool of myself at that thing in Stockholm. Not gonna do that again."

They all laughed and Dawson squirmed uncomfortably.

"It's not funny," he said, eyes downcast.

"Tommy, don't worry," Kate assured him. "I'll personally help you find the best tailor in New Mexico. When he's done you'll be spiffed up like a U.S. Senator."

"Really?"

"Yes, really."

"And you'll all be there?"

"Absolutely."

"Well, okay then." Dawson looked relieved. He stood and began picking up the plates and silverware. He carried them into the kitchen. The others looked at each other. Reeves cocked an eyebrow and whispered to Kate, "What have you done to Tommy?" She smiled and said nothing.

Dawson returned with a six-pack of Cokes, set them in the middle of the table and helped himself to one. He opened it and took a drink.

Reeves took one as well. As he fumbled with the pop-top lid he looked curiously across the table. Chris was sitting beside Kate, apparently lost in thought.

"Can I ask you something, Chris?"

"Sure."

"About what happened back there at Muntz's house. How did you do that exactly? I mean, when you typed the password it just seemed right, logical—but you were so certain. How did you know?"

Chris shrugged and glanced at Kate.

"Well, let's just say I had a stroke of intuition."

"Intuition, eh? Some intuition. It was as if an angel whispered the answer in your ear."

Chris laughed and Kate joined in.

"Well, I can't explain it really..."

"Hell, Fisher, I can," interjected Dawson. "You focused your mind on that totem of yours and it let you tap into the extra dimensions. I do it all the time. There's more to human senses than science understands."

Reeves stared at Dawson, then turned his eyes back to Chris. Kate was beaming and he saw that she was fingering her Thunderbird pendant.

"You guys really believe that?" he asked. Chris merely smiled. "Well, okay, I can't argue with results. But never mind. We need to prepare for the press conference."

"Yes, what are you going to tell them?" Dawson asked.

Everyone looked at the acting director. He thought for a moment before replying.

"I'll give them some of the background on the quantum computer and the troubles we've had to deal with," he told Dawson. "Then I'm going to say that I'm proud to introduce my friend, the world's greatest physicist and—turn the lectern over to you to announce whatever you decide to do."

Dawson laughed nervously.

"No, I mean really."

Reeves smiled and reached over to put a hand on Dawson's shoulder.

"I'm not kidding, Tommy. This is your discovery. Only you have the right to make any announcement."

Dawson stood up and walked around in a little circle. For a moment they thought he was about to start a manic dance, but he remained calm. He stopped, picked up the Coke and took another sip.

"Okay," he said matter-of-factly. "But I'm not going to try to explain my theory to a bunch of ignorant journalists." He glanced at Kate and reddened slightly. "Present company excepted, of course."

"Thank you, Tommy. No, I wouldn't recommend that you do."

She reached in her bag, extracted a printout and pushed it across to Dawson.

"I've written this backgrounder that we can use as a handout, just the bare bones on what you've achieved. If we decide to go public with your time theory, it'll give the press some red meat without throwing them the whole steer. Read it over and let me know if it needs any revisions."

Looking relieved, Dawson snatched up the papers. He sat back down and turned to Chris.

"Well, then, Fisher, we need to decide how we're going to handle this. I've listened to a lot of ideas over the past few days. We all agree that the ability to view past time will change our understanding of history. I always thought of it as a positive thing, but some of you have raised questions about that. Good, valid points. As we've learned, there are secrets that some people will kill to protect."

"No question about that," Reeves remarked. "Time viewing will change the world in ways that we probably can't even imagine." The others nodded agreement.

"Well, I've been giving this a lot of thought," Dawson informed them. "You know what I said about 'If you want peace, prepare for peace'? Well, I didn't realize it at the time, but now I think it can work."

"What do you mean?" Reeves asked.

Dawson looked around the table and thought for a moment before responding.

"The old maxim about preparing for war in order to attain peace worked because only force could counter force and thus maintain a status quo. We saw that in the nuclear standoff of the Cold War.

"But there's something else that can enforce peace—the truth. When space-time viewing comes into widespread use, truth will be the order of the day. I think we'll learn it's more powerful than any military force. No one will need to live in fear under the threat of violence and death. We've demonstrated that already, just in the last few days."

Nobody spoke for a long time as they pondered his words. Finally Kate spoke up.

"Tommy, I think you're right. Your discovery *will* be a force for good."

"Bravo," Bonner chimed in. "You're dead on, Kate. I've been thinking about the ethical issues we've raised, and it would be a tragic mistake not to let the world share in Tommy's brilliant discovery. I'm just a tired old detective, but it sounds to me like the moral thing to do."

"Your judgment is worth a lot," Chris assured him, "and we're all in agreement. This is your decision to make, Tommy, and it sounds like you're on the right track."

Dawson looked around the table.

"It's a lot to put on my shoulders," he murmured, "but I know it's right. I've got a couple of hours to think about what to say." He looked over at Kate. "I trust your judgment, so go ahead and have copies of your fact sheets ready."

Everyone sat still for a moment, heads nodding.

"Great, Tommy," Reeves spoke up at last. "I'll do the intro, and you can take it from there. I know you'll find the right words."

"Sure, but I need to get a clean shirt from my house," Dawson said, looking down at the stained Grateful Dead T-shirt he'd worn for several days. "And maybe a fresh pair of jeans," he added, noticing some prominent stains where he had wiped his hands on his pants leg.

He took off his Quantum Cowboy cap and examined it closely. He shook it and a little cloud of dust flew out.

"This'll do fine," he announced placing it back on his head.

Chapter 64

Los Alamos
Saturday, 3:47 p.m. MST

The high school auditorium was filling up fast. Every major television outlet was represented and cameras and lights were everywhere. The podium had sprouted a forest of microphones. There were camera crews not only from the American networks and cable news channels but also the BBC, Deutche Welle, two Japanese networks and a host of others including French, Italian, Spanish, Brazilian and Korean channels.

Familiar faces from the media dotted the room, including prominent anchors, investigative reporters and columnists. Tabloid writers, their notoriety measured more by the outrageous unlikelihood of their stories than reliance on facts, were present in force. Several dozen still photographers were huddled on the floor near the stage, festooned with cameras. Even so-called Paparazzi who usually spent their time stalking celebrities were on hand, carrying their huge telephoto lenses like hunters stalking dangerous game.

There was a general sense of impatience and even a hint of anger among the members of the press, resentful at being kept in the dark about unexplained and perhaps important events. Now, like a hungry tiger too long denied food, the international press corps was poised to pounce.

Peering out from behind the stage, Dawson studied this mob with indignation.

"I can't talk to those people," he declared to Kate. "Look at them. They're idiots and maniacs."

She smiled sardonically.

"Yes, that's a good characterization for most of them. But they think they're something special. In fact, most of them believe they're actually the story. It's narcissism. They've caught the celebrity bug. They're in love with themselves and they feed off of each other. That's the main reason why I left the mainstream media."

"But how can they do their jobs?" Dawson asked her. "To report the news, I mean."

"Oh, their job isn't to report the news, not really. They're entertainers, the modern day version of court jesters. The news is just the medium on which they feed. If there isn't any news, they make some up."

Dawson stared at her in amazement.

"It's really that bad?" he murmured, taking another peek at the hubbub in front. "Then we'll never get our story out."

Kate shook her head in denial.

"Deep down, most ordinary people understand that most of these journalists are self-centered clowns. The real stories come through for those intelligent enough to recognize them. There are still a few serious journalists left and the perceptive readers and viewers know the difference."

"Oh, then it's like with an undergrad class," Dawson responded. "Hopeless morons, most of them, but a small minority who will actually learn something."

"Yeah, like that," Chris chimed in. He was keeping an eye on his SEAL unit, gathered behind the stage.

"Well, then, I know how to talk to them," Dawson exclaimed. "That old KISS rule: Keep It Simple, Stupid."

"There you go."

Stuart Reeves walked up to them looking at his watch.

"It's nearly time to start and people are still milling around like cattle out there," he complained."

"They *are* like cattle," Kate told him with a chuckle. "You need to get out your bullwhip and round 'em up. Don't expect them to be cooperative."

"Yeah, I guess so." Reeves walked onto the stage and stepped behind the lectern. He stared at the massive display of microphones, placed there by producers and sound engineers from a gaggle of media outlets. He could barely see over them. The swarming members of the press ignored him. He tapped tentatively on some of the mikes. The only result was that several producers wearing headphones looked up from their recorders and glared at him.

Kate walked up beside him. She reached down to a shelf inside the lectern and pulled out a wireless mike.

"Tommy asked for this," she explained, handing it to Reeves. "It's hooked into the PA system."

"Ah!" Reeves snatched the mike and turned on the power switch. He strode to the front of the stage.

"Please take your seats," he bellowed and his amplified voice rang from every corner of the auditorium. For a moment no one seemed to notice, but attention began to come his way. "We're going to start in thirty seconds."

That got people moving a bit faster and before long most of them were taking their seats. Like the patter of raindrops in a sudden Spring shower came the sound of camera shutters as the gang of still photographers began to shoot. Producers donned headsets and red lights flared on more than a dozen television cameras. Slowly a semblance of silence fell across the room.

"Okay, thank you. My name is Stuart Reeves and I'm acting director of the Los Alamos National Laboratory."

"Speak into the mikes!" someone from the audience shouted. Other voices joined in.

Reeves lowered the hand-held microphone and glared around the room, waiting for quiet. When he thought he could be heard he raised the mike to his lips.

"Ladies and gentlemen, I'm holding the only microphone I plan to use. My words are coming clearly through the speakers around this room and your microphones can capture my voice from there. I do not intend to hide behind that lectern with its monstrous growth of electronic gadgetry. If that doesn't suit you, it's your problem."

Mutterings rippled through the room, but there was some sympathetic laughter from members of the print media. Reeves waited patiently for things to settle down.

"We've called this press conference as a courtesy to you, the members of the press, but more importantly to the world at large. As you've surmised there have been some unusual events here at Los Alamos. The details haven't been announced. There were good reasons for that, reasons which even today I'm not free to share with you in full."

"Why not?" a voice shouted. "It's our right!" cried a second. Others joined in.

Once more Reeves lowered the mike and glared around the room. Gradually the outbursts subsided. He raised the mike but looked around with a steady gaze before speaking.

"Ladies and gentlemen, I have witnessed your behavior many times in the past at televised press conferences. I must say that I am not impressed. We are here today not to become victims to your arrogant fantasies of self-worth, but to communicate some very important information to the world."

He turned and gestured back stage. Leonard Perry stepped into view followed by the other SEALs. They were wearing camouflage battle dress with Kevlar helmets. Automatic pistols were in the holsters at their waists and each carried a long black crowd control baton. They took up position, three on each side of the stage.

Simultaneously two squads of elite Marines appeared at the back of the auditorium, similarly armed. They spread out around the room and took positions along the outer walls. While the men took their places, Reeves continued to speak.

"Members of the press, you are here as representatives of your employers, but you owe a more important allegiance. I mean to the people of the Earth. The Los Alamos National Laboratory is a high-security research facility of the United States Government. Much of what we will share with you today has been classified Top Secret. As such, we have no obligation to share this information with you. However, due to unusual circumstances we've received special dispensation from the highest sources to waive that shield of secrecy, at least in part."

He paused and examined the room. The SEAL and Marine presence had provided a dampening effect on the press. He smiled and continued.

"No matter how much it may go against your training and instincts, I'm asking you for once to behave like mature, civilized human beings and listen to what we have to say without further interruptions. I invite any of you who feel you cannot contain yourselves to leave at this time."

There were gasps of indignation, then a few snickers. Finally someone laughed out loud and a little wave of restrained hilarity spread through the room and quickly died out.

A respected network anchor stood up in the front row and politely held up his hand. Reeves gestured to him to proceed.

"Doctor Reeves, I believe that I can speak for my fellow journalists when I say that we of the media have never before been treated in such a way. I will add that it is quite refreshing." There was more laughter. "I presume that by the presence of these bold fighting men that you are prepared to eject anyone who steps out of line…?"

Reeves grinned.

The anchor flashed his own familiar smile and turned toward the audience, holding up both arms for attention.

"I think we're about to get a big story here," he bellowed, projecting his voice to fill the room. "As the text-messagers say, let's just STFU and listen for once."

There was a low rumble of agreement. Even the chief reporter for a notorious supermarket tabloid nodded in agreement.

"Doctor Reeves, please proceed." The anchor sat down.

Behind the stage Dawson turned to Kate with a grin.

"Told them, didn't he?" She returned the smile.

Chris chuckled appreciatively. "It remains to be seen whether he'll get away with it," he remarked.

Reeves began to talk, describing the development of the quantum computer and the events such as the Hummer attack and Black Hawk helicopter episode. These were already known to the public, if only in general terms. He did not mention Dawson's theory of six-dimensional space-time, the rogue CIA involvement, or the contributions of Chris and Kate.

"As you can obviously see, there's more to this story," he concluded. "Some will be revealed at a press briefing from the White House tomorrow morning. The President and his key intelligence officials will have some news that you will find quite interesting."

The crowd had remained relatively quiet but at the mention of the President a press swarm broke out. Dozens of reporters leaped to their feet, shouting for attention and yelling questions.

Master Chief Leonard Perry took two steps forward, stamped his booted feet twice on the hardwood stage and issued a command.

"Platoon! Att'en...*Hut!*"

Honed through years on the parade ground his order rolled across the cacophony of the squabbling journalists. Around the room, as one the SEALs and Marines snapped to attention.

"Pre...sent...*Harms!*"

The warriors snapped their batons into position in front of their chests.

The room fell silent. Reporters who a moment before were shouting and waving their arms were suddenly sitting down trying to look innocent.

Reeves looked at the senior anchor. Running his fingers through his abundant silver hair the man stood and turned to address the mob once more.

"You guys embarrass me," he said with a mournful look. "Shame on you. Shame on us." He turned back to the stage. "Please continue, Doctor Reeves."

Perry's voice rang out once again.

"Platoon! Stand at...*Ease!*"

Reeves looked around with satisfaction.

"Thank you," he said at last. "Now, as I was saying, new information will be released from the White House tomorrow. I am not in a position to upstage the President of the United States, and do not intend to. Your questions will be answered in due course, but not at this time and not by me.

"However, we've prepared some briefing notes that will be useful to you in putting into perspective what we've revealed here today. Copies will be passed out at the conclusion of this event.

"But now, I'm pleased to introduce the man behind the quantum computer, one of the world's greatest scientists and my friend, the Nobel laureate and leading physicist Doctor Thomas Dawson."

He turned as Dawson stepped into view and approached the front of the stage. He was wearing a new T-shirt, this time featuring the portraits of Galileo, Newton and Einstein. He wore his Quantum Cowboy cap centered firmly on his head. He reached for the mike and held it down to his side. He glanced around the room, assuring himself that everyone was going to remain on good behavior.

After a moment he raised the mike to his lips and began to speak in a calm, matter-of-fact tone.

"Ladies and gentlemen of the press, thank you for your attention. I noticed my friend Stuart was a little hard on you. I want to apologize in advance if I should equally offend."

He paused after this punch-line, smiling slightly at the little ripple of nervous laughter.

"You may wonder at my casual attire." He looked down at the T-shirt. "These three men were scientists like me. They are honored for their contributions, for ideas that changed the world."

He pointed at the drawing of Galileo.

"This man, Galileo Galilei, began to overthrow two thousand years of Aristotelian natural philosophy, ideas that placed the Earth at the center of the Universe. He could be called the father of modern physics. Physics is my bag, too, so he's one of my personal heroes."

He pointed at the image of Newton.

"Isaac Newton was the next to shake the tree of knowledge. As we all know that caused an apple to fall and hit him on the head."

Like a seasoned standup comic he paused for the punch-line and was rewarded with a few chuckles.

"Sir Isaac gave us the theory of gravity, the explanation of how apples, planets and galaxies are bound together by a single force in three dimensions of space."

He pointed at the image of Einstein.

"But poor Newton's theories lasted only a couple hundred years before this galoot came along and overturned his apple cart. This is Albert Einstein, the man who discovered that everything is relative. He conceived of what he called a four-dimensional space-time continuum."

He began to walk across the stage, letting his eyes travel over the packed auditorium. The watching journalists were keeping quiet but showing signs of impatience.

"What I'm here to report in no way detracts from these great men," Dawson continued. "However, just as Newton made discoveries that surpassed the insights of Galileo, and just as Einstein's theory made Newtonian physics obsolete, our team of physicists and engineers has made a breakthrough that could be of importance equal to or even greater than those past achievements."

Now he had their full attention. He smiled and took another turn around the stage.

"I cannot even begin to explain the details to you in a way that you could understand. Could Albert back in 1905 have informed the press of that day about his new theory? No. In fact, for some time after he published his paper it was said that only a handful of physicists and mathematicians understood relativity. So I will attempt to give you the 'My Weekly Reader' version. It's the best I can do."

Heads were nodding with eager understanding. The community of the press may have an inflated sense of its own wisdom and knowledge, but when it comes to science and math it knows it's in way over its head.

"Don't try too hard to follow what I say. There'll be a second fact sheet to take with you that gives a general outline of our discovery. I'm preparing a series of papers that will appear in scientific journals but it will take a long time before the full details will be available. That's the way of science. It's a slow, methodical process.

"Before I go on, I should inform you that one purpose for building the quantum computer that Doctor Reeves described was to test a new theory about the nature of the Universe. The theory predicts that Einstein's concept of a four-dimensional space-time continuum is incomplete."

Dawson paused for effect. He reached the far end of the stage, turned and began to walk slowly back.

"We set out to demonstrate that Einstein was two dimensions short—and we believe that we have succeeded. We've discovered solid evidence that we live in a Universe of six dimensions, a space-time continuum with three dimensions of space...and three dimensions of time."

A ripple of murmurs passed through the crowd. Dawson raised his left hand to signal for calm.

"Yes, three dimensions of time. You heard me right. And what does that mean? Well, it means many things but one of those is that with the power of quantum computing that Doctor Reeves told you about—vastly more powerful than the old chips-and-dips type of silicon-based computers—we can actually access those additional dimensions of time."

He reached the center of the stage, stopped and faced the audience.

"We have in fact done that," he told them. "We've seen into the past, and even into the near future."

Pandemonium broke out as the unnatural restraint of the press corps broke in the face of this statement. Dawson let the microphone dangle from his hand and waited patiently for order to return. From his position at the side of the stage Perry glared meaningfully at the audience. After a minute the clamor subsided and Dawson began to speak again.

"Now I bet you don't realize just how big a deal that is," he told the reporters. "I know I didn't, not at first." He shrugged. "For a long time I thought, 'Oh, we'll be able to look back in time and see how things really happened. Isn't that nice.' But what I didn't realize is that it might not seem nice to everyone."

He wheeled around and continued to stroll toward the other end of the stage, gesturing with his free hand as he continued to speak.

"Now so that you can understand what being able to view events in time actually means, let me put it in the simplest way I can, by example."

He pointed at one of the reporters in the front row and asked him to stand.

"Let's say just as a hypothetical for-instance that last Christmas this man dismembered his grandmother, ran her remains through a wood chipper and fed the result to wild pigs, all in order to get title to her farm."

A ripple of laughter passed through the audience. The reporter turned to the crowd and took a comic bow, causing more laughter.

"You covered your tracks pretty well," Dawson continued, "so there's no way anyone could ever prove what you did. Granny just went missing, probably ran away with her hired hand or was abducted by aliens, right?"

Dawson leered at the reporter, then gestured for him to sit down again.

"Now, of course I know this fine man didn't do that terrible thing..." He paused for effect, stopping and turning directly toward the audience. "Or at least I hope not, because if he did he should be very worried, because there's a new sheriff in town. With the use of a quantum computer and some software we've developed, Johnny Law can look back in time. He can see our friend here doing in poor old Granny, document his actions in perfect detail, and wrap up the case against him with irrefutable evidence that no jury could ignore."

Again the mob began to stir but Dawson raised his hand and they settled back. He noticed many of the reporters were furiously writing notes.

"So you see, the ability to visualize the past can change a lot of things, solve long-standing mysteries if you will. What happened to Jimmy Hoffa? Judge Crater? Was there someone on the grassy knoll in Dallas? That sort of thing. Interesting questions."

One of the reporters stood up and raised his hand. For a moment Dawson looked at him skeptically then nodded for him to go ahead.

"Sir, excuse me but what you say seems too fantastic to believe. What proof do you have to back up these claims?"

"Ah, proof is it? Proof! Yes, you must remember that I am a scientist and proof lies at the very heart of science. We have proof. Doctor Reeves outlined some of the things that happened here at the quantum center, but he didn't tell you the full story behind those events. Quite assuredly, sir, if we had not had the ability to see into the past and the future none of us would be here today. For the duration of the siege, and I think it would be fair to use that term to describe what we have gone through, we used the secret weapon of time viewing to foil our attackers."

The reporter was still standing.

"Can you give us some examples?"

"Gladly, but let me introduce someone who played a key role in this." Dawson walked toward the back of the stage and gestured for Chris to come forward.

"This is Lieutenant-Commander Chris Fisher of the United States Navy," he said, placing a hand possessively on Chris's shoulder. "Doctor Fisher was one of my graduate assistants at Cal-Tech and now he's a distinguished full professor at Stanford, the youngest ever. He was called to active duty to help us in the quantum project, which is primarily funded by the Navy. Doctor Fisher, tell the nice journalists some of the things we achieved." He handed over the mike.

Chris took a moment to collect his thoughts. He started with a brief description of how past viewing had identified Jalaly, the spy who infiltrated the center and murdered two security officers. The members of the press listened with rapt attention as he described the high-resolution image of the spy's retina as the key to solving the mystery. He told how they had used past viewing to trace the Iranian to his cabin hideout. Next, skipping over the details, he described how future viewing had given them warning of the coming attacks on the center from the Hummer and Black Hawk.

"I can't tell you more right now because there are some on-going investigations," he concluded. "You'll learn more as we are able to release it. But be assured, Doctor Dawson's theory has already been proven many times over, and in real-life situations. As he told you, we wouldn't have survived without the ability to see forward and backwards in time."

He handed back the mike and stepped aside.

"I'm not going to say much more about the details now," Dawson said, "but I do want to share one more thing. When I first conceived of the extra time dimensions and realized that they might allow us to view past and future events, I thought only of the positive benefits. I didn't consider how much of the world's history is based on myths, assumptions, misdirection—well, we might as well call them for what they are, bald-faced lies. I realized the history books would have to be re-written, but it never occurred to me that revealing the true past could shake things up pretty badly."

He stopped and gazed around the room.

"After my friends pointed out the downside of this, it really upset me. For a while I thought I might have done something bad, coming up with this theory. I started to believe that maybe revealing all those hidden secrets could undermine the very foundations of our civilization. That's a pretty heavy thought, isn't it?"

He looked inquiringly around the room and saw thoughtful nods of agreement.

"But after seeing what time viewing can actually do, the way it saved our lives, I had an epiphany. I realized that the only people to be threatened by the truth are those who have things to hide. We've seen proof of that—and learned that the truth can trump lies."

Dawson stopped in the middle of the stage and looked around the auditorium. He seemed to become distant, his eyes rising toward the ceiling.

"I started to think once again about the positive side, about what kind of world we could have if dishonest people could no longer get away with their deceptions." He was speaking softly now in an almost wistful tone. Then his gaze snapped back down to the gathered journalists and he spoke with authority.

"Well, we're going to find out what kind of world that is, because as you must realize by now I've decided to release my theory to the world at large. Time viewing is going to become a new force, for better or for worse. How ironic it would be if such a thing, with its very ability to uncover secrets, were itself to be kept under wraps."

He took several steps forward to the edge of the stage and fixed the mass of journalists with a determined gaze. For a long moment he stood there, gazing implacably around the room. The crowd was silent, waiting to learn more. After a moment he began to speak again, once more in a soft, wistful tone.

"When I faced my doubts, I felt like Pandora must have. You all know her legend. She was created by the ancient Greek gods. Each gave her a special gift, one of which was curiosity. Now according to the myth, at that time there was a sealed box that was said to contain important secrets. Pandora opened that box—and all of the evils of mankind were inadvertently released into the world."

Dawson waved his arms to simulate dark forces scattering like bats or locusts in every direction.

"It was tragic because she could never put them back. Well, because of the potential of my discovery, I was afraid that, like Pandora, I was about to release bad things into the world. It was a very sad idea."

For a long moment he gazed steadily around at the sea of faces, each turned toward him with unwavering attention. Then he smiled, beaming like an oversized cherub. It was like the Sun coming up.

"Well, ladies and gentlemen, I thought it through and at last I came to an opposite conclusion. I will leave you with this thought. My discovery, the ability to view events in time, will cast the light of truth on evils and help us do what Pandora never could—to put them back where they belong.

"I thank you for your patience. I'm not going to answer questions and I earnestly beseech you not to ask any. There will be much more to reveal, all in its own good time—no pun intended."

With that Dawson turned and strode off the stage, leaving the press corps in silent awe.

There was no sound of trumpets, no roll of thunder, but in that fleeting instant the future of human civilization began to change forever.

The End

Author's Notes

The concept of multiple time dimensions as related in this novel is purely the product of my imagination. However, some cutting edge "theories of everything," ideas that attempt to connect the quantum world and Einsteinian space-time, predict the existence of more than four dimensions. Most seem to assume that the extra dimensions are of space, not time, but at least one theorist, Itzhak Bars of the University of Southern California in Los Angeles, has stated that there is a second dimension of time.

The mysteries of quantum dynamics, entanglement, and the uncertainty principle are generally as I have described, within the limits of my understanding of these esoteric subjects.

The potential of quantum computers is real and some simple working models have been built. It's extremely unlikely that if and when a fully functioning quantum computer is built that it will equal the performance of the one described here.

As mentioned in this story, Albert Einstein is reported to have said that he thought in images rather than words. He is accurately quoted as saying that entanglement is "spooky action at a distance."

The description of Little Fox being able to "see" a hidden fish and catch it in his hand is based on an actual event that I witnessed as a boy. Although taking place under different circumstances, the real event occurred exactly as described in the fictional version. Many people have reported having similar experiences, hinting at a mysterious sixth sense, something which science generally refuses to recognize and for which it has no explanation.

The description of the complex number used to test the quantum computer, a product of two primes, is based on fact. As described, a 167-digit cryptographic key was factored in 1997 by Dr. Samuel Wagstaff at Purdue University. The task required about 100,000 hours of parallel computing time. Since then, even larger numbers have been factored, containing up to 600+ digits, but requiring immense amounts of computing power.

The Los Alamos National Laboratory is real but no attempt has been made to accurately describe its facilities, organization, or infrastructure. My research on the LANL consisted of driving past the entrance gate with a friend who works there. I asked him if he could take me inside, and he said: "Yes, once."

A reliable source, when asked whether a nuclear warhead could be rigged to override its safety features as described in this book, told me that only an expert could, but yes, it's possible. Tom Clancy described something similar in his novel *The Sum of All Fears*.

The W-87 nuclear warhead exists more or less as described, and was designed at the Lawrence Livermore National Lab in California, a fact which provided an important clue in the resolution of the plot. The similar W-88 warhead now used in the Navy's Trident ballistic missiles was designed at Los Alamos.

All conventional weapons described in the course of the book exist and their features and performance are accurately described to the best of my ability.

The New Mexico Institute does not exist although it was inspired by the nearby Santa Fe Institute, an important scientific think tank.

The Naval Intelligence Agency exists and is headquartered in Suitland, Maryland. Its methods and activities are unknown to me and those I describe are purely fictitious. Its emblem and motto are as described. It is the oldest of all U.S. intelligence agencies.

The linear accelerator program at the SLAC facility in California with which Chris Fisher is alleged to work actually exists.

All references made to myths and legends related to time are historically accurate.

As a private joke I have planted a tiny clue connecting this story with *The Hitchhiker's Guide to the Galaxy* by Douglas Adams in which a gigantic computer labors for centuries to answer the question "What is the meaning of life, the universe and everything?" only to yield the answer "Forty-two." I assigned the number 42 to the bus station locker where Admiral Chase had stashed an emergency kit of guns and equipment.

This book was a lot of fun to write and I became quite fond of the characters. I hope you enjoyed reading about them and their adventures. I offer my special thanks to friends who helped me in the creation of this novel, particularly Val Germann, David Ponton, Barbara Deputy and my wife Patricia. — *David L. Brown*

About the Author

David L. Brown has been a newspaper reporter, columnist and editor. He blogs at StarPhoenixBase.com and has written two other books: *After Calamity*, a novel about the dangers of the near future, and *Dead End Path: How Industrial Agriculture Has Stolen Our Future*, an extended essay and report on the fate of civilization. He is currently vice president of the Rio Grande Chapter of the Society of Professional Journalists and was named "2010 Citizen Journalist of the Year" by SPJ's Region 9 covering New Mexico, Colorado, Utah and Wyoming. He lives in Rio Rancho, New Mexico.

CPSIA information can be obtained at www.ICGtesting.com
Printed in the USA
BVOW05s1222030314

346422BV00004B/43/P

9 781609 104658